Reactivity in Organic Chemistry

GERHARD W. KLUMPP
Free University Amsterdam
The Netherlands

Translated from the German by
Ludmila Birladeanu, Department of Chemistry,
Harvard University, Cambridge, Massachusetts

A WILEY-INTERSCIENCE PUBLICATION
JOHN WILEY & SONS
New York Chichester Brisbane Toronto Singapore

Originally published as *Reaktivität in der organischen Chemie I: Produkte, Geschwindigkeiten; II: Übergangzustände*

© 1977, 1978 by Georg Thieme Verlag, Stuttgart

Copyright © 1982 by John Wiley & Sons, Inc.

Library of Congress Cataloging in Publication Data:

Klumpp, G. W.
  Reactivity in organic chemistry.

  Translation of: Reaktivität in der organischen Chemie.
  "A Wiley-Interscience publication."
  Bibliography: p.
  Includes index.
  1. Chemistry, Physical organic. 2. Reactivity (Chemistry) I. Title.

QD476.K5413   547.1'3   81-16437
ISBN 0-471-06285-5   AACR2

Printed in the United States of America

10 9 8 7 6 5 4 3 2

Preface to the English Edition

It is gratifying to see "*Reaktivität in der organischen Chemie, I, Produkte, Geschwindigkeiten,* and *II, Übergangszustände,* translated into the language in which most of the important contributions to the field appear (and which is also the language of many of my friends in the profession).

The translation also gave me a welcome opportunity to make corrections and revisions and to introduce a considerable amount of new material including work published in 1980.

Upon completion of the undertaking I express my gratitude foremost to Dr. Ludmila Birladeanu, of the Department of Chemistry, Harvard University, for her fine work in translating my German and otherwise helping in the preparation and publication of the book.

I am also deeply indebted to Prof. W. von E. Doering for his enthusiastic support of the English version, and I thank A. H. W. van den Berg for several new drawings.

GERHARD W. KLUMPP

Amsterdam
September 1981

Preface

This book is based on lectures I gave at Free University Amsterdam. Assuming knowledge of material commonly treated in textbooks of organic chemistry, I aim to give an overview of essential aspects of present-day organic chemistry. No attempt has been made to be comprehensive or to adhere to historical order.

The central theme, *reactivity*, is illustrated by examples taken from the areas of synthetically important reactions and properties and behavior of reactive intermediates and covers such practical questions as "Which products are to be expected from certain starting materials?" and "What is the reaction time needed for a given substrate?"

The treatment is three-layered: an introductory description of the multiple reactivity of organic compounds is followed by a discussion of the effects of reaction conditions (thermodynamic control, various kinetic systems) on the ratios of products formed. Relationships (mostly quantitative) between structure of substrate and reaction rates are then described. Finally, to get an insight into systems whose reactivity is amenable to qualitative treatment only (and which represent the majority of situations commonly encountered in the laboratory), properties of transition states and models for their understanding are presented.

The didactic strategy used is that of repetition at various stages: nucleophilic reactions, reactions of carbonyl compounds and olefins, cycloadditions, Cope and Claisen rearrangements, aromatic substitutions, and formation and reactions of reactive intermediates are leitmotives recurring throughout the text and are treated at all three levels.

A main concern is to present material that because of the rapid development of organic chemistry is not treated in available textbooks. The novice experimenter starting out alone is offered a vast array of data taken from practice. Along with the fact that such a text does not exist in German it is hoped that these two elements will make *Reactivity in Organic Chemistry, I, Products, Rates,* and *II, Transition States* a useful and stimulating source not only for advanced students but also for colleagues and industrial chemists.

Having reached the end of the road, it is a pleasure to express my gratitude to all who contributed directly or indirectly to the accomplishment of the task: to my family, to my students, co-workers and colleagues for help, stimulation, and criticism, to Mrs. P. M. M. Zwebe-Böeseken for typing the initial draft of the manuscript [in German] and preparing many drawings, and to Georg Thieme Verlag and its staff, in particular Dr. G. Zartner, for unrestricted support in setting up the final form of the German edition.

GERHARD W. KLUMPP

Amsterdam
September 1977

Contents

TRANSITION STATES

**Reactivity in
Organic Chemistry**

Introduction

"Populate the earth and subdue it" (Genesis 1:26). Results of modern organic chemistry seem to indicate that in this field the biblical command has been fulfilled. An arsenal of instrumental and theoretical methods makes it possible to isolate and determine the structure of the most complicated natural products on vanishingly small amounts of material, to perform their total syntheses, and to solve other synthetic problems that have intrigued chemists for decades. One of the many examples is the structure proof by B. Witkop et al.[1] of the extremely toxic cardiotoxin *batrachotoxin* (1) (used as arrow poison), available as one of the four components of a mixture isolated from 5000 Colombian frogs in the amount of only 110 mg.

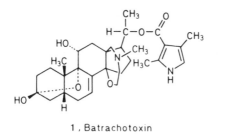

1 , Batrachotoxin

3 Cyclobutadiene

Up to now the ultimate triumph in organic synthesis may be the synthesis of *vitamin B*$_{12}$ (2) by the joint efforts of R. B. Woodward and A. Eschenmoser and their collaborators.[2] One should also mention the recent synthesis and characterization of *cyclobutadiene* (3) in several laboratories, a problem that first emerged in 1900.[3, /4/*] It is not surprising, therefore, to encounter statements such as "[it is] the modern chemist's attitude that he can very probably make anything he wants."[5]

---

*The symbol // indicates references containing comments along with literature sources.

1

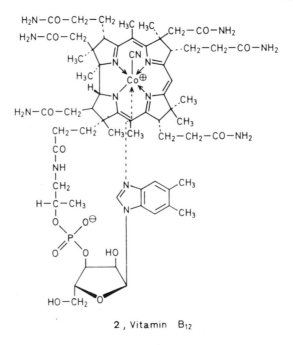

2 , Vitamin B$_{12}$

However, even someone who has spent only a short time in an organic research laboratory knows that the conquest is far from total and that battles must be fought almost daily. Only in relatively few cases (e.g., the epoxidation of cyclohexene) does the reaction result in a single product. But if instead of the simplest, symmetric substances one wishes to use polyfunctional compounds as starting materials, the formation of a single product is rather the exception while the formation of several products is the rule. For instance, the epoxidation of 3-(*cis*-1-propenyl)-cyclopentene (*1*) yields eight epoxides as racemic mixtures. Even worse results are obtained when more complex substrates and highly reactive reagents are used. In the extreme only undesired products may be formed. In such instances, rules that teach what changes should occur in a molecule and in what part of it do not give a clear-cut answer. Only *experiment* can provide a solution, as well as hints for more successful approaches.

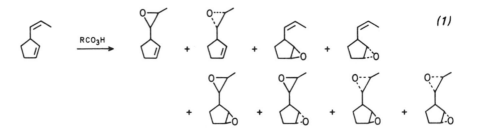

(1)

The more complex and unusual the substrate, the greater the danger of occurrence of unexpected or undesired reactions. In multistep syntheses, where the amount of available material diminishes with every step and thus becomes more precious [e.g., see (2)], satisfactory total yields can be obtained without too much loss of material and time only if one takes into consideration all possible reactions and their rates and chooses appropriate reaction conditions accordingly.

$$A \xrightarrow{80\%} B \xrightarrow{60\%} C \xrightarrow{90\%} D \xrightarrow{40\%} E \quad 17\% \text{ total} \qquad (2)$$

Before success was achieved in the synthesis of vitamin B_{12} (2), the interplay of theoretical considerations and experiment required 11 years of work by 99 people.

In kinetic and mechanistic studies the formation of several products is often desirable, since the elucidation of their structure and relative concentrations may give valuable information with respect to the *reaction mechanism* and the *properties of potential intermediates*.

The concern about products is thus the most elementary and compelling aspect in any chemical reaction. In planning new reactions the following three questions must be asked:

1 Which primary *products* are possible under a given set of conditions?
2 What are the *relative amounts* of the possible products, and are they *stable* under the reaction conditions?
3 How can one *influence* the outcome in 1 and 2? In the following the attempt is made to:
   (a) Correlate *reactivity* and *composition* of reaction mixtures.
   (b) Review the quantitative empirical and semiempirical *structure-reactivity* relationships and their application.
   (c) Present the methods of the *transition state theory* and evaluate the possibilities of influencing product composition and reactivities.

Examples used as illustrations of general principles should familiarize the reader with topics such as "*(synthetically) important classes of compounds and their transformations*" and "*(important) properties of reactive intermediates.*" Various aspects of a given topic often appear in a different context. It is assumed that the reader is familiar with the fundamental aspects of the transition state theory, reaction mechanisms, and conformational analysis.

PRODUCTS

CHAPTER 1

Multiple Reactivity
of Organic Compounds

1.1 TRANSFORMATION OF POLYFUNCTIONAL COMPOUNDS

The simultaneous and independent occurrence of several reactions, for example,

$$A + B \longrightarrow \begin{array}{c} C \\ D \end{array} \qquad \qquad (3)$$

may be due to several causes. Obviously, compounds containing several identical or similar groups can be attacked by a reagent at any of these groups, leading to a mixture of *structural isomers*. [See, e.g., the epoxidation of propenylcyclopentene in *(1)*. Since the molecule is asymmetric, diastereomers can arise as well.]

Similar situations are encountered in compounds with two functional groups, one of which is not compatible with the reagent that is supposed to interact with the other—for example, the interaction of a halogenated ester with organolithium reagents. Such problems are very important in planning a synthesis. In these instances one may try to achieve selectivity by a judicious choice of reagents able to sense small, but ever-present differences between substitutents (e.g., of steric or electronic nature), or even to enhance these differences. Better still is the "timely" introduction of *protecting groups*, or their timely *modification*. An impressive example is the synthesis of vitamin B_{12} *(2)* via cobyrinic acid *(4)*, which contains six amide groups in addition to a carboxy group. Since it is impossible to selectively transform one of seven similar groups in an immediate precursor, the differentiation between the carboxy and the amide groups had to be built in at the beginning of the synthesis.

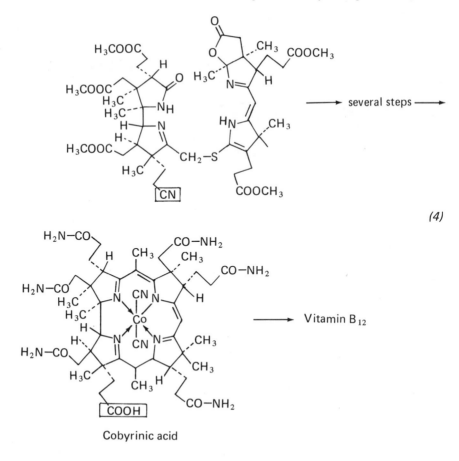

(4)

Cobyrinic acid

1.2 SEVERAL REACTIONS AT THE SAME GROUP

1.2.1 Nucleophilic Substitution and Elimination

The best known example of one group having the ability to undergo different reactions is the simultaneous occurrence of nucleophilic substitution and elimination.[6] In *bimolecular reactions* (*5a*) the ratio of substitution to elimination for a given combination of base or nucleophile and solvent depends on the *starting material*. In *unimolecular reactions* (*5b*) the ratio is determined by the nature of the intermediate *carbonium ion** or *ion pair*. Some examples are given in Table 1.

---

*Tricoordinate carbocations are habitually called carbonium ions. Olah has suggested the use of different terms for pentacoordinate carbocations (*carbonium ions*) and for tricoordinate carbocations (*carbenium ions*): G. A. Olah, J. Am. Chem. Soc., 94, 808 (1972).

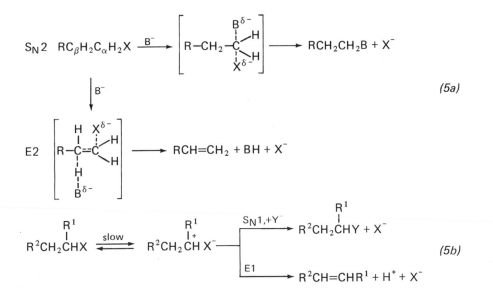

$$\text{S}_\text{N}2 \quad RC_\beta H_2 C_\alpha H_2 X \xrightarrow{\;B^-\;} \left[R-CH_2-\overset{\overset{B^{\delta -}}{|}}{\underset{\underset{X^{\delta -}}{|}}{C}}\overset{H}{\underset{H}{\diagdown}} \right] \longrightarrow RCH_2CH_2B + X^-$$

$$\qquad\qquad\qquad (5a)$$

$$\text{E2} \quad \left[R-\overset{\overset{H}{|}}{\underset{\underset{B^{\delta -}}{|}}{C}}=\overset{X^{\delta -}}{\underset{\underset{H}{}}{C}}\overset{H}{\diagdown} \right] \longrightarrow RCH=CH_2 + BH + X^-$$

$$\underset{R^2CH_2CHX}{\overset{R^1}{|}} \underset{\text{slow}}{\rightleftarrows} \underset{R^2CH_2\overset{+}{C}H\,X^-}{\overset{R^1}{|}} \left\{ \begin{array}{l} \xrightarrow{\text{S}_\text{N}1,+Y^-} \underset{R^2CH_2\overset{}{C}HY}{\overset{\overset{R^1}{|}}{}} + X^- \\[2ex] \xrightarrow{\text{E1}} R^2CH=CHR^1 + H^+ + X^- \end{array} \right.$$

$$\qquad\qquad\qquad (5b)$$

TABLE 1 Effect of Structure and Reaction Conditions on Yields of Olefins[6]

Substrate	Type of Reaction	Temperature (°C)	Reaction Conditions	%E
$(H_3C)_2CHBr$	$E2/S_N2$	50	a	58
		80	a	61
		100	a	66.5
		55	b	79
C_2H_5Br	$E2/S_N2$	55	b	1
$(H_3C)_3CBr$	$E2/S_N1$	55	b	100
$H_5C_2-\overset{+}{S}(CH_3)_2$	$E2/S_N2$	45	b	12
$(H_3C)_2CH-\overset{+}{S}(CH_3)_2$	$E2/S_N2$	45	b	61
$(H_3C)_3C-\overset{+}{S}(CH_3)_2$	$E2/S_N2$	45	b	97-100
$(H_3C)_3CCl$	$E1/S_N1$	25	c	17
		50	c	24
		65	c	36
$(H_3C)_2\overset{\overset{C_2H_5}{\diagup}}{\underset{\underset{Cl}{\diagdown}}{C}}$	$E1/S_N1$	50	c	40
$(H_3C)_2\overset{\overset{C_2H_5}{\diagup}}{\underset{\underset{\overset{+}{S}(CH_3)_2}{\diagdown}}{C}}$	$E1/S_N1$	50	c	48

^{a} 80% aqueous ethanol containing the conjugate base.
^{b} Ethanol-ethoxide.
^{c} 80% aqueous ethanol.

The following factors govern the ratio of substitution to elimination in bimolecular reactions and thus may affect their outcome:

1 **Reaction Temperature.** Since elimination is associated with greater changes in bonding than substitution, its activation energy is higher and the relative amount of elimination will increase with temperature.

2 **Structural Effects and Conformation.** Branching in the starting material at C_α and/or C_β results in *steric hindrance* and affects the rate of substitution more than that of elimination, since in the latter attack of the reagent takes place more at the periphery of the molecule.[6a] In addition, elimination is favored if *more stable* higher substituted olefins can be formed (see also p. 32 and 44).

In cyclic compounds *conformational effects* may favor elimination as well as substitution. In cyclohexane systems with equatorial leaving groups substitution may be subject to strong steric hindrance (due to axial hydrogen atoms). However, elimination may still suffer more from the impossibility of achieving the *trans*-coplanar arrangement of β-proton and leaving group. Also, in mobile systems the tendency of larger leaving groups to adopt the equatorial position can make transition states of elimination less favorable.

The requirement of a *trans*-diaxial orientation of the β-proton and the *leaving group* is particularly striking in (4-*t*-butylcyclohexyl)trimethylammonium hydroxides (**4a, 4b**)[7]:

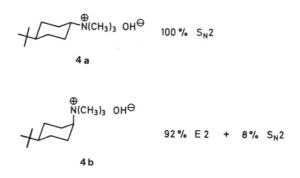

4a

4b

3 **Reagents and Solvents.** Typical bases, like OR^- and OH^- favor elimination, whereas carbon nucleophiles like $H_5C_6S^-$, N_3^-, CN^-, or malonate favor substitution (see p. 147). Conjugate bases of acids with pK_a's under 11 rarely induce elimination in protic media. However, in dipolar aprotic media elimination can be induced even by halide ions (see p. 188). In general, yields of olefins increase in the order

$$H_2O < CH_3CH_2OH < (CH_3)_2CHOH < (CH_3)_3COH$$

Such protic solvents also contain their conjugate bases, and a change in solvent results in a simultaneous change in base. In the series above steric factors and basicity increase, while polarity decreases. The latter factor also seems to be somewhat responsible for the larger increase in elimination rate, since charge delocalization in the transition state is larger for elimination than for substitution.

Most factors mentioned above operate in the same sense in *unimolecular reactions*. (Since the intermediate carbonium ions are often prone to rearrangement, such reactions are only seldom used in synthetic work.) Bulky alkyl groups oppose olefin formation (bond angle 120°) less than the formation of the substitution product (tetrahedral bond angle). Obviously, in polar media, the bulky leaving group has little influence on the substitution-elimination ratio, which ratio depends on the carbonium ion. In less polar solvents, however, the anion is locked in the *ion pair*. In this situation substitution is hampered and the amount of elimination increases.

1.2.2 Reactions of Carbonyl Compounds

Multiple reactivity plays a major role in reactions of carbonyl compounds essential for C—C bond formation. The *ambident* behavior of *enolates* in alkylation reactions (6) is presented more fully on page 156 and in Section 5.2.3.2.

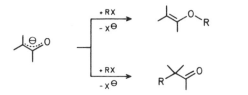

(6)

$$O$$
$$\|$$
TABLE 2   Effect of Metal on the Reaction $H_5C_6-C-CH_3 + C_6H_5M$

Metal	Ratio of Enolization to Addition
K	10:1
Na	2:1
Li	1:23
MgX	1:∞

On treatment of carbonyl derivatives with organometallic compounds, the desired 1,2-additions are sometimes accompanied by *enolization* and *reduction* [(7); Tables 2 and 3].[8] In conjugated

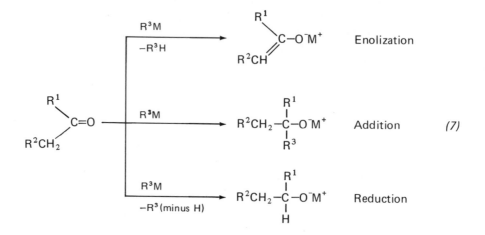

TABLE 3   Effect of Alkyl Groups on the Reaction of Alkyl-Grignard Compounds with 2,4-Dimethyl-3-oxopentane

R in RMgBr	% Enolization	% Reduction	% Addition
Methyl	0	0	95
Ethyl	2	21	77
Propyl	2	60	36
2-Propyl	29	65	0 (!)
2-Methylpropyl	11	78	8
2,2-Dimethylpropyl	90	0 (!)	4

carbonyl compounds there is the additional possibility of the synthetically desirable *1,4-addition* [(8); Table 4].[8] The ionic character of the phenyl-metal bond favors proton abstraction from simple ketones (Table 2) and increases the relative amount of 1,2-addition of α,β-unsaturated ketones (Table 4).

$$H_5C_6-CH=CH-CO-C_6H_5 \quad + \quad C_6H_5M$$

$$\xrightarrow{1,2} \quad H_5C_6-CH=CH-\overset{O^{\ominus}M^{\oplus}}{\underset{}{C}}(C_6H_5)_2$$

$$\xrightarrow{1,4} \quad (H_5C_6)_2CH-CH=\overset{O^{\ominus}M^{\oplus}}{\underset{}{C}}-C_6H_5 \qquad (8)$$

TABLE 4   Effect of Metals on the Ratio of 1,2:1,4-Addition in the
Reaction of Benzalacetophenone with Phenyl-Metal Compounds

Metal	% 1,2-Addition	% 1,4-Addition
K	67	—
Na	60	14
Li	75	14
MgX	—	94

An increase in relative amount of reduction (Table 3), on treatment of 2,4-dimethyl-3-oxopentane with Grignard reagents corresponds to the increase in elimination on treatment of branched halides with bases (peripheral reaction being favored by increase of bulk).

In reactions involving *Grignard reagents* undesired reduction and enolization leading to loss of material can be inhibited by addition of magnesium bromide, thus promoting addition to the carbonyl double bond[9]. In the case of α,β-unsaturated carbonyl compounds, addition of Cu(I) favors 1,4-addition.[10]

1.2.3   Aromatic Substitution

Electrophilic aromatic substitution is one of the best studied organic reactions.[11] The ratio of possible position isomers formed depends primarily on the *electrophilicity* of the reagent. Increase in electrophilic character promoted by electron-withdrawing substituents [e.g., in the Friedel-Crafts acetylation of toluene (9)] or by reduction of the nucleophilicity of the medium [in sulfonation of toluene (9)] results in increase in *ortho* product at the expense of the *para* product (Tables 5 and 6).

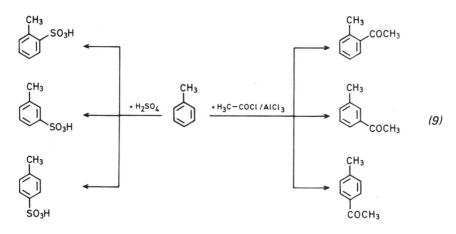

(9)

TABLE 5   Distribution of Isomers in the Acetylation of Toluene Catalyzed by AlCl$_3$[11]

Acid Chloride	% ortho	% meta	% para
$H_3C-C{\displaystyle {\overset{O}{\underset{Cl}{}}}}$	2.5	2.0	95.5
$(H_3C)_2CH-C{\displaystyle {\overset{O}{\underset{Cl}{}}}}$	3.2	2.4	94.4
$ClH_2C-C{\displaystyle {\overset{O}{\underset{Cl}{}}}}$	11.1	2.3	86.6
$Cl_2HC-C{\displaystyle {\overset{O}{\underset{Cl}{}}}}$	17.3	3.2	79.5

TABLE 6   Sulfonation of Toluene with Sulfuric Acid of Varying Strength[11]

% H$_2$SO$_4$	% ortho	% meta	% para
77,8	21,2	2,1	76,7
84,3	38,8	2,6	58,6
99–100 in C$_6$H$_5$NO$_2$	50,2	4,9	44,9

1.2.4 Short-Lived Reactive Intermediates

The simultaneous (and consecutive) occurrence of several reactions is most ubiquitous in transformations of *reactive* and thus *short-lived intermediates of low selectivity* (e.g., free radicals, carbonium ions, carbenes).

Let us consider as an example the behavior of *carbalkoxycarbene*; this species is synthetically important for the formation of carbon frameworks. Simple thermolysis of diazoacetates in cyclohexene (*10*) yields a mixture of addition and insertion products.[12]

Less complicated product mixtures may be obtained on addition of copper salts or copper dust, which transform the reactive species into more stable, thus more selective, carbene complexes (carbenoids), so that insertion reactions become much less important.

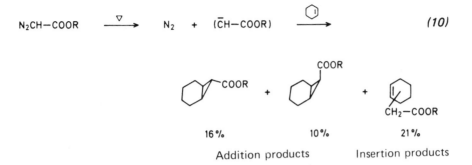

$$N_2CH-COOR \xrightarrow{\nabla} N_2 + (\bar{C}H-COOR) \xrightarrow{\hspace{1cm}} \tag{10}$$

16%   10%   21%

Addition products   Insertion products

The reaction with 1,4-cyclohexadiene (*11*) then proceeds as follows[13]:

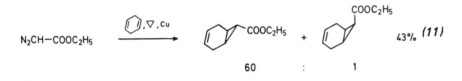

$$N_2CH-COOC_2H_5 \xrightarrow{\nabla, Cu} \quad + \quad 43\% \tag{11}$$

60   :   1

1.3 ORIENTATION PHENOMENA: STEREOSELECTIVITY AND REGIOSELECTIVITY*

The epoxidation of 3-(*cis*-propenyl)-cyclopentene (*1*) and the addition of carbalkoxycarbenes to cyclohexenes are examples of reactions in which the attack of the reagent takes place at the same site, but from

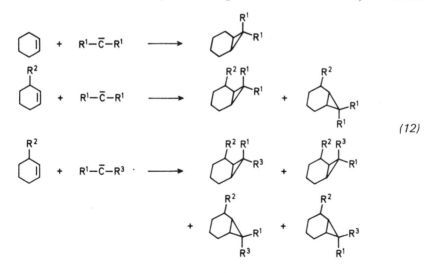

$$\tag{12}$$

*A more extensive scheme of selectivity nomenclature has recently been proposed by Seebach [D. Seebach, *Angew. Chem. Int. Ed.* **18**, 239 (1979)].

two different directions, resulting in mixtures of *diastereomers*. The number of possible diastereomers increases when the symmetry of the reagents decreases (*12*).

The addition of *unsymmetric* reagents may further result in the formation of mixtures of *structural isomers* (*13*):

$$RCH{=}CH_2 + X{-}Y \rightarrow \underset{Y \quad X}{RCH{-}CH_2} \quad \text{and/or} \quad \underset{X \quad Y}{RCH{-}CH_2} \quad \textit{(13)}$$

Generally accepted terms for situations represented by (*12*) and (*13*) are *stereoselectivity* and *stereospecificity* and *regioselectivity* and *regiospecificity*, respectively.

One speaks of *stereoselectivity* when a given starting material yields one diastereomer in larger amount than the other, or, conversely, when in a given transformation one diastereomer reacts faster than the other. In the extreme case, when one stereoisomer gives only one diastereomer, whereas another stereoisomer gives exclusively another diastereomer, one speaks of *stereospecificity*.

For cases of a given starting material that yields predominantly one from among all possible structural isomers, Hassner[14] proposed the term *regioselectivity*. When the structural isomer is formed exclusively one speaks of *regiospecificity*.

It must be pointed out that the specificity of a given reaction is a quality that depends on the ability to detect experimentally potential by-products and may be lost when detection methods improve.

1.3.1 Some Stereospecific or Stereoselective Reactions

1.3.1.1 Addition of Bromine to Olefins

The addition of bromine to the isomeric 2-butenes via *bromonium ions*[15] is stereospecific (*14a, 14b*).

The corresponding reaction of the isomeric 1-phenylpropenes (*15a, 15b*), is, on the other hand, only stereoselective. In carbon tetrachloride, the *cis* isomer gives *predominantly* the *threo*-dibromide, (corresponding to the *d,l* form) whereas the *trans* isomers gives predominantly the *erythro* dibromide, corresponding to the *meso* form.[16a, /16d/, /16e/]

The rate constants of ring-substituted derivatives show a strong dependence ($\rho = -4.5$) on σ^+ (see p. 108), leading to the conclusion that in these reactions the intermediates are not the loosely bridged *bromonium ions* (**6**) with partial charge transfer to the bromine atom and restricted rotation, but rather "open" carbonium ions [**5**, (*15a*)] in which the positive charge is delocalized exclusively into the phenyl ring.[17, /17a/]

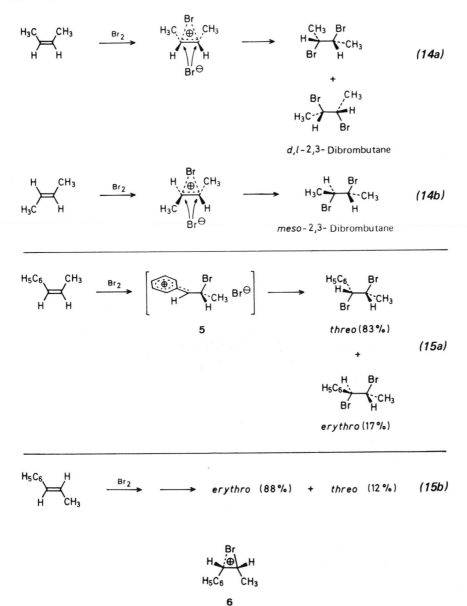

The similarity of the *threo-erythro* ratio of dibromides from the *cis* olefin and the erythro-threo ratio of dibromides from the *trans* olefin is rationalized by assuming that the ratios of rate constants for internal rotation versus nucleophilic attack by Br⁻ (path rot-*cis* and rot-*trans* versus path *trans*) are very similar for the enantiomeric intermediates **5a** and **5b** in this solvent (Scheme 1). This interpretation is

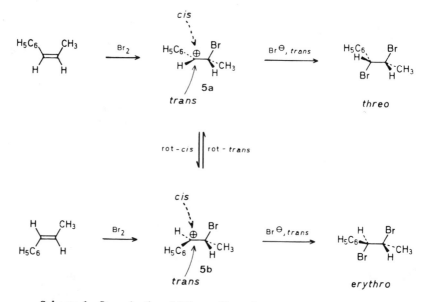

Scheme 1   Steps in the addition of bromine to 1-phenylpropene.

supported by the fact that the stereoselectivity of addition to the *cis* olefin decreases with increasing solvent polarity—in acetic acid and nitrobenzene it disappears practically completely; on the other hand, the stereoselectivity of addition to the *trans* isomer is only slightly affected by solvent.[16b] Obviously, the intermediates have longer life-times in polar media, and so the tendency of **5a** to undergo internal rotation (driving force: relief of partially eclipsed interaction between the phenyl and methyl groups) can manifest itself more strongly. An alternative interpretation, competing nucleophilic attack from the *cis* side, is less likely, since the final products would be initially formed in the less favorable eclipsed conformation (Scheme 1). If the diastereo-meric cations were to interconvert by internal rotation faster than they reacted with bromine, both isomeric olefins should yield the same mixture of *threo* and *erythro* products[16c] (see Section 4.2.2).

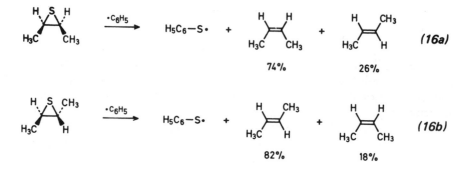

1.3.1.2   Free Radical Desulfurization of Episulfides

The fact that the desulfurization of episulfides with phenyl radicals is only *stereoselective*, not stereospecific (*16a, 16b*) rules out mechanism (*17*) and requires as immediate precursor of the end products a species able to undergo internal rotation (e.g., 7).[18]

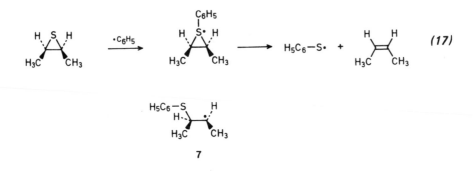

(17)

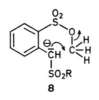

7

1.3.1.3   The S_N2 Reaction

The attack of the bromonium ion by the bromide ion, proceeding with inversion (*14*), exemplifies the most important stereospecific reaction of organic chemistry, S_N2 *substitution* at the tetravalent carbon, extensively studied by Ingold and Hughes.[19] It was recently shown by Eschenmoser[20] that even under the most favorable circumstances (see 8), namely, the possibility of an *intramolecular substitution of a "good" leaving group by a strong nucleophile at a primary carbon*, the reaction does *not* proceed with retention of configuration at the substitution center, as shown by arrows in 8. Instead, the carbanionic part of one molecule of 8 attacks *intermolecularly* the methyl group of a second molecule.

8

The accepted fact that S_N2 reactions proceed stereospecifically with inversion plays an important role in establishing *stereochemical correlations*. In turn, such correlations may provide insight into mechanisms of other reactions:

Example[21]

Assuming that the nucleophilic substitution 10 → 11 proceeds with 100% inversion, formation of the known (optically active) acid 13

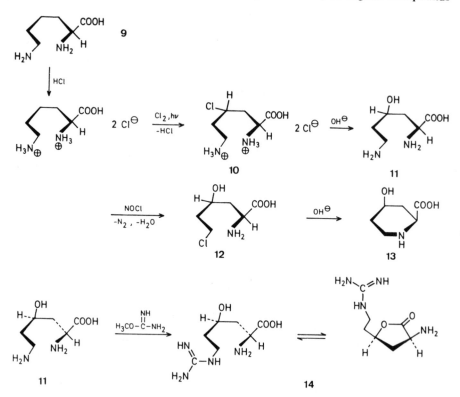

shows that the novel photochlorination of L-lysine 9 in strong acid proceeds not only regiospecifically but also stereospecifically, resulting in the formation of *erythro*-γ-chloro-L-lysine: [Such behavior may be due to polar effects. The doubly protonated substrate undergoes electrophilic attack by chlorine atoms (see Section 4.2.4.2) at the C—H bond furthest removed from the two positive centers. Since 11 (of now known absolute configuration) could be transformed into a new naturally occurring amino acid 14 (at that time of unknown absolute configuration), reactions 9 → 10 → 11 and the correlation of 11 and 13 enabled the establishment of the absolute configuration of 14 by total synthesis from starting materials of known configuration.]

1.3.1.4 Multicenter Reactions

Stereospecificity is often encountered in *multicenter reactions*. In these cases the transformation of reactants into products takes place *directly* (i.e., without occurrence of intermediate steps), with *simultaneous change in bonding at several centers* (concerted reactions).[22] For instance, retention of relative configurations at adjacent carbon atoms in epoxidation or addition of singlet carbenes (e.g., CCl_2) to olefins can

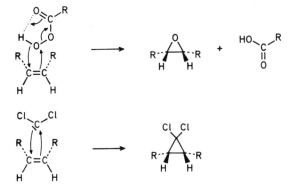

Scheme 2 Stereospecific addition of peracids and singlet carbenes (e.g., CCl₂) to olefins.

be easily explained if one assumes that the two new bonds are formed simultaneously (Scheme 2).

Triplet carbenes (e.g., fluorenylidene) react with olefins in a *stepwise* fashion because of the spin conservation requirement. The intermediate 1,3-diradical undergoes internal rotation followed by spin inversion, yielding a mixture of cis and trans cyclopropanes. (Scheme 3).

Well-known stereospecific cycloadditions are the *Diels-Alder reaction* (*18*) and *1,3-dipolar additions* (*19*).[24a]

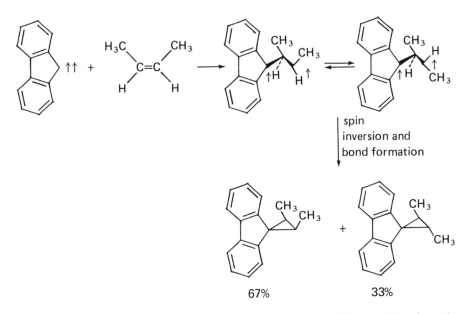

Scheme 3 Nonstereospecific addition of a triplet carbene (fluorenylidene) to *cis*-2-butene.

In many Diels-Alder reactions and in some 1,3-dipolar additions one observes not only *stereospecificity* with respect to the configuration of diene and dienophile, or dipolarophile, respectively, but also *stereoselective* formation of adducts (15, 16a) such that, unsaturated groups of both partners (here C_6H_5 and C=O) become bound on the *same* side of the newly formed ring (see Section 5.2.4.6).

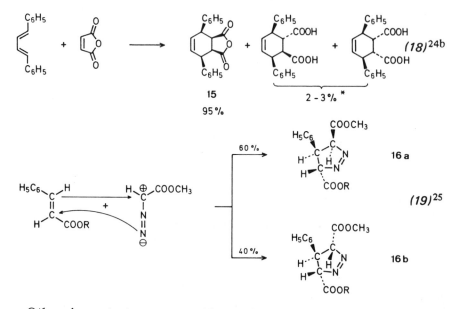

Other important *stereospecific* multicenter reactions are *electrocyclic reactions* [cyclization of an *n*-π electron system to a $(n-2)\pi + 2\sigma$ electron system (20)] and *sigmatropic rearrangements* [shift of a σ bond in conjugated systems, e.g., H in (21)].

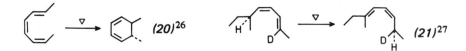

The reason for this behavior is the prevalence of *rigorous selection rules* that relate to the properties of the molecular orbitals involved in these transformations (Section 5.2.4).

1.3.1.5   Addition Reactions of Cyclic and Polycyclic Substrates

Addition reactions to many cyclic and polycyclic compounds lead quite often to that stereoisomer whose formation involves the least *steric*

---

*These are probably artifacts caused by hydrolysis and epimerization of the main product.

interactions and *conformational changes*. Thus, *cis*-bicyclo[3.3.0]oc-
tanes **16c** and **16d** are attacked from the exo side with rather high
degrees of stereoselectivity.[45]

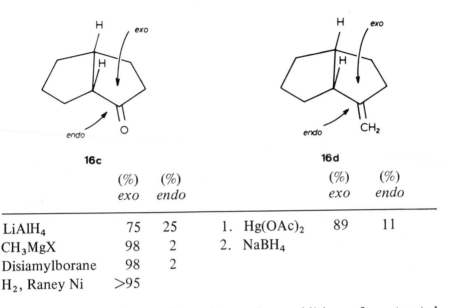

| | (%) | (%) | | (%) | (%) |
	exo	*endo*		*exo*	*endo*
LiAlH₄	75	25	1. Hg(OAc)₂	89	11
CH₃MgX	98	2	2. NaBH₄		
Disiamylborane	98	2			
H₂, Raney Ni	>95				

Special attention has been directed toward *exo* additions of unsaturated
bicyclo[2.2.1]heptane derivatives, where the attack of the reagent
takes place preferentially from the side of the shorter one of the two
bridges surrounding the reaction center. Different explanations have
been offered for this behavior. The simplest view is that reagents
approaching norbornene from the *exo* side encounter less steric hindrance from the hydrogen at C-7 than from hydrogen atoms at C-5 and
C-6 in *endo addition* (22).

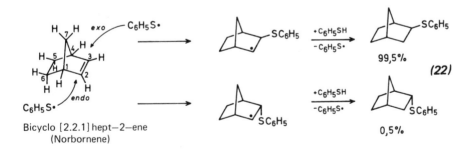

(22)

Alternative interpretations for the favored *exo* addition (which is
also observed with bicyclo[2.2.1]hepta-2,5-dienes lacking *endo* hydrogens at C-5 and C-6) are based on *torsion effects* and on the *principle of*

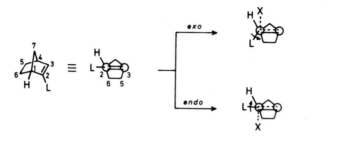

(23)

least nuclear motion (see Section 5.2.5): the motion of ligand L at C-2 in *endo* addition takes place in the sense of increasing eclipsing interaction with C-1—H, whereas in the case of *exo* addition this unfavorable interaction is less severe *(23)*.[28a] In addition, it can be shown that the sum of all changes in position of atoms involved both in reactant and product is smaller in *exo* reactions (rearrangements, additions and their reverse, dissociations) than in *endo* processes.[29] According to the *principle of least nuclear motion* (see Section 5.2.5) *exo* processes should thus be favored.

The preference for *exo* attack shown by electrophiles in reactions with norbornene and related hydrocarbons has also been ascribed to the electron density of the π orbital being greater on the *exo* face than on the *endo* face.[29a]

The effect of replacing the *syn* hydrogen at C-7 by a methyl group on addition reactions of norbornene may be a diagnostic tool for the mechanism and geometry of addition.[30]

One-center reactions in which a proton or a thiophenyl radical approaches C-2 or C-3 sidewise proceed preferentially by *exo* attack, even in the methylated substrate, and their rate is only slightly reduced *(24)*[30a]:

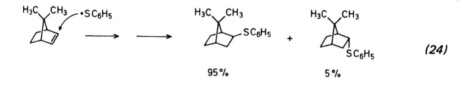

95 %   5 %

(24)

On the other hand, rates of *multicenter reactions*, primarily those involving three- and four-membered rings (e.g., epoxidation and hydroboration) are reduced a thousandfold. Addition of dichlorocarbene and 1,3-dipolar addition of phenylazide are completely inhibited by the presence of a *syn*-7-methyl group, because hindrance on both sides of the double bond is too great in the tight cyclic transition states. Such great rate differences in reactions of norbornene and 7,7-dimethylnorbornene may be used as criteria for the occurrence of cyclic transition states.

syn-7-Methyl groups in 2-oxo-norbornane guide one-center attacks by bulkier reagents (e.g., aggregat Grignard reagents, or complex hydrides) in the *endo* direction. In this instance hindrance by *endo*-hydrogen atoms at C-5 and C-6 is obviously smaller than hindrance by the methyl group (*25*):

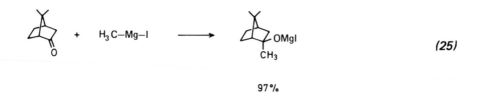

$$\text{(25)}$$

97%

Preferencial formation of axial alcohol 18a on reduction of 17 [(*26*)] is due to steric hindrance of axially approaching reducing agent by the axial methyl group.[31,462a]

Normally reduction of simple cyclohexanones (e.g., **19**) yields exclusively equatorial alcohols **20** (*27*); see also ref. 165c. Electronic factors may be responsible for this outcome (see Section 5.2.4.9).

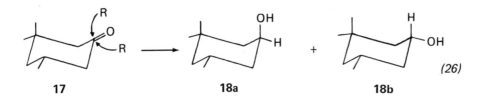

$$\text{(26)}$$

17   **18a**   **18b**

R	18a:18b
H_2/Pt	83:17
LiAlH(OR)$_3$	98: 2
LiAlH$_4$	55:45

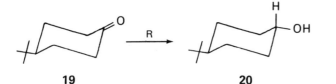

$$\text{(27)}$$

19   **20**

Stereoselective formation of *trans-diaxial dibromides* on bromination of rigid cyclohexenes is a consequence of the fact that of the two competitive S_N2 reactions on the bromonium ion (formed in a steroid at the sterically more accessible α-side$^{/16d/}$), the one leading to the more stable chair form (*a*) proceeds at a faster rate (*28*)*:

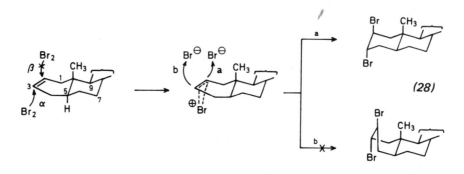

$$(28)$$

Similarly, opening of the corresponding epoxide proceeds both stereo- and regioselectively to give the diaxial 2β-bromo-3α-hydroxy derivative. Selective formation of the chair form is also observed in additions of nucleophiles to rigid γ-lacto dioxolenium ions (p. 2472 of ref. 33a).

1.3.1.6 Formation and Reactions of Delocalized Systems; The Stereoelectronic Factor

Any transformation leading to stabilizing electron delocalization is favored. Certain spatial arrangements of reaction partners (or a certain diastereomer) allow optimum overlap of the interacting orbitals (and thus maximum delocalization) and, as a consequence, result in faster reaction. Such a reaction is stereoselective or regioselective. The outcome may be described both in electronic terms (electron delocalization) and in steric terms (spatial arrangement) and is known as the *stereoelectronic factor*.

Using deuterated rigid cyclohexanones [(*29*), D instead of H_{axial} (H_a) or $H_{equatorial}$ (H_e)] Corey and Sneen[32] were able to show that on acid-catalyzed enolization *axial* hydrogen atoms are removed much faster than *equatorial* ones and that on *protonation of enols* axial hydrogen atoms are accepted faster than equatorial ones:

---

*Contrary to usual practice *a* and *b* in (*28*) point from the *electron-poor* to the *electron-rich* reaction partner; this is done to emphasize more clearly the change in position of the carbon atoms involved.

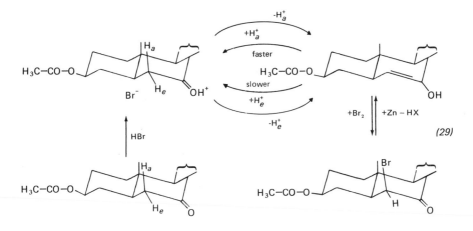

$$(29)$$

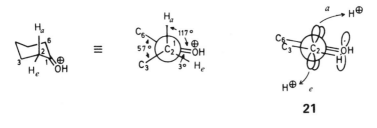

The reason for this behavior can be understood from the Newman projection of a protonated cyclohexanone (Fig. 1). The C-2—H_a bond is almost parallel to the axes of the p_z orbitals of the neighboring carbonyl group. When this bond begins to break (and increasingly so in later stages of the reaction) stabilization of the enol becomes possible, leading to an increased reaction rate. To remove H_e, energy is required to first rotate the C-2—H_e bond parallel to the axes of the p_z orbitals* and only then can H_e depart as described above for C-2—H_a.

Similarly, axial protonation of enol 21 does not require great changes in position of C-3 and H and the three orbitals involved remain largely parallel; equatorial protonation requires more substantial changes (H → H_a) and thus early loss of orbital overlap or formation of the unfavorable "twist" conformation.

Preferential abstraction of axial α-protons was recently proved[33] in base-catalyzed hydrogen-deuterium exchange in 4-*tert*-butylcyclo-hexanone.

21

Fig. 1 Newman projection of protonated cyclohexanone, its enol 21, and the two possible reaction paths for protonation of 21.

*The "twist-boat" conformation of the cyclohexane ring is thus achieved.

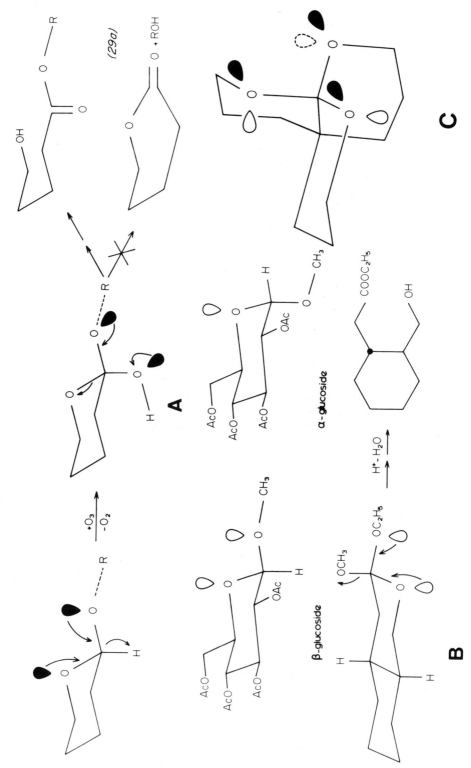

Stereoelectronic control has been recognized[33a] as a major factor in reactions proceeding through cleavage of a tetrahedral species in which *two* heteroatoms (oxygen and/or nitrogen, preferably negatively charged) and a leaving group are linked to the same carbon. Antiperiplanar orientation of the leaving group and a lone pair orbital *on each* of the heteroatoms is thought to provide a powerful driving force for expelling the leaving group at the same time causing this process to be stereospecific. Thus, in the reaction of acetals with ozone (*29a*), which is thought to occur via electrophilic attack of ozone on the acetal hydrogen (= leaving group) to give an intermediate like A, β-glucosides, which fulfill the orbital requirement, react readily, whereas α-glucosides, in which the lone pair orbitals of the ring oxygen can never be antiperiplanar to C−H, are inert. Tetrahydropyranyl ethers specifically lead to hydroxyesters, suggesting that decomposition of A is faster (because of the favorable orbital arrangement) than ring inversion.

The same stereoelectronic factor can be invoked to explain the specific hydrolysis of the rigid mixed orthoester B. Taking into account rotational preferences of the two alkoxy groups the first step must be removal of the methoxy group. Orthoester C whose C−O cannot be antiperiplanar to a lone pair orbital on each one of the two remaining oxygen atoms, is inert under these conditions. Other reactions besides those mentioned where this factor is thought to operate include the haloform reaction, the retro-Claisen condensation, the benzilic acid rearrangement, the Cannizzaro reaction, the Favorski rearrangement, and the Haller-Bauer reaction.[33b]

The following examples show the effect of the stereoelectronic factor on carbonium ion formation.

Of the two conformers of 9-chlorotetraphenyltribenzocycloheptatriene 22 and 23 (X = Cl), only the "axial" conformer reacts with nucleophiles, probably through the intermediacy of the corresponding tropylium ion 24 (*30*), whose seven-membered ring could adopt a shallow boat conformation.[34] [The bulky phenyl groups in the 1,4-position would prevent total planarization of the ion, just as they would prevent the flipping (22 ⇄ 23)].

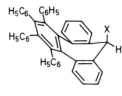

22

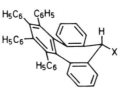

23

24

$$\textbf{22 (X = Cl)} \xrightarrow{\text{CH}_3\text{OH}} \textbf{24} + \text{Cl}^\ominus \xrightarrow{\text{CH}_3\text{OH}} \textbf{22 (X = OCH}_3\text{)} + \text{HCl} \qquad \textit{(30)}$$

The stereospecificity of the ionization and of the reaction of **24** with nucleophiles may be due to the fact that in **23** the axial C—X bond is almost parallel to the p_z orbitals of the benzosubstituents, whereas in **23** the C—X bond is almost *orthogonal* to the p_z orbitals. Preferred reaction of the axial bond of C-9 is also observed in free radical and anionic reactions of appropriate compounds **22** and **23**.[34a]

Triphenylchloromethane **25** undergoes hydrolysis 4×10^6 times faster than *t*-butylchloride. The reason is extensive electron delocalization in the triphenylmethyl cation and the possibility for the C—Cl bond to adopt an orientation parallel to the p_z orbitals in **25**. In contrast, in bromotriptycene **26** the C—Br bond and the p_z orbitals are orthogonal and electron delocalization is not possible either in the step leading to the formation of the ion, or in the ion itself. The consequence is total lack of reactivity in hydrolysis. Compound **26** does not undergo ionization even with alcoholic silver nitrate.

25　　　　　　　　　　　　　　**26**

Unlike **26**, the cyclopropane derivative **27** undergoes ionization very readily (10^2 times faster than *t*-butylchloride see also p. 290).[35] In **27** deviation from the stereoelectronically favorable conformation **28** is much smaller (60°, cf. **29**), compared to the C^+—C_6H_4 situation in **26**, and the cyclopropane rings can contribute significantly to the stabilization of the positively charged carbon atom in ion **29**, thus facilitating its formation.

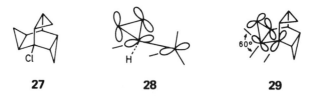

27　　　　　　　　**28**　　　　　　　　**29**

Enhanced tendency toward ionization in the *anti*-7-norbornenyl derivative **32**, compared to the *syn* isomer **30** and the saturated compound **31** is due to a favorable stereoelectronic situation. *Bishomocyclopropenyl* cations **33**[35a] formed as a result of *neighboring group participation* react stereospecifically with nucleophiles yielding a mixture of *anti*-7-norbornenyl derivative (attack of nucleophile from *a*) and *endo*-6-tricyclo[3.2.0.0^{2,7}]heptyl-derivative: **34** (attack from *e*) [(*31*); see also p. 160].

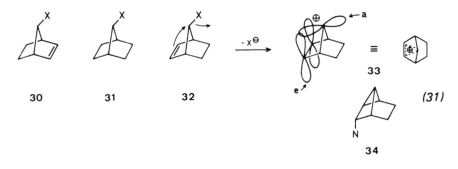

30   31   32   33

34   (31)

Preferential abstraction of the *exo*-4-proton in bicyclo[3.2.1]octa-2,6-diene 35 by base may be interpreted in the same way[36]: the incipient p_z orbital can interact even during its formation with the lower lobe of the p_z orbital at C-6, leading to the bishomocyclopentadienyl anion 36 [(*32*)].

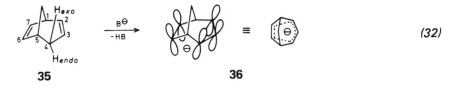

35   36   (32)

The reason for *exo selectivity* in the solvolysis of 2-norbornyl derivatives is controversial. The *exo*-2-tosylate undergoes acetolysis about 300 times faster than the *endo* isomer. According to Winstein this is due to the neighboring group participation of the C-1—C-6 σ bond in the ionization of the *exo* derivative resulting in the formation of the stabilized nonclassical 2-norbornyl cation 37. Recent results[37] show that this ion is best viewed as a "corner-protonated" cyclopropane (*33*).

In reversal of its formation nucleophiles can then attack 37 at C-1 or C-2 by S_N2 substitution of the partial (···) bond only from the *exo* side.

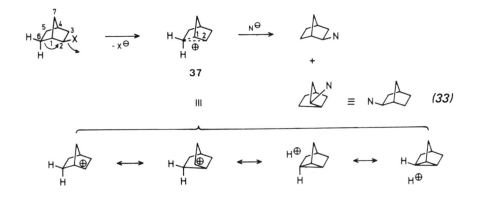

37   (33)

The intramolecular substitution of X by the C-1—C-6 bond is not possible in *endo* derivatives. These compounds must slowly form the less stable classical ion **38**, which undergoes a fast rearrangement to **37** [(*34*); see, however, p. 219].

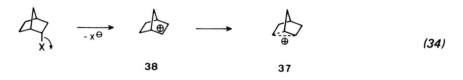

$$-X^\ominus$$

(34)

38   **37**

According to H. C. Brown, *exo* selectivity of 2-norbornyl derivatives in solvolysis as well as in other (e.g., nonionic) reactions is due to normal steric effects (see Section 1.3.1.5 and p. 136). Both approach and departure of groups from the *endo* side should be inhibited by the *endo*-hydrogen atoms at C-5 and C-6, whereas *exo* reactions proceed at a normal rate.[38]

1.3.2 Some Regioselective Reactions

Markovnikov addition of HX to unsymmetric olefins proceeds in such a way as to form the *more stable* cations on protonation of the olefin.

Hydrogen atoms in boranes have hydride ion character and accordingly, the stereospecific *cis* addition of boranes to unsymmetric olefins (hydroboration) proceeds *regioselectively* in the *anti-Markovnikov* sense (*35*). The trialkylboranes are formed in practically quanti-

(35)

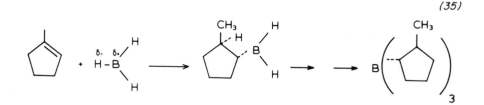

tative yield and can undergo further transformations into a variety of compounds (Fig. 2).[39] Elimination of HX (X = Hal or OTos) on treatment of halides or tosylates with base (*36*) proceeds in such a way as to form predominantly the *higher substituted* olefins ("Saytzev rule"). The stabilizing effect of alkyl groups on double bonds increases the rate of formation of the more highly substituted unsaturated system:

$$H_3C-CH_2-\underset{\underset{Br}{|}}{CH}-CH_3 \xrightarrow{\ OH^\ominus\ } H_3C-CH=CH-CH_3 \ + \ H_3C-CH_2-CH=CH_2$$

(36)

81%   19%

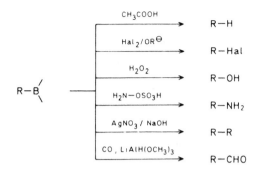

Fig. 2   Transformations of trialkylboranes.

On the other hand, *polar effects* are probably responsible for the *regioselective* formation of *less substituted terminal* olefins (*37*) on decomposition of tetraalkylammonium and trialkylsulfonium salts in the presence of base ("Hofmann rule"):

$$H_3C-CH_2-\underset{\underset{CH_3}{|}}{CH}-\overset{\oplus}{S}(CH_3)_2 \quad \xrightarrow{\ominus OC_2H_5} \quad H_3C-CH_2-CH=CH_2 \quad + \quad H_3C-CH=CH-CH_3 \qquad (37)$$

$$74\% \qquad\qquad\qquad 26\%$$

The polar effect of the positively charged heteroatom facilitates the abstraction of a proton by base; the abstraction precedes the elimination of the "bad" leaving group and the formation of the double bond. The transition state for double bond formation therefore has a carbanionicity that is better accommodated by carbon atoms without alkyl substituents.

Increased steric demand of bases used in elimination [e.g., $KO-C(CH_3)_3$] yields even in "Saytzev" substrates *larger* amounts of terminal olefins.

According to the theory of *variable E2 transition states*[40] Saytzev and Hofmann transition states represent different regions of the "spectrum" of E2 transition states, encompassing at one extreme the *deprotonation of carbocations* and on the other the *elimination of anions from carbanions* (Fig. 3).

Most interesting from a synthetic point of view are *regioselective alkylations, acylations*, and *condensations* of unsymmetric ketones. Since these objectives can often be achieved via equilibrium reactions, the methods used are described in the next chapter. At this point we mention only the cleavage of cyclopropylketones by alkali metals in liquid ammonia, as an example of regioselective enol formation. (Scheme 4). The C—C bond (*b*) more nearly parallel to the axes of the CO *p*-orbitals is preferentially broken,[41] since the delocalized enol

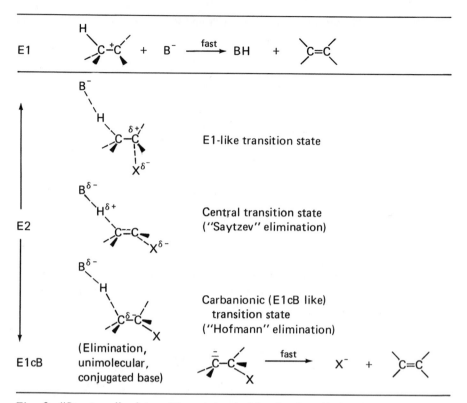

Fig. 3 "Spectrum" of transition states in bimolecular eliminations, from E1 to E1cB. There are E1cB reactions where elimination of X⁻ is the slowest step and the formation of the carbanion is reversible.

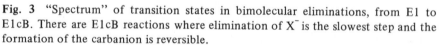

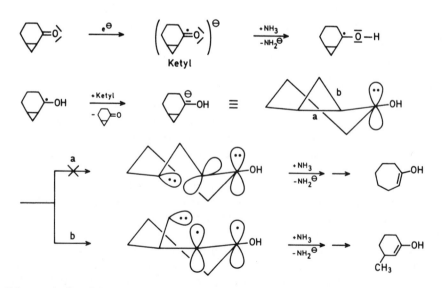

Scheme 4 Possible mechanism and stereoelectronic factor in the reductive (enol-forming) cleavage of cyclopropyl ketones.

system can be formed directly. This would not be the case if bond (*a*) were to break first.

Other regioselective reactions are transformations of ambident nucleophiles and the large class of aromatic substitution reactions that can be *ortho-*, *meta-*, or *para-*regioselective.

CHAPTER 2

Product Ratios

2.1 EQUILIBRIUM REACTIONS; THERMODYNAMIC CONTROL

Questions regarding *ratios* in which several possible products are formed or ratios in which starting materials and products are present at the end of an experiment can be answered most easily if one can be sure that *equilibrium* has been established between various products and/or between product and starting material. In such cases one speaks about *thermodynamic control*. (Proof that a reaction mixture is at equilibrium is obtained if the same mixture is formed starting with either the original reactants or any one of the products formed.)

Equilibrium constants may be calculated from (free) enthalpies and entropies of reaction (e.g., *38a, 38b*):

$$RT \ln K = -\Delta G = -\Delta H + T \Delta S \qquad\qquad (38a)$$

$$\frac{\mathrm{d}\ln K}{\mathrm{d}(\frac{1}{T})} = \frac{-\Delta H}{R} \qquad\qquad (38b)$$

It follows from (*38a*) that even if ΔG between two isomers in equilibrium (A $\rightleftarrows$ B) is only 20 kJ/mol, the favored isomer will be present to the extent of 99.9% (Table 7).

So if one wants to synthesize the *more stable* isomer, one must choose reaction conditions favoring *thermodynamic control*.

In natural products asymmetric centers very often have the more stable of the two possible configurations. One can take advantage of this fact in their synthesis; for example, the usual method of construction of the B-C-D system in steroids consists of base-catalyzed closure of ring B (Scheme 5). The reaction conditions allow enolization and reprotonation, and the final product will adopt the more stable con-

TABLE 7 Relationship Between ΔG (at 25 and 80°C), the Equilibrium Constant K, and Percent of the More Stable Isomer Present at Equilibrium (A $\rightleftarrows$ B)

More Stable Isomer (%)	K	ΔG^{298} (J/mol)	ΔG^{353} (J/mol)
50	1.00	0	0
55	1.22	499	591
60	1.50	1,006	1,194
65	1.86	1,538	1,818
70	2.33	2,103	2,493
75	3.00	2,728	3,230
80	4.00	3,440	4,077
85	5.67	4,307	5,099
90	9.00	5,455	6,461
95	19.00	7,311	8,656
98	49.00	9,662	11,443
99	99.00	11,409	13,513
99.9	999.00	17,145	20,309
99.99	9999.00	22,865	27,084

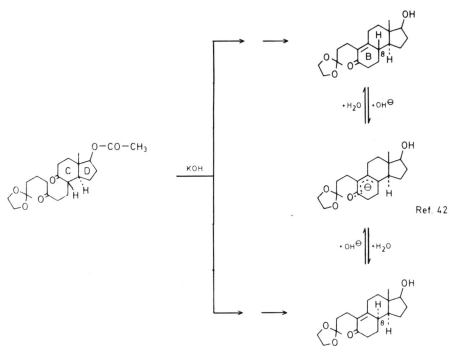

Scheme 5 Formation of the B ring of steroids by base-catalyzed condensation. The intermediate dienolate enables equilibration and accumulation of the more stable stereoisomer.

37

figuration, with the C-8 hydrogen in the β-position, even if the initially formed product were the α-isomer.

2.1.1 Calculation and Estimation of Equilibrium Constants[42a]; Relative Stabilities of Organic Compounds

Values of thermodynamic functions of many simple organic compounds can be found in tables or calculated from increments given in tables.[43] This allows the calculation of free energies of reaction and equilibrium constants.

In reactions of complicated organic molecules all one can usually do is to *estimate* reaction enthalpies ΔH (from changes in *bond energies, strain, conformation* and *delocalization energies*, and—in polar systems—*association and solvation energies*). [see, e.g., p. 448, Section 5.2.6.3)].

A short survey of such data is given in Tables 8-10.

TABLE 8 Bond Energies (kJ/mol)

H–H	436	C–H	415	C–C	348	C–F	486
O=O	499	N–H	390	C=C	612	C–Cl	339
N≡N	947	O–H	465	C≡C	838	C–Br	285
C≡O	1073	S–H	348	C–N	306	C–J	214
F–F	155	P–H	318	C=N	616	C–S	272
Cl–Cl	243	N–N	163	C≡N	894		
Br–Br	193	N=N	419	C–O	356		
J–J	151	O–O	147	C=O	746		
H–F	566	S–S	226				
H–Cl	432						
H–Br	364						
H–J	297						

TABLE 9 Delocalization Energies (kJ/mol)[a]

Benzene 151	Butadiene 15	Allyl Radical 52

[a]For a collection of delocalization energies of many more radicals and ions, see ref. 43g; for those of cyclic conjugated hydrocarbons, see ref. 43h.

To detect "special factors" (e.g., strain, resonance stabilization), it is advantageous to determine heats of reaction (from experimental and/or calculated heats of formation of the compounds involved) of hypothetical *isodesmic reactions*.[43d] These are bond separation reac-

TABLE 10 Steric Energies[a]

Compound	Energy (kJ/mol)
Strain energies[163a]	
Cyclopropane	115
Oxirane	115
Aziridine	116
Thiirane	74
Cyclobutane	110
Oxetane	110
Azetidine	110
Thietane	81
Cyclopentane	26
Tetrahydrofuran	28
Tetrahydropyrrole	28
Tetrahydrothiophene	7
Cyclopentanone	25
Succinic anhydride	19
Maleic anhydride	19
Succinimide	26
Cyclohexane	0
Tetrahydropyran	9
Piperidine	4
Tetrahydrothiopyran	0
Cyclohexanone	14
Glutaric anhydride	6
Cycloheptane	27
Cyclooctane	41
"Twist" conformation of cyclohexane	23

Free energy differences

1 Between axial and equatorial conformers of primary and secondary alkylcyclohexanes: *ca.* 8 kJ/mol (review: J. A. Hirsch, *Top Stereochem* **1**, 199 (1967)].

2 Between staggered and eclipsed conformations in acyclic groups $-\overset{\overset{\displaystyle CH_3}{|}}{\underset{\underset{\displaystyle R}{|}}{C}}-\overset{}{C}-$ (R = alkyl and aryl): *ca.* 8 kJ/mol.

[a]Strain energies of many alicyclic compounds may be found in E. M. Engler, J. D. Andose, and P. v. R. Schleyer, *J. Amer. Chem. Soc.* **95**, 8005 (1973).

tions in which a molecule is transformed into a set of simpler molecules, each containing one of the types of bond whose interaction in
the starting material is to be studied. To keep the stoichiometric
balance, the required number of the simple hydrides (CH_4, H_2O, NH_3)
is added. In general isodesmic reactions measure deviations from the
additivity of bond energies.

Example[43e]
Calculation of the interaction energy between vinyl group and cyclopropane ring in vinylcyclopropane.

1 Bond separation by addition of methane:

2 Isodesmic reaction with (quantum mechanically) calculated heats of
   formation*:

| -191.60520 | -39.72686 | -115.66616 | -115.66030 |

3 Stabilizing interaction energy: $\Delta H = 15$ kJ/mol.

In the same way the interaction energy between two vinyl groups in
1,3-butadiene was calculated as 24 kJ/mol. Although quantum
mechanical caculations cannot reproduce experimental data absolutely,
they are able to indicate differences qualitatively. In the present case,
the interaction of a vinyl group with a cyclopropane ring is similar but
smaller than its interaction with a second vinyl group.
   Obviously, *differences in entropy* can also affect the position of the
equilibrium. Basically, *association reactions* (a special intramolecular
version of these are *cyclization reactions*) have negative entropies of
reaction because of loss of translational and rotational freedom. Data
for gas phase processes may be found in tables in ref. 43. Scheme 5A[43f]
is illustrative.
   Relationships are more complicated in polar systems in condensed
phases. The association of two oppositely charged ions to give a neutral
molecule leads in this instance to an *increase* in entropy, since solvent
molecules are set free from their attachment to ions in the solvation

*In hartrees; 1 hartree = 2629 kJ/mol.

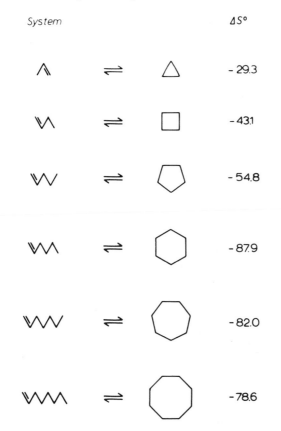

| System | | | $\Delta S°$ |

Scheme 5A   Entropy changes accompanying cyclization at 298 K (kJ/mol)

shell. Schleyer[44] determined the influence of entropy on the equilibrium between isomeric bicyclooctanes:

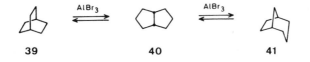

The order of relative stabilities is **41** > **40** > **39** at all temperatures (Table 11). Compounds **39** and **41** have practically identical enthalpies; the higher stability of **41** is the result of its higher entropy; **39** and **41** are both less strained than **40** (ΔH = − 8 kJ/mol), but this factor is compensated by the high entropy of **40**. Consequently, above 378 K **40** must be the stablest of the three isomers. The reason for the high entropy of **40** is common to all *cyclopentane* derivatives, namely, their *flexibility*, resulting in a large number of possible conformations.

TABLE 11   Amounts of Isomeric Bicyclooctanes (%)

Temperature (°K)	Bicyclo[2.2.2]- octane, 39	cis-Bicyclo[3.3.0]- octane, 40	Bicyclo[3.2.1]- octane, 41
296.8	3.66	32.95	63.35
317.4	3.41	37.20	59.35
321.8	3.36	37.85	58.78
331.7	3.33	39.78	56.86
345.4	3.11	42.34	54.50

The difference in entropy between **39** and **41** is a consequence of the higher symmetry of **39**. Symmetry contributions to entropy are $-R \ln \sigma$, where σ is the symmetry number.* Compound **39** has the symmetry number 6, whereas σ for **41** is 1. For this reason alone $\Delta S_{39 \to 41}$ should be 15 J/mol (Table 12). Substitution products of **39** have lower symmetry and thus are more favored relative to derivatives of **41** [(*38*)].

30%   70%   *(38)*

TABLE 12   Thermodynamic Parameters for the Isomerization of Bicyclooctanes

Reactant	Product	ΔG^{298}(kJ/mol)	ΔH(kJ/mol)	ΔS^{298}(J/K mol)
39	40	−5.5 ±0.4	+8.1 ±1.9	+45.7 ±6.7
40	41	−1.6 ±0.2	−7.9 ±1.0	−21.0 ±3.3
39	41	−7.0 ±0.5	+0.2 ±1.9	+23.0 ±10.5

It must be pointed out that the formation of derivatives of bicyclo-[3.3.0]octane is observed in many other eactions (Scheme 6).

The role played by entropy effects in these reactions is unknown. An important factor is undoubtedly the disappearance of unfavorable *transannular* interactions common to eight-membered rings. The transformation of various substrates into the 1-bicyclo[3.3.0]octyl cation in equilibrating "superacid" media may occur because it is the *only strain-free* tertiary cation possible in this series of carbon skeletons.

---

*The value of σ is equal to the number of undistinguishable positions adopted by the molecule (considered rigid) by simple rotations.

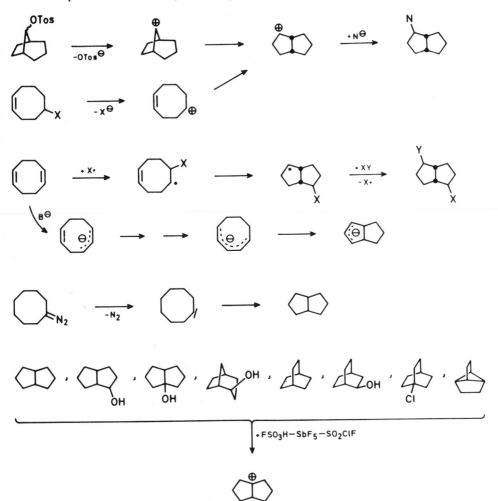

Scheme 6   Formation of the bicyclo[3.3.0]octane system.

A well-known example of differences in equilibrium composition due to entropy is the dissociation of the enol forms of acetyl acetone (*39*) and of dihydroresorcinol (*40*). Loss of entropy due to planarization to give the mesomeric anion does not occur in the cyclic com-

$$(\text{pK} = -\lg K = \frac{\Delta G}{2,3\,RT}; \quad (\Delta\text{pK})_{\Delta H} = \text{const.} = -\frac{\Delta\Delta S}{2,3\,R}$$

(39)

$$\text{pK} = 8,24; \quad \Delta S = -72,49 \text{ J/K} \cdot \text{mol}$$

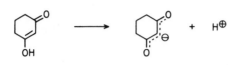

$$pK = 5{,}26; \quad \Delta S = -43{,}58 \text{ J/K} \cdot \text{mol}$$

pound. The resulting entropy difference contributes 1.5 units to the lowering of pK.[46].

The high precision of modern analytical methods makes it possible to determine the composition of equilibrium mixtures with great accuracy and thus to measure very small effects of *structure, substituents*, and *medium* on the free energy of molecules; this would be very difficult to do by thermochemical methods. For example, using $Fe(CO)_5$ as catalyst it was possible to equilibrate the isomeric pentenyl-methyl ethers[47]:

$$H_3C-O-(CH_2)_3-CH=CH_2 \quad : \quad H_3C-O-(CH_2)_2-CH=CH-CH_3 \quad :$$

$$H_3C-O-CH_2-CH=CH-CH_2-CH_3 \quad : \quad H_3C-O-CH=CH-(CH_2)_2-CH_3$$

$$= \quad 1 \quad : \quad 9 \quad : \quad 5 \quad : \quad 85$$

From the ratio (9:85) of the 3-pentenyl ether to the 1-pentenyl ether it follows that the latter is 6.3 kJ/mol more stable. Since it is known that an alkyl group stabilizes a double bond by about 10.5 kJ/mol, the stabilizing effect of an alkoxy group must be about 16.8 kJ/mol. The composition of the equilibrium mixture of 3-methoxy-1-phenylpropene (**42**) and 1-methoxy-3-phenylpropene (**43a, 43b**), obtained, for example, on treatment with $KO-t-C_4H_9$-DMSO shows that the stabilizing effect of a methoxy group on a C=C bond is slightly greater than that of a phenyl group.[48]

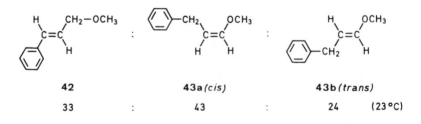

42	:	**43a** *(cis)*	:	**43b** *(trans)*
33	:	43	:	24 (23 °C)

Remarkably, the equilibrium between **42** and **43a** is established in a few hours, whereas the achievement of the equilibrium concentration of **43b** requires several days. This means that the more stable *cis*-1-methoxyphenylallyl anion (**44a**) is preferentially formed from **42** (Scheme 7).

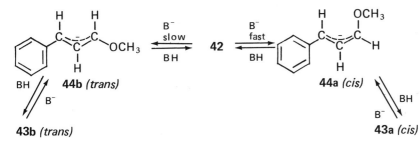

Scheme 7 Kinetic effects in the base-catalyzed isomerization of 3-methoxy-1-phenylpropene 42.

The *cis* isomer is also more stable in 1-alkyl allyl anions.[48a] Consequently, 1-*alkenes* form preferentially *cis-2-alkenes* on isomerization with $KO-t-C_4H_9$-DMSO.

Under the same conditions *cis-2-alkenes* form stereospecifically the rearranged *trans-alkenes*, and *trans-2-alkenes* rearrange stereospecifically to *cis-alkenes* (Scheme 8). The reactions may be explained by assuming the *cis,trans-1,3-dialkylallyl* anion as intermediate. Unfavorable steric effects prevent one of the two alkyl groups from assuming the favored *cis* position in the allylic system.

Statistical factors must be taken into consideration in the analysis of the equilibrium mixture of hexalines 45-48 obtained via the isomeric pentadienyl anions (Fig. 4).[49a] Independent of values of the rate con-

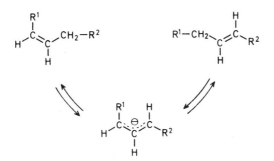

Scheme 8 Stereospecific double bond migration on treatment of 2-alkenes with base.

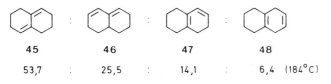

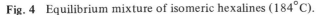

45		46		47		48	
53,7	:	25,5	:	14,1	:	6,4	(184°C)

Fig. 4 Equilibrium mixture of isomeric hexalines (184°C).

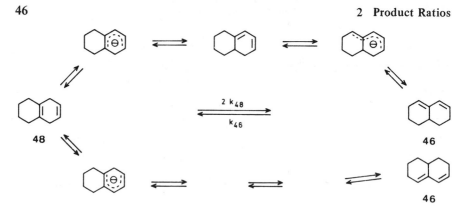

Scheme 9 Statistical effect in the equilibration of decalines **46** and **48**.

stants, the probability of attack of molecule **48** leading to **46** is twice as large as the probability of attack of **46** leading to **48** (Scheme 9). Consequently, at equilibrium:

$$[46] \cdot k_{46} = [48] \cdot 2 k_{48}$$

$$\frac{[46]}{[48]} = \frac{2 k_{48}}{k_{46}} = 2 K_{46-48}$$

To compute the contribution $K_{46\text{-}48}$ caused by stability differences the effective equilibrium constant $K = [46]/[48]$ must be divided by 2. The same is true for **46** and **47** in pairs 45-46, 45-47, and 47-48.

After making these corrections, the results show that **45**, with a more highly alkylated π system, is 5.5 kJ/mol more stable than **46** and 8 kJ/mol more stable than **47** and **48**, respectively.

In addition, since other possible isomers (e.g., **49**, **50**) could have been detected even if present in amounts as small as 1%, it follows that the planar *transoid* 1,3-diene **45** is more stable than the undetected identically substituted planar *cis*-1,3-diene **49** by at least 15 kJ/mol; (similarly, **46** must be more stable than **50** by at least 12 kJ/mol). This *cis* effect was already known for acyclic butadienes. The results obtained in the bicyclic system show that it is not due solely to the steric effect between the inner protons at C-1 and C-4 in *s-cis-butadiene* **51**.

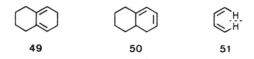

49　　　　　　　**50**　　　　　　　**51**

Another possible cause may be the transannular *van der Waals* repulsion between the π-electron clouds of the two *cis*-oriented double bonds.[50a]

TABLE 13 Equilibration of Cycloalkenones[51] (at Reflux Temperatures of the Isomer Mixtures)

Ring Size	Time Required to Establish Equilibrium (h)	$\%\Delta^2$	$\%\Delta^3$
6	Not determined	99	1
7	1	73	27
8	2	20	80
9	22	<0.3	<99.7

The stabilities of **47** and **48** (as well as those of the parent compounds 1,3- and 1,4-cyclohexadiene[49b]) differ very little. This seems to indicate that in six-membered rings the *cis* effect is strong enough to completely wipe out stabilization due to conjugation (*ca.* 15 kJ/mol[50b, 49c]).

A scale of base-solvent systems of different strengths for olefin isomerizations was set up by Bank.[50c]

Acid-catalyzed equilibration of cycloalkenones [(*41*), Table 13] shows that in eight- and nine-membered ("medium-size") rings *conformational effects* are stronger than *conjugative effects*:

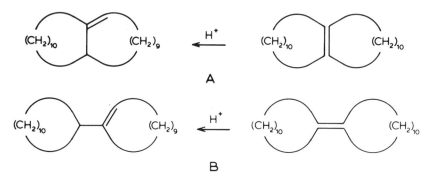

Reversal of the usual stability order is also observed when simple carbon-carbon double bonds are incorporated into medium-size rings: In the olefin pairs **A** and **B** trialkylated double bonds are preferred over tetraalkylated ones.[51a]

2.1.2   Further Examples of Thermodynamically Controlled Reactions

The transformation of allyl alcohols (e.g., 52) into cyclopropyl carbinols (53) and the rearrangement of the latter via cyclopropyl carbinyl cations (54) into homoallyl derivatives (*42*) is an important *homologation reaction*, best performed under themodynamic control, for example[52]:

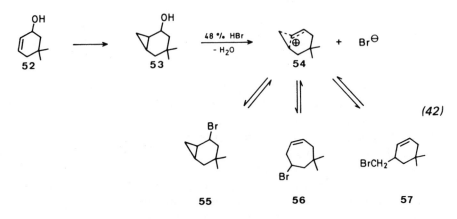

$$(42)$$

The desired compound 56 is obtained as the major product after 6 h (after 10 min the major product is 55, and after 3 days the only product is the most stable isomer 57).

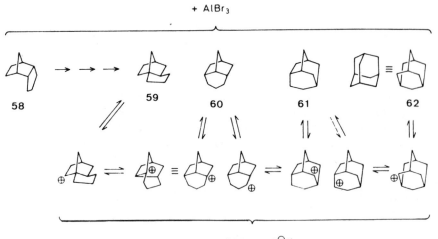

Scheme 10 Mechanism of the $AlBr_3$-catalyzed transformation of tetrahydrodicyclopentadiene into adamantane. The driving force of the reaction is the transformation of secondary carbocations formed by hydride abstraction from the hydrocarbons into isomeric cations of more favorable conformation via Wagner-Meerwein rearrangement.

One of the most remarkable equilibrations is the formation of *adamantanes* from a host of organic compounds in the presence of *aluminum halides* (Scheme 10). The reaction was first observed in the case of tetrahydrodicyclopentadiene (58), and some of the possible intermediate transformations (59 → 60 → 61 → 62) were recently investigated experimentally.[53]

Under similar conditions, such compounds as cholesterol, nujol, cedrene, caryophyllene, cyclohexane, and abietic acid, all yield *poly-alkyladamantanes*.[54] The statement "adamantane may be conceived of as a bottomless pit, into which all rearranging molecules may irreversibly fall"[55] is certainly true here: under the destructive reaction conditions all other compounds are transformed into polymers, tar, or volatile products. Only the very stable adamantanes survive and can be easily isolated, since no similar compounds are present.

Regiospecific alkylations, acylations, and condensations of *unsymmetric ketones* are of great importance in synthesis. A complicating factor is the ease of equilibration of various isomeric enolates. For instance, on addition of 2-methylcyclohexanone 63 to excess solution of triphenylmethyllithium, the proton from the unsubstituted α-position is abstracted preferentially,* yielding irreversibly a mixture of enolates 64 and 65 in the ratio 86:14.[57, /57a/] On addition of 10% 63 as proton transfer catalyst (in dimethoxyethane, 85°C) equilibrium is achieved after 3 h (composition of the equilibrium mixture 10% 64 + 90% 65).[57†]

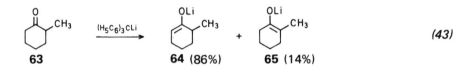

$$63 \quad \xrightarrow{(H_5C_6)_3CLi} \quad 64 \ (86\%) \quad + \quad 65 \ (14\%) \tag{43}$$

On methylation of a mixture of enolates from 63, the role of proton transfer catalyst is assumed by the primary products 66 and 67; this results in the formation of the enolates of 66 and 67, which in turn undergo methylation. Thus, the end product is a very complicated mixture[58, /58a/] (44).

Regioselective transformations may be achieved in such instances by means of the important methods of *selective activation* or *selective blocking*.[56] Such methods are based on the very high stability of *anions of 1,3-diketones* obtained under thermodynamic control. In the case of

---

*For steric and polar reasons.
†The strong coordination of lithium and oxygen makes the charge effect of the methyl group less important and its stabilizing effect on the double bond (p. 44) takes over.

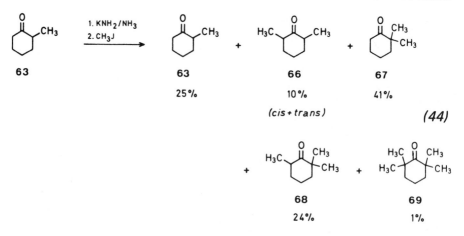

(44)

63 such an anion (70) can be formed only at the unsubstituted CH_2 group (Scheme 11).

After alkylation of 70, the *activating* group can be removed (a); alternatively 70 can be transformed into the dianion 71, which will be alkylated regiospecifically at the site of the less stable enolate (b).[59] Finally, 70 may be transformed into derivatives 72.[56] The CH_2 group of 63 is now *blocked*, and enolate formation and alkylation is possible only at the methylated site (c). The elimination of blocking groups is sometimes difficult to achieve.

As a last example of equilibration involving enolates we mention the epimerization of alcohol 73 to the sterically favored *trans*-substituted cyclopentane derivative 74[60] by means of the *retro-aldol condensation* (75 → 77) (Scheme 12).

2.2 IRREVERSIBLE PARALLEL REACTIONS; KINETIC CONTROL

Irreversible reactions are termed to be under *kinetic control*. In these instances product ratios are determined by the ratios of rate constants of simultaneously occurring reactions. Empirical and semiempirical relationships and qualitative rules concerning relative rate constants are discussed in the following chapters.

This section deals with formal relationships between relative rate constants and ratios of products or unreacted starting materials, the determination of relative reactivities based on these relationships, and the identification of reactive intermediates from characteristic product ratios.

2.2.1 Intramolecular Competition

The simplest situation occurs when a *single* reactant gives rise to *several* stable products via *irreversible* concurrent (parallel) reactions (45).

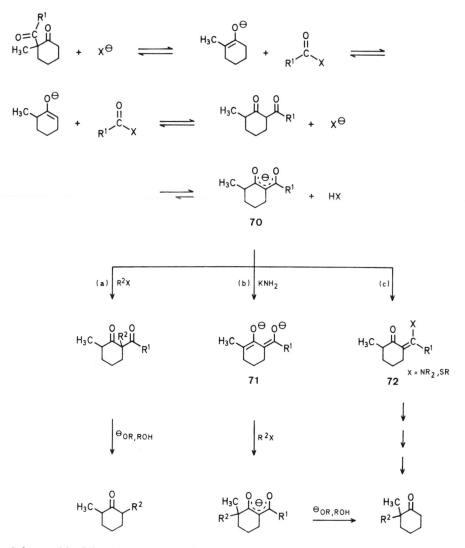

Scheme 11 Selective activation *(a)* and blocking *(b)*, *(c)* of the CH$_2$ group of ketones of general structure R^{1}R^{2}CHCOCH$_2$R^3 (e.g., **63**) via formation of anions of 1,3-diketones (e.g., **70**), which are strongly favored thermodynamically.

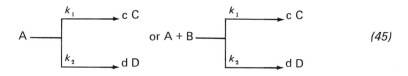

(45)

(Since one is dealing with concurrent reactions occurring at different sites of the same compound, one can speak about intramolecular competition.)

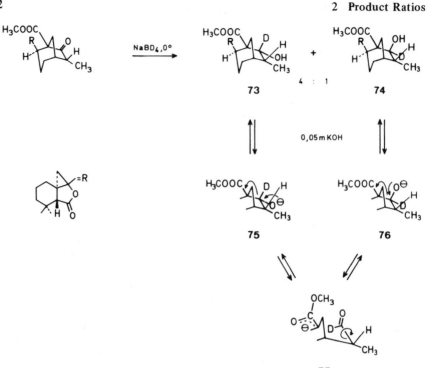

Scheme 12 Equilibrating epimerization of a β-hydroxyester via *retro*-aldol condensation.

$$\frac{d[C]}{dt} = ck_1 [A] \text{ resp. } \frac{d[C]}{dt} = ck_1 [A][B]$$

Since at any given time the same values of [A] and [B] are active in both concurrent reactions, one can divide by the corresponding equation for [D] and integrate. For $[C]_0 = [D]_0 = 0$:

$$\frac{[C]}{[D]} = \frac{ck_1}{dk_2}$$

This relationship allows us to determine *at any given time* the ratios of rate constants from product ratios, provided the products are stable under the reaction conditions.

The great advantage of such a procedure is obvious: instead of performing multiple measurements of concentrations at various times to obtain the absolute rates of formation of various individual products, one can perform a *single analysis* of the product mixture (which often can be done with great accuracy) and determine the relative rate

constants of their formation. For establishing structure-reactivity relationships, these values are as suitable as the absolute ones. This is
especially true for reactions where the determination of absolute rate
constants is difficult: very fast reactions and reactions with low repro-
ducibility (e.g., heterogeneous reactions, reactions involving very sensitive reagents). More details are given later.

By determining a single absolute rate constant, a whole series of
relative rate constants can be transformed into *absolute* ones.

Examples

1. *Partial reaction rate factors for reactions of substituted aromatic
   compounds (46):*

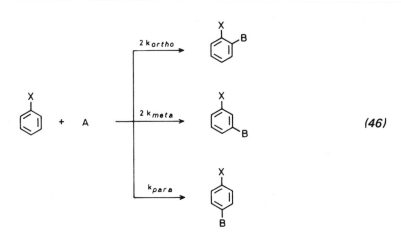

(46)

The ratio of isomeric substitution products gives the ratio of rate
constants:

$$\%\text{-}ortho:\%\text{-}meta:\%\text{-}para:2k_{ortho}:2k_{meta}:k_{para}$$

The *total rate constant* being

$$k_{C_6H_5X} \ (= 2k_{ortho} + 2k_{meta} + k_{para})$$

$$\frac{\%\text{-}ortho}{100} = \frac{2k_{ortho}}{k_{C_6H_5X}} \quad \text{and} \quad k_{ortho} = \frac{\%\ ortho \cdot k_{C_6H_5X}}{200}$$

For the same reactions of benzene (47):

$$\bigcirc + A \xrightarrow{\ 6\ k_H\ } \bigcirc^B \ ; \quad k_{C_6H_6} = 6\ k_H$$

(47)

The reactivity of *one ortho* position in C_6H_5X *versus* the reactivity of a *single* position in benzene can then be expressed by the so-called *partial rate factor* f^x_{ortho}:

$$f^x_{ortho} = \frac{k_{ortho}}{(k_{C_6H_6}/6)} = \frac{3 \cdot \%\text{-}ortho}{100} \cdot \frac{k_{C_6H_5X}}{k_{C_6H_6}}$$

Partial rate factors are larger (smaller) than 1 if in a given reaction a substituent activates (deactivates) the corresponding position. Their computation involves only the determination of *the ratio of isomers* formed and of the *relative reactivity* $k_{C_6H_5X}/k_{C_6H_6}$. The latter is easily obtained from product ratios of *intermolecular* concurrent reactions (see below).

A collection of partial rate factors for many electrophilic substitution reactions can be found in Stock and Brown[61] and Marino[133] (heterocycles).

Table 14 gives values for some reactions of toluene.

TABLE 14   Partial Rate Factors for Some Electrophilic Substitution Reactions of Toluene

	$f^{CH_3}_o$	$f^{CH_3}_m$	$f^{CH_3}_p$
Bromination			
(Br_2, CH_3COOH–H_2O, 25°C)	600	5.5	2420
Chlorination			
(Cl_2, CH_3COOH, 25°C)	617	4.95	820
Deuteration			
(D_2O–CF_3COOH, 70°C)	253	3.8	420
Acetylation			
(CH_3COCl, $AlCl_3$, $C_2H_4Cl_2$, 25°C)	4.5	4.8	749
Benzoylation			
(C_6H_5COCl, $AlCl_3$, $C_6H_5NO_2$, 25°)	32.6	5.0	831
Mercuration			
($Hg(OOC$–$CH_3)_2$, CH_3COOH, 50°C)	4.60	1.98	16.8
Sulfonation			
(14.8 M H_2SO_4, 25°C)	63.4	5.7	258
t-Butylation			
(t-C_4H_9Br, $SnCl_4$, CH_3NO_2, 25°C)	0	3.2	93.2

By multiplying the corresponding $f^x_{o(m,p)}$ values one can compute the reactivities of polysubstituted aromatic rings. Results based on this method (which corresponds to the summation of Hammett's σ constants) (see p. 106) are available primarily for polymethylbenzenes.[62]

For example to calculate the reactivity of hemimellitene (1,2,3-trimethylbenzene) 78 on chlorination in acetic acid at 25°C, one uses the following relation:

 78

Ratio of total reaction rates

$$\frac{k_{\text{hemimellitene}}}{k_{\text{benzene}}} = \frac{2f_A^{\text{hemimellitene}} + f_B^{\text{hemimellitene}}}{6}$$

The *partial rate factors* for A and B (see 78) are obtained by multiplying the f^{CH_3} values for this reaction corresponding to individual methyl groups (see Table 14).

$$f_A^{\text{hemimell}} = f_o^{CH_3} \cdot f_m^{CH_3} \cdot f_p^{CH_3}$$
$$f_B^{\text{hemimell}} = f_m^{CH_3} \cdot f_m^{CH_3} \cdot f_p^{CH_3}$$

The calculated value $k_{\text{hemimellitene}}/k_{\text{benzene}} = 8.37 \times 10^5$ is slightly higher than the experimental one of 4.58×10^5. The discrepancy is probably steric in origin.

The simplest model for electrophilic aromatic substitution, which thus has great theoretic importance, is the protodetritiation reaction (*48*):

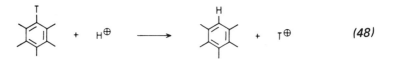 *(48)*

The literature lists more than 200 partial rate factors for this reaction.[63]

2. *Determination of migratory aptitudes of substituents.* Carbocations, other cations, and carbenes readily undergo 1,2-shifts of substituents and it is important to have a scale of *migratory aptitudes* for various groups.

If several groups migrate irreversibly, as it is the case in the *pinacol rearrangement*, relative migratory aptitudes can be determined from product ratios and total rate constants (Scheme 13).[64]:

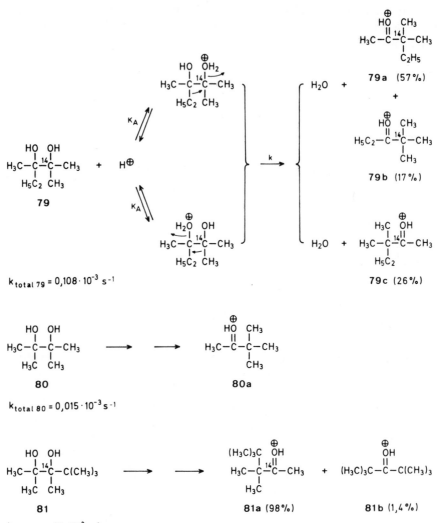

$k_{\text{total 79}} = 0{,}108 \cdot 10^{-3}\ s^{-1}$

$k_{\text{total 80}} = 0{,}015 \cdot 10^{-3}\ s^{-1}$

$k_{\text{total 81}} > 15 \cdot 10^{-3}\ s^{-1}$

Scheme 13 Mechanism, products, and rate constants in the pinacol rearrangement.

Partial rate constants, k_p for migration of individual groups can be determined from

$$k_p = \frac{\%\ \text{migration product}}{100} \cdot k$$

where k is the acidity-independent rate constant of protonated glycols in the total reaction; k is related to the experimentally determined *total rate constant* k_{total} for **79** through the *equilibrium constant for proto-nation* K_A (see "Stationary States")

$$k = \frac{k_{\text{total}}}{K_A[H^+]}$$

Since the values for K_A are not known, values for k_{total} obtained at identical acid strengths can be used only to determine the *relative migratory aptitudes* (assuming that K_A of all glycols studied are the same).

Taking into account the statistical advantage for methyl migration in **80**, we can write

$$\frac{k_p^{C_2H_5}}{k_p^{CH_3}} = \frac{\%\text{-}79a \cdot k_{\text{total}}\,79}{\%\text{-}80a \cdot k_{\text{total}}\,80/4} = 16$$

$$\frac{k_p^{t\text{-}C_4H_9}}{k_p^{CH_3}} > 4000$$

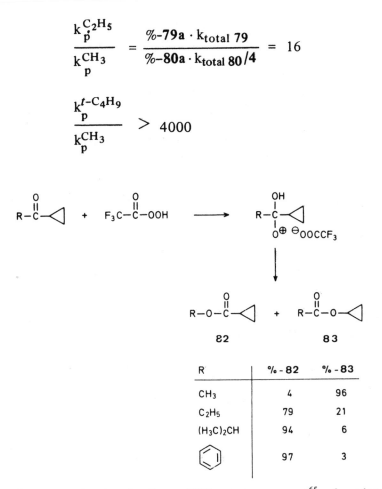

R	%-82	%-83
CH_3	4	96
C_2H_5	79	21
$(H_3C)_2CH$	94	6
⬡	97	3

Fig. 5 Product ratios in the Baeyer-Villiger rearrangement[65] of cyclopropyl ketones.

The increase in migratory aptitude in the order methyl $<$ ethyl $<$ *sec-* or *t*-alkyl[†] is the same as that observed in the *Baeyer-Villiger rearrangement* (Fig. 5). It is certainly due, at least in part, to the increased ability of the groups to bear a positive charge and agrees with the formulation of the transition states for these rearrangements as in *(49)*.

$$\begin{array}{c} R \\ \diagup \\ C-X^+ \end{array} \longleftrightarrow \begin{array}{c} R^+ \\ \diagdown \\ C=X \end{array} \longleftrightarrow \begin{array}{c} R \\ \diagdown \\ \overset{+}{C}-X \end{array} \tag{49}$$

$$X = C\diagup \quad \text{resp. O}$$

However, ratios of relative migratory aptitudes may be influenced by other factors (depending on the systems under consideration). This can be clearly seen on examination of percentages of products **80a**, **79b**, and **81b** (Scheme 13), which give the rates of migration of the methyl group in the pinacol rearrangement (Table 15).

TABLE 15   Rate of Migration of the Methyl Group in the Pinacol Rearrangement

Relative Rate of Migration of CH_3	Product Formed
1	**80a**
5	**79b**
>56	**81b**

In each case the methyl group migrates toward a $\overset{+}{C}(CH_3)_2$ group. However, in the first case its neighbor at the migration origin is a methyl group, in the second case an ethyl group, and in the third a *t*-butyl group. Obviously, these differences in rate are steric. The increasingly bulkier alkyl groups repel the methyl groups to an increasingly greater extent and oblige them to migrate.*

Other important factors may affect the tendency toward rearrangement[66]:

1 *Preponderant population* of certain conformations in the starting material. This factor is particularly important in *very fast* rearrange-

---

*A careful analysis of such a situation was recently performed by J. E. Dubois and P. Bauer, *J. Amer. Chem. Soc.* **98**, 6993, 6999 (1976).
[†] See, however: E. Wistuba and C. Rüchardt, Tetrahedron Lett. 4069 (1981).

ments when the migrating and leaving groups must adopt a *trans*-coplanar (antiperiplanar) arrangement (stereoelectronic factor).

2 *Steric effects* in transition states.

3 *Relative stabilities* of the unrearranged and rearranged cation[67] (see also Section 5.2.2.2).

In general, one can assume the sequence aryl > H > alkyl. However, depending on the dominating factor, relative migratory aptitudes may vary with the nature of the reaction or even from one compound to the next.

Turro[68] discussed the influence of steric factors on the ring enlargement of trimethylcyclopropanone with diazoalkanes. On addition of diazoethane four diastereomeric diazonium betaines may be formed (84, 86, 88, 90). They are represented in Scheme 14, each one along with one of its rotamers.

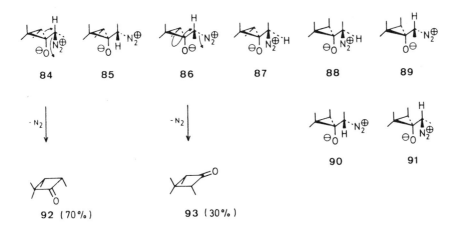

Scheme 14 Formation of cyclobutanones from trimethylcyclopropanone and diazoethane via diazonium betaines.

The products are cyclobutanones 92 and 93 formed in yields of 70% and 30%, respectively, which means that the secondary alkyl group migrates in preference to the tertiary one. A possible interpretation is as follows:

1 Rearrangement and heterolysis of the $C-N_2^+$ bond are concerted and the C—C bond *trans* coplanar to the leaving group is the one to undergo migration.

2 The diastereomeric pairs of rotamers 88, 89 and 90, 91, which would be formed if diazoethane were to add at the face carrying the two *cis* methyl groups, are not being formed. The same is true for

85 and 87, which have an unfavorable 1,3-diaxial methyl-methyl interaction. This interaction would also prevent the buildup of significant equilibrium concentrations of **85** and **87** (from **84** and **86**) in case internal rotation were to compete with the decomposition of the diazonium betaines.

3 The unfavorable interaction of the methyl group in diazoethane and the *gem*-dimethyl group in cyclopropanone causes **86** to be formed more slowly than **84**. Consequently, the ratio 92:93 must then be equal to the ratio of rate constants for the formation of **84** and **86**.

Differences between **84** and **87** or **85** and **86** disappear when the reagent is diazomethane (*50*). The rotamers corresponding to **84** and **86** become nearly equivalent and the greater tendency toward rearrangement of the *t*-alkyl substituent can now manifest itself.

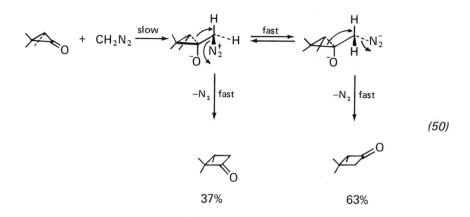

(50)

37%   63%

However, if the decomposition of rotameric diazonium betaines is very much faster than their interconversion by internal rotation, the outcome again will reflect only the ratios of rate constants for the formation of the two rotamers from ketone and diazomethane.

The ideal system for determining the characteristic migratory aptitudes of substituents R is thought to be the *dienone-phenol rearrangement*[69] (*51*).

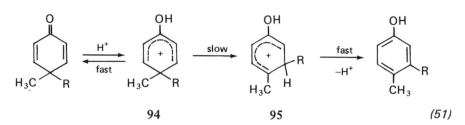

94   95   (51)

Differences in population of various conformations cannot occur in the planar rigid cation **94**; in addition, steric effects cannot be very important in **94** or in the irreversible rearrangement **94** → **95**.

In this system migration of the ethyl group takes place 25 times faster than that of the methyl group.

(Also, the 4-ethoxycarbonyl group migrates *faster* than the 4-methyl group. This could be the result of participation of additional electron pairs of the ester group, but a clear interpretation of this behavior is not yet available.[69a])

2.2.2 Intermolecular Competition

When two reaction partners (A and B) compete for a third (C) (*52*),

$$A + C \xrightarrow{\ k_1\ } D + \ldots\ldots$$

$$B + C \xrightarrow{\ k_2\ } E + \ldots\ldots \qquad (52)$$

[C] has again the same value for both reactions at any given time. When A and B are used in large excess, such that

$$[A] = [A]_o = \text{const.}\ (\approx 7 \cdot [C])\ ,$$

$$[B] = [B]_o = \text{const.}\ (\approx 7 \cdot [C])\ ,$$

product formation is given by

$$\frac{d[D]}{d[E]} = \frac{k_1[A]_o}{k_2[B]_o} \qquad (53)$$

and (for $[D]_0 = [E]_0 = 0$) the ratio of products at any given time is

$$\frac{[D]}{[E]} = \frac{k_1[A]_o}{k_2[B]_o}$$

If the concentrations of A and B are changing during the course of the reaction, the ratio of the two competing starting materials is given by

$$\frac{d[A]}{d[B]} = \frac{k_1[A]}{k_2[B]} \quad , \quad \frac{k_1}{k_2} = \frac{\lg \dfrac{[A]}{[A]_o}}{\lg \dfrac{[B]}{[B]_o}} \tag{54}$$

The quotient k_1/k_2 is called *competition constant*.

For clarity we specifically state the conditions for which equations (53) and (54) are valid:*

1 It is essential that the reaction orders with which A and B enter into competition for C be known and that the reaction order of C be the same with A and B (it is not unlikely that similar compounds would react by the same mechanism).[69b]

2 The two reaction products should not undergo further reactions (including back reactions) under the given reaction conditions.

3 The reaction rates should not be greater than the diffusion-controlled rate in solution. If they are greater, two situations can arise: in a *closed* system (e.g., C is a reactive intermediate formed *in situ*, see p. 67), A and B will be consumed equally fast (at a rate at which C is formed and diffuses to the reaction partners). If C is *added dropwise* to (A + B), then in the region where the drop touches the solution, the more reactive compound will be depleted and the less reactive one will be attacked *before* the normal ratio A:B can be reestablished by diffusion. Consequently, the difference in reactivity between A and B will appear to be smaller than it really is.

Such a situation seems to occur on nitration of benzene-toluene mixtures with nitrylium tetrafluoborate.[70] Whereas under normal conditions (HNO_3 + H_2SO_4) toluene reacts about 20 times faster than benzene, on dropwise addition of a solution of nitrylium fluoborate in sulfolane to the benzene-toluene mixture, toluene appears to react only 1.6 times faster than benzene.

Similarly, reaction of one mole of nitrylium tetrafluoborate with two moles of bibenzyl (as a model for toluene) does not give a statistical mixture of 37.5% mononitro derivative, 6.25% dinitro derivatives, and 56.25% bibenzyl[†] but substantially more dinitro derivatives, probably because the mononitro derivative reacts faster than new dibenzyl can be brought to the reaction site.

---

*An important amplification of (54) has been described by V. S. Martin et al., J. Amer. Chem. Soc., **103**, 6237 (1981).

[†]Probabilities: Attack at a single phenyl group: 0.25; dinitration: $0.25 \cdot 0.25 = 0.0625$; mononitration: $2 \cdot 0.25 - 2 \cdot 0.0625 = 0.375$.

The reliability of the competition method may be checked in the following way: first one determines experimentally the relative reactivities of two pairs of reagents (F,H and G,H) toward C (Scheme 15). From this one can compute the reactivity ratio of F and G and check this value against that obtained in a direct competition experiment of F and G with C. Also, on working with different initial concentrations one must obtain identical values for the competition constants. The greatest precision is achieved on working with compounds of similar reactivities.

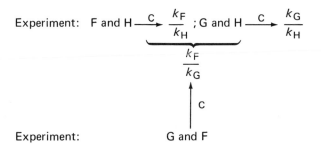

Scheme 15   Check on results in competition experiments: comparisons of directly and indirectly obtained values of competition constants.

The intermolecular competition method for determining relative reactivities of A and B versus C allows one not only to save time, but also to deal with the following situations:

1   Reactions are so *fast* that it is difficult, if not impossible to determine absolute rate constants by normal means. Examples are the addition of chlorine to olefins (*55*),[71] the addition of mercuric acetate to olefins (oxymercuration) (*56*),[72] and the reduction of aromatic compounds with solutions of alkali metals in liquid ammonia ("Birch reduction") (*57*).[73]

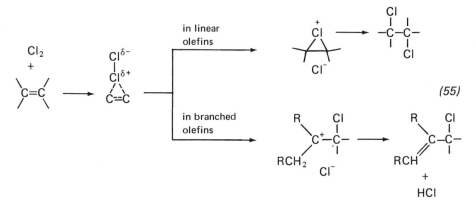

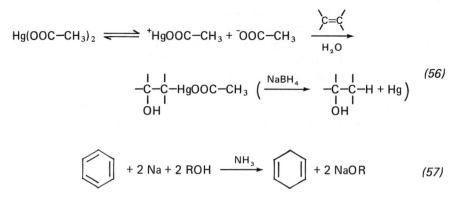

$$Hg(OOC-CH_3)_2 \rightleftharpoons {}^+HgOOC-CH_3 + {}^-OOC-CH_3 \xrightarrow[H_2O]{\overset{\backslash\ \ /}{\underset{/\ \ \backslash}{C=C}}}$$

$$\underset{OH}{-\overset{|}{C}-\overset{|}{C}-HgOOC-CH_3} \left(\xrightarrow{NaBH_4} \underset{OH}{-\overset{|}{C}-\overset{|}{C}-H} + Hg \right) \qquad (56)$$

$$\bigcirc\!\!| + 2\ Na + 2\ ROH \xrightarrow{NH_3} \bigcirc\!\!|\,| + 2\ NaOR \qquad (57)$$

Relative reactivities of olefins in chlorination (*55*) and oxymer-curation (*56*) (see Table 16), show that both reactions involve electrophilic attack on the double bond. In chlorination each additional alkyl group increases the reactivity by a factor of 10^1-10^2. The effect of alkyl groups is smaller in oxymercuration; in this instance unfavorable steric effects probably oppose favorable electronic effects. In both cases higher energy *cis* olefins react slightly faster than their *trans* isomers (see Section 4.2.3). In combination

TABLE 16　Relative Reactivities of Olefins

Chlorination	$k/k_{\bigcirc}$	Oxymercuration	$k/k_{\bigcirc}$
	0.02		6.6
	0.023		0.15
	32		48
	1		1
	1.23		0.56
	1		0.17
	220		1.24
	8600		0.061

TABLE 17   Relative Reactivities in Birch Reduction

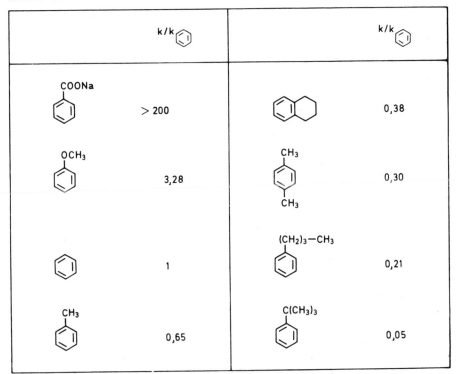

with the very easy reduction of β-hydroxyorganomercuric compounds, oxymercuration constitutes a synthetically very important method for *Markovnikov hydration* of olefins.

Birch reduction [(57)], involving the formation of delocalized radical anions, is favored by electron-withdrawing groups (COO⁻ and OCH₃), which facilitate the primary electron transfer, whereas alkyl groups have an unfavorable effect (see Table 17). The powerful effect of large alkyl groups may be due to steric inhibition of solvation (see p. 254, Section 4.2.3).

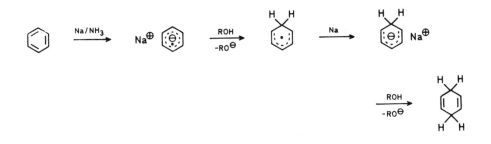

2 Conditions for reaction (A+C) *cannot be exactly reproduced* for reaction (B+C). This is, for instance, the case in the heterogeneous oxidation of alcohols with silver carbonate on Celite (Fetizon reagent) [(58) and (59)]. The following competition reactions were performed to determine the *kinetic deuterium isotope effect* k_H/k_D in this reaction[74] (see Section 4.2.4):

$$R^1 - \overset{\overset{\displaystyle H}{|}}{\underset{\underset{\displaystyle R^2}{|}}{\overset{14}{C}}} - OH \quad + \quad Ag_2CO_3 \quad \xrightarrow[-CO_2;-H_2O]{k_H} \quad \overset{\overset{\displaystyle R^1}{\diagdown}}{\underset{\underset{\displaystyle R^2}{\diagup}}{\overset{14}{C}}} = O \quad + \quad 2 \ Ag \qquad (58)$$

$$\text{(A)} \qquad\qquad\qquad \text{(C)}$$

$$R^1 - \overset{\overset{\displaystyle D}{|}}{\underset{\underset{\displaystyle R^2}{|}}{C}} - OH \quad + \quad Ag_2CO_3 \quad \xrightarrow[-CO_2;-DHO]{k_D} \quad \overset{\overset{\displaystyle R^1}{\diagdown}}{\underset{\underset{\displaystyle R^2}{\diagup}}{C}} = O \quad + \quad 2 \ Ag \qquad (59)$$

$$\text{(B)} \qquad\qquad\qquad \text{(C)}$$

Since the reaction product is the same in both competing reactions, (54) must be used. Its two unknowns [A] and [B] necessitate two independent measurements, which is achieved by radioactive labeling (^{14}C) of one of the reactants and measuring the specific activities of both the mixture of still unreacted reactants

$$\text{spec act} \ = \ \frac{[A]}{[A] + [B]}$$

and the mixture of products present at the same moment

$$\text{spec act} \ = \ \frac{[A]_0 - [A]}{[A]_0 - [A] + [B]_0 - [B]}$$

(The isotope effect due to ^{14}C is very small in comparison with the deuterium isotope effect and can be neglected.)

The determination of *kinetic acidities* of hydrocarbons by measuring rates of isotopic exchange with cyclohexylamine catalyzed by cesium cyclohexylamide could be achieved only in a *relative sense*, by comparison with the *rate of exchange* of added benzene, since this acid-base system is exceedingly reactive and sensitive [(60) and (61)][75]:

$$R - \overset{|}{\underset{|}{C}} - T \quad + \quad \langle \ \rangle - NH_2 \quad \underset{\overset{\displaystyle \langle \ \rangle - NH-Cs}{\rightleftharpoons}}{} \quad R - \overset{|}{\underset{|}{C}} - H \quad + \quad \langle \ \rangle - NH-T \qquad (60)$$

$$(61)$$

3   The determination of competition constants is the only method that provides quantitative data on reactivities in systems involving non-isolable *reactive intermediates* whose *stationary* concentration (p. 96) cannot be measured. A classic example is the demonstration of the *electrophilic character* of *dichlorocarbene* produced *in situ* (62) in the presence of a series of olefins (Table 18).[76] In this instance (63) all mixtures of (cyclopropane)$_1$ + (cyclopropane)$_2$ are formed at a maximum rate (i.e., in the absence of side reactions of CCl_2 this rate is equal to the rate of formation of dichlorocarbene). However, the *ratio* of the rates of formation of the two cyclopropane derivatives is given by (53).

$$HCCl_3 + {}^-O-C(CH_3)_3 \xrightarrow{\text{fast}} {}^-CCl_3 + HO-C(CH_3)_3$$

$$(62)$$

$${}^-CCl_3 \xrightarrow{\text{slow}} CCl_2 + Cl^-$$

TABLE 18   Relative Reactivities of Olefins versus CCl_2 (lg k_{rel} at 15 ±5°C)

	lg k_{rel}		lg k_{rel}
	1,73		0,27
	1,37		0,21
	0,92		0 (Definition)
	0,74		-0,86
	0,33		

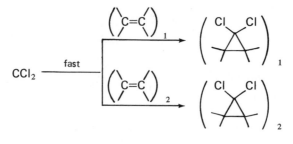

(63)

2.2.3 Characterization of Reactive Intermediates by Means of Competition Constants

Product mixtures of *thermodynamically* controlled reactions cannot yield any information with respect to the mechanism and possible occurrence of reactive intermediates. On the other hand, product ratios of *irreversible* reactions can yield important information.

Just as *stable*, isolable compounds can be identified by their *melting points*, *spectra*, and so on, *unstable*, nonisolable intermediates may be identified by their *competition constants* (valid only under certain conditions).

Thus, Huisgen and Knorr[77] reacted four different precursors of benzyne with mixtures of furan and 1,3-cyclohexadiene. Their results show that these Diels-Alder reactions must involve the same product-forming intermediate, namely, benzyne (Scheme 16).

Important examples may be found in the field of *carbenes* and *carbenoids* (carbenoids are intermediates that by themselves are not *free* carbenes, but possess similar reactivities).

Skell and Cholod[78] synthesized *free* dichlorocarbene by high vacuum pyrolysis of chloroform at 1400 K. Competition constants

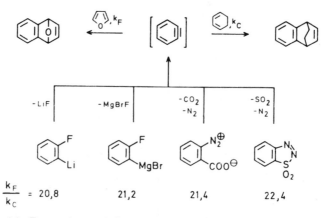

Scheme 16 Formation and characterization of benzyne (dehydrobenzene).

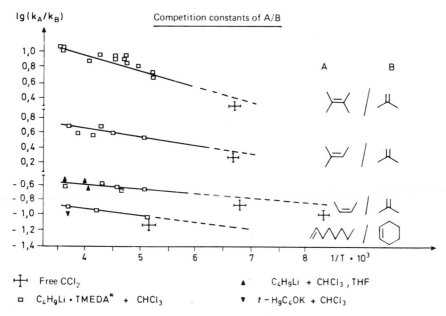

Fig. 6 Competition constants of CCl_2 generated from various precursors toward olefins A (B) (at different temperatures). [P. S. Skell and M. S. Cholod, *J. Amer. Chem. Soc.* 91, 6035 (1969) (ref. 78).] *N,N,N′,N′-Tetramethylethylenediamine

obtained in reactions of this species with pairs of olefins (at −150 to −78°C) lay practically on the line representing the temperature dependence of competition constants (Fig. 6) obtained at different temperatures from CCl_2 generated in various ways.

Finally, based on the identity of competition constants, it was possible to postulate that all methods of preparation listed below involve free dichlorocarbene.

1. $HCCl_3$ $\xrightarrow{1400\ °K}$ H + Cl + CCl_2 (64)

2. $KO-C(CH_3)_3$ + $HCCl_3$ $\longrightarrow$ $\longrightarrow$ CCl_2 (65)

3. RLi + $HCCl_3$ $\longrightarrow$ $\longrightarrow$ CCl_2 (66)

4. (structure) $\xrightarrow{h\nu, 195°K}$ (structure) C_6H_5 + CCl_2 (67)

5. $C_6H_5HgCCl_2Br$ $\rightleftharpoons$ C_6H_5HgBr + CCl_2 (68)

6. NaOH + $HCCl_3$ $\xrightarrow{NR_4 X}$ $\longrightarrow$ CCl_2 [79]   (69)

Dichlorocarbene generated by *phase transfer catalysis* [(*69*) and Scheme 17,][79] reacts with olefins that prove to be unreactive toward dichlorocarbene generated in other ways [e.g., the monodichlorocyclopropanation products of polyenes, which are deactivated by the inductive effect of the chlorine atoms (*70*)].

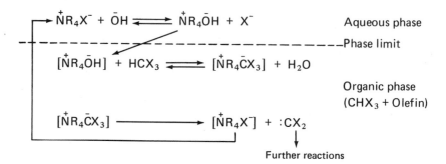

Scheme 17   Phase transfer catalysis. Ion pairs are in square brackets.

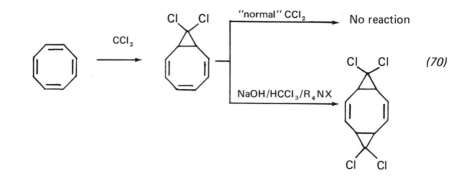

This does not mean that phase transfer catalysis involves a different species, with different reactivity, but rather that under these conditions CCl_2 shows an enhanced "net reactivity."[79] In phase transfer catalysis CCl_2 is generated in large excess at the site where it will undergo further reaction. Side reactions (e.g., with solvent or carbene precursors) play a much less important role relative to the reaction with the olefin, and as a consequence even unreactive olefins can be made to react.

On the other hand, the synthetically very important *CH_2 transfer reactions* behave differently. Using as an example the intramolecular competition of the two double bonds in 4-vinylcyclohexene [(*71*) and Table 19), Huisgen and Burger have shown that different systems yield different carbenoids $M-CH_2-Hal$ (M = metal).[80]

TABLE 19 Competition Constants of "CH_2"

Carbenoid Square	Solvent	Temperature °C	k^{96}/k^{97}
$C_4H_9Li + BrClCH_2$	Pentane	-50	1.15
$t\text{-}C_4H_9Li + BrClCH_2$	Pentane	-50	1.14
$t\text{-}C_4H_9Li + Br_2CH_2$	Pentane	-50	1.11
$t\text{-}C_4H_9Li + I_2CH_2$	Pentane	-50	1.10
$H_5C_2\text{-}AlCl_2 + CH_2N_2$	Pentane	-50	0.11
$Zn/Cu + I_2CH_2$[a]	Ether	35	0.51, 0.49, 0.50
$ZnI_2 + CH_2N_2$	Ether	35	0.51, 0.50, 0.50
$Zn/Cu + Br_2CH_2$	Ether	35	0.46, 0.44, 0.45
$ZnBr_2 + CH_2N_2$	Ether	35	0.45, 0.44, 0.44

[a]The Simmons-Smith reagent.

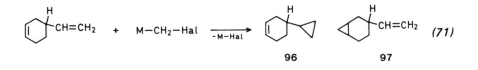

$$(71)$$

96   **97**

Various carbenoid precursors of the exceedingly unstable CH_2 are probably reactive enough to interact with the olefins, possibly by means of the transition state shown in Fig. 7.

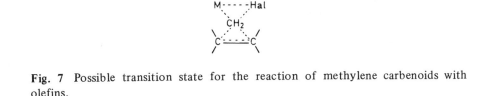

Fig. 7 Possible transition state for the reaction of methylene carbenoids with olefins.

On the other hand, dichlorocarbene is so highly stabilized by the mesomeric effect of the two chlorine atoms that only the free carbene is capable of reaction with olefins.

An example from carbonium ion chemistry is the acetolysis of isomeric chlorides 98-101 (Scheme 18). The reaction products 102 and 103 are formed in different ratios. This means that there must be subtle differences between the 9-decalyl cations generated from different precursors.[82]

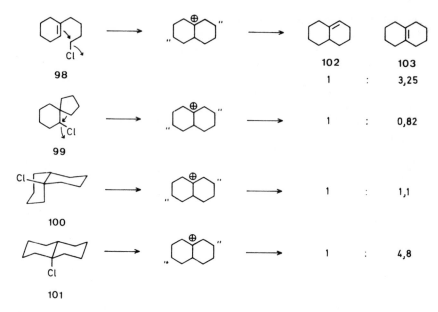

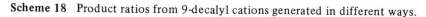

Scheme 18   Product ratios from 9-decalyl cations generated in different ways.

Basically one must take into account differences in conformation and in environment (position of the counterion, asymmetric solvation; see Scheme 19).

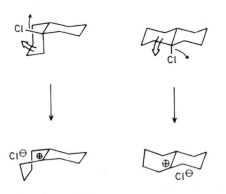

Scheme 19   Possible influences of the reactant on the conformation of the 9-decalyl cation.

This means that the elimination of the proton must be faster than the equilibration of such differences.[82a] Berson has observed similar effects, which he calls "memory effects," with cage compounds. A clear discussion of this work is given in ref. 82b.

Other examples where the method was applied are:

1 The work of Roth and Grimme et al.,[83] who showed that compounds **104-108** yield on thermolysis compounds **110** and **111** in the same ratio of 2:1, implying the intermediacy of the same diradical **109** (Scheme 20).

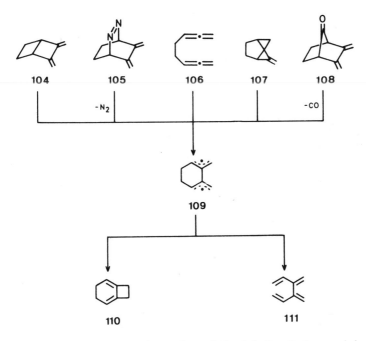

Scheme 20 Formation and transformation of the 2,3-dimethylenecyclohexadiyl-1,4 diradical (**109**).

2 Proof of occurrence of *singlet oxygen* both in the dye-sensitized photooxygenation of olefins and in their reactions with hypochlorite-H_2O_2 (Scheme 21).[83a]

3 Postulation of a *new radical chain mechanism* (*72*) for allylic and benzylic bromination with *N-bromosuccinimide* (NBS) based on the great similarity of relative reactivities of arylalkanes toward bromine and NBS[84]:

$$Br\cdot \ + \ \langle\!\!\langle\ \rangle\!\!\rangle\text{--}CHR_2 \longrightarrow HBr \ + \ \left[\langle\!\!\langle\ \rangle\!\!\rangle\text{---}CR_2\right]^{\cdot}$$

$$\left[\langle\!\!\langle\ \rangle\!\!\rangle\text{---}CR_2\right]^{\cdot} + Br_2 \longrightarrow Br\cdot \ + \ \langle\!\!\langle\ \rangle\!\!\rangle\text{--}CBrR_2$$

(*72*)

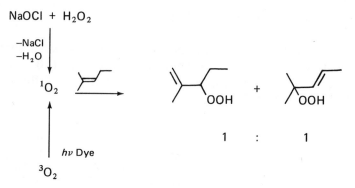

Scheme 21 Modes of formation of singlet oxygen and its reaction with olefins.

The role of NBS is to maintain a low stationary concentration of bromine (*73*), not to provide the succinimide radical for chain propagation (*74*), as previously believed.

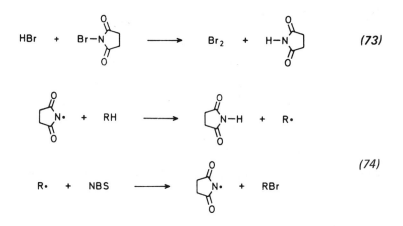

There is, of course, the possibility that two different species may coincidentally show identical competition constants, which would lead to false results. This becomes less likely when competition constants are determined under different sets of conditions. For instance, three different alcohols (R = CH_3, C_6H_5, C_6H_{11}, Scheme 22) were treated with Ag_2O/Br_2 to yield an intermediate in which the ratio in which diastereotopic[16d] hydrogen atoms H_a and H_b (distinguishable by deuteration) are abstracted by radicalic oxygen is identical to the ratio obtained on treatment of the alcohols with $Pb(OOC-CH_3)_4$, where the free radical character of the intermediate had been firmly established[85] (Scheme 22).

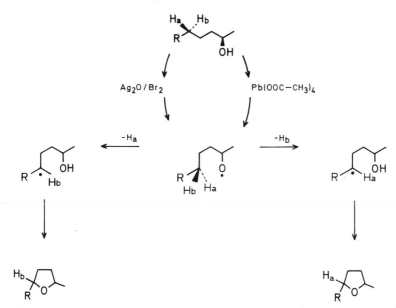

Scheme 22 Two methods for generation of oxyradicals from alcohols. The carbon radicals resulting from the 1,5-hydrogen shift are oxidized by the reagents to the corresponding cations. Nucleophilic attack of the cation by the HO group yields tetrahydrofuran.

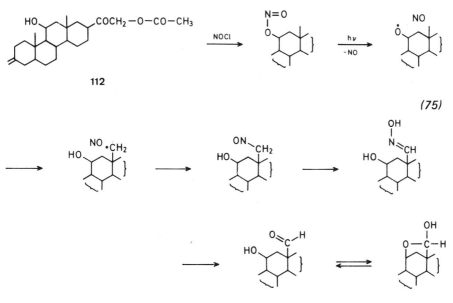

(75)

The same technique was used to substantiate the occurrence of the same intermediate ($R = CH_3$) on elimination of water from 2-hexanol in the mass spectrograph.

Abstraction of hydrogen atoms in δ-position by oxyradicals is synthetically important as a tool to functionalize unactivated C—H bonds. The first step in the "Barton reaction" is the transformation of alcohols into nitrite esters. Photolysis of these esters gives oxyradicals and NO; NO traps the carbon radical formed as a result of hydrogen transfer to give a nitroso compound, which tautomerizes to an oxime. An example is the synthesis of aldosterone-21-acetate (113) from corticosterone acetate (112) (75):

Comparison of competition constants obtained at different temperatures is less reliable, since the temperature dependence of these constants is generally small, (e.g., Fig. 6; see, however, Fig. 98). Consequently their determination requires very precise analytical methods. The temperature dependence of competition constants was used, for example in the determination of C—H dissociation energies D(R—H) (76)

$$R-H \rightarrow R\cdot + H\cdot \tag{76}$$

by means of competitive photobromination of mixtures of alkanes[86] (Scheme 23).

$$Br_2 \xrightarrow{\ h\nu\ } 2\,Br\cdot$$

$$Br\cdot \begin{cases} + & R^1H \xrightarrow{\ k_1\ } R^1\cdot + HBr \\[2mm] + & R^2H \xrightarrow{\ k_2\ } R^2\cdot + HBr \end{cases}$$

$$R^1\cdot (R^2\cdot) + Br_2 \xrightarrow{\ fast\ } R^1Br(R^2Br) + Br\cdot$$

Scheme 23 Chain reactions in the competitive photobromination of two alkanes $R^{1(2)}H$.

Formulation of rate constants by means of the Arrhenius equation [(77), see Section 4.1]

$$k = A \cdot e^{-E/RT} \tag{77}$$

gives

$$\left(\frac{[R^1Br]}{[R^2Br]}\right)_{T_1} = \left(\frac{k_1}{k_2}\right)_{T_1} = \frac{A_1}{A_2}\, e^{\frac{E_{R^2H}-E_{R^1H}}{RT_1}} \tag{78}$$

$$\left(\frac{[R^1Br]}{[R^2Br]}\right)_{T_2} = \left(\frac{k_1}{k_2}\right)_{T_2} = \frac{A_1}{A_2} e^{\frac{E_{R^2H}-E_{R^1H}}{RT_2}} \tag{79}$$

Using the *known* Arrhenius parameters A and E for butane one can obtain the Arrhenius parameters for hydrogen abstraction from cyclo-alkanes by bromine atoms. Values for D(R—H) were then obtained from the *empirical* relation (cf. ref. 86a)

$$E_{RH} = 0.87[D(R-H) - 347] \tag{80}$$

and C—H dissociation energies D(R—H) (kJ/mol) were found, as follows:

Cyclobutane   404
Cyclopentane   395
Cyclohexane   400
Cycloheptane   388

The higher values of D(R—H) for cyclobutane and cyclohexane may be associated with increase in ring strain on transformation of these hydro-carbons into their radicals. On the other hand, ring strain is reduced when radicals are formed from cyclopentane or cycloheptane.

2.2.4 Competition Between Internal and External Reactions; Formation and Trapping of Radicals; Absolute Values of Rate Constants of Very Fast Reactions

Other important kinetic situations arise when there is competition be-tween rearrangement (A → A') and *direct reaction* of A with a reactant B (Scheme 24).

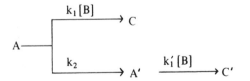

Scheme 24 Competition between rearrangement (to A') and bimolecular reaction (with B to C).

Under normal conditions (C and C' are stable, $[B] = [B]_0 = $ con-stant, reactions with B are not diffusion controlled), one can write for reactive species A and A' (when $k_1'[B] > k_2$):

$$\frac{[C]}{[C']} = \frac{k_1[B]_0}{k_2} \qquad\qquad (81)$$

The ratio of unrearranged product to rearranged product can be changed by varying $[B]_0$. Reactions of radicals able to undergo rearrangement (e.g., 114-116) (Scheme 25) offer the opportunity to describe some important ways for the formation and trapping of radicals.

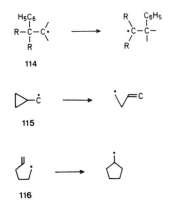

114

115

116

Scheme 25 Some radicals whose rearrangement can compete with trapping.

In each case rearrangement yields more stable radicals. Examples for the formation of radicals 114 are given in Scheme 26.

In pure 117 the ratio 122:120 = 1.3. On dilution of 117 with chlorobenzene (lowering of [117]) the ratio 122:120 becomes 4:1.

In the presence of trapping agents more powerful than 117 (higher values for k_1) 119 is no longer able to rearrange to 121. This is the case when the decarbonylation of 117 takes place in the presence of thiols,[88] in the peroxide-catalyzed chlorination of 123[89] (where Cl_2 appears to be a very effective trapping agent), and on reduction of 124 with tin hydride.[90]

In the analogous reduction of 125 rearrangement is observed.[91] The reason may be that rearrangement to the more stable radical 126 is faster than rearrangement to 121.

Similar reactions are known also in the cyclopropylcarbinyl system (Schemes 27-31).

When favorable substitutents are present (Scheme 29), or under the influence of conformational properties of bicyclic systems (Scheme 31), the rearrangement can go the other way—the allylcarbinyl system may rearrange to the cyclopropylcarbinyl system:

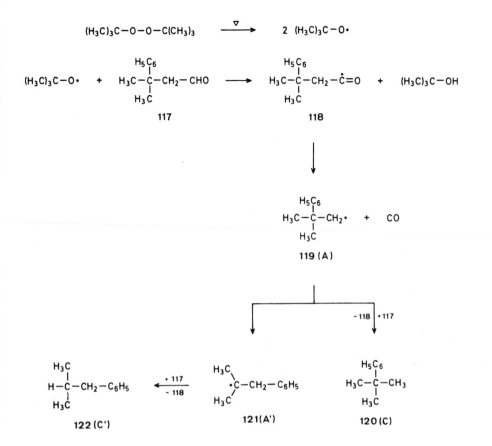

Scheme 26 Peroxide-initiated decarbonylation of 3-methyl-3-phenylbutanal 117.[87]

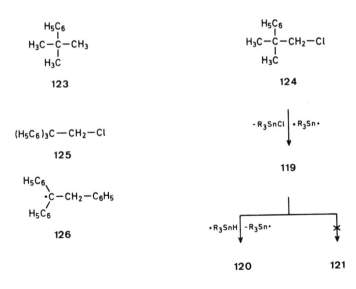

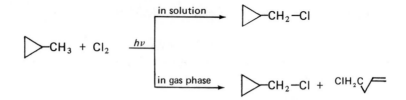

Scheme 27 Chlorination of methylcyclopropane in solution and in gas phase.[93]

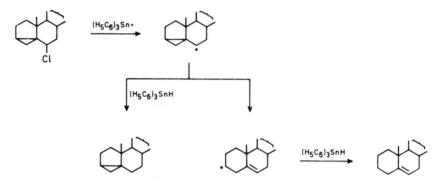

Scheme 28 Competition between hydrogen abstraction and ring opening in the tin hydride reduction of a steroid cyclopropylcarbinyl chloride.[94]

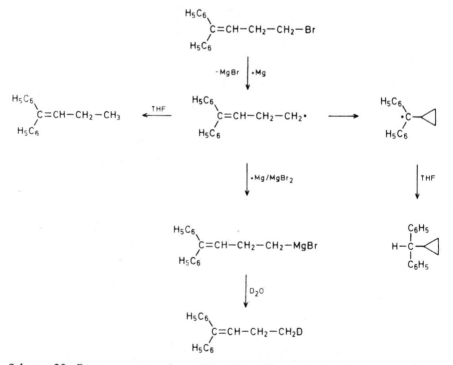

Scheme 29 Rearrangement of an allylcarbinyl (homoallyl) radical to a cyclopropylcarbinyl radical stabilized by two phenyl groups.[95]

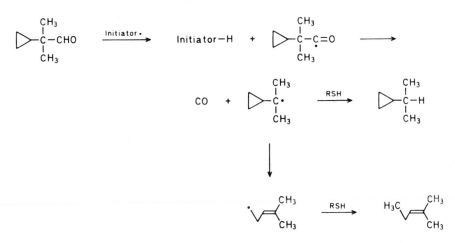

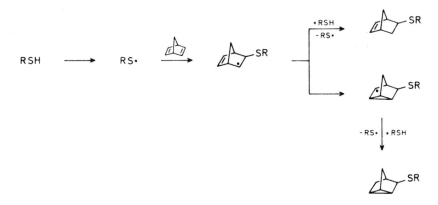

Scheme 30 Competition between hydrogen abstraction and ring opening of the dimethylcyclopropylcarbinyl radical (obtained by decarbonylation of 2-cyclopropyl-2-methylpropanal).[92]

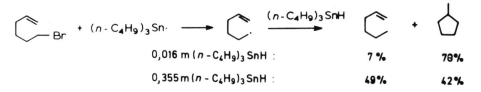

Scheme 31 In the bicyclo[2.2.1]heptane system cyclopropylcarbinyl isomers can be formed from the isomeric allylcarbinyl (homoallyl) radicals.[96]

A typical reaction of the 5-hexenyl radical is[97]:

$$
\underset{Br}{\diagup}\!\!\!\diagdown + (n\text{-}C_4H_9)_3Sn\cdot \longrightarrow \diagup\!\!\!\diagdown \xrightarrow{(n\text{-}C_4H_9)_3SnH} \diagup\!\!\!\diagdown + \bigcirc
$$

0,016 m (n-C$_4$H$_9$)$_3$SnH :	7 %	78%
0,355 m (n-C$_4$H$_9$)$_3$SnH :	49%	42%

So far it is not known definitively why the 5-hexenyl radical does not cyclize to the cyclohexyl radical, which should be more stable than the cyclopentylmethyl radical[98] both on steric grounds and because of its

Attack at tetrahedral carbon

Favored   *3 to 7-exo-tet*, e.g.,

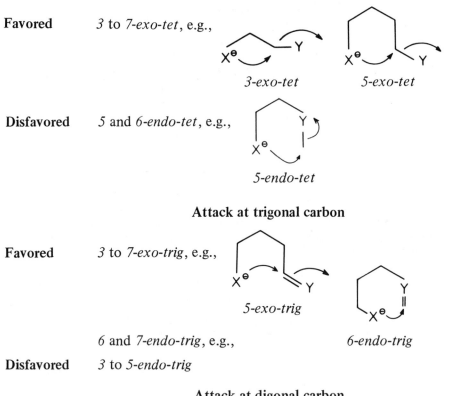

3-exo-tet   *5-exo-tet*

Disfavored   *5 and 6-endo-tet*, e.g.,

5-endo-tet

Attack at trigonal carbon

Favored   *3 to 7-exo-trig*, e.g.,

5-exo-trig

6 and 7-endo-trig, e.g.,   *6-endo-trig*

Disfavored   *3 to 5-endo-trig*

Attack at digonal carbon

Favored   *5 to 7-exo-dig*, e.g.,

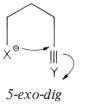

5-exo-dig

3 to 7-endo-dig, e.g.,   *4-endo-dig*

Disfavored   *3 and 4-exo-dig*, e.g.,

4-exo-dig

Scheme 31A Baldwin's rules for ring closure. The numeral prefix indicates the ring size; *exo* and *endo*, respectively, indicate whether the bond broken is exocyclic or endocyclic to the ring formed.

secondary nature. Possible reasons for kinetic favoring of the less stable radical are [99a, /99b/]:

1 A more favorable *entropy of activation* (see Section 4.1 and p. 41] for the formation of cyclopentanes.

2 Unfavorable *steric interactions* on formation of the chair form of the cyclohexyl radical;

3 If one assumes that the radical center at the end of the chain interacts with the π^* orbital of the double bond, examination of models shows that both orbital overlap and steric interactions are more favorable in the formation of the five-membered (**127**) than the six-membered ring (**128**) (the double arrow shows hydrogen atoms that repel each other). Structure **129** is totally unfavorable, since bonding interactions are canceled by antibonding ones.

4 It has been suggested that favorable interaction of the highest occupied molecular orbital of the carbon sigma skeleton with the lowest unoccupied molecular orbital (see Section 5.2.4.3) of the two electrons involved in bond formation is possible only for five-membered ring formation.[99c]

The favored mode of ring closure of the 5-hexenyl radical exemplifies one case covered by *Baldwin's rules for ring closure*.[99d] These rules were arrived at by considering how well a carbon chain can accommodate the ideal transition state geometry. They are shown in Scheme 31A. Here X must be a first row element. The rules apply not only to X⁻ as shown in Scheme 31A but likewise to X· and X⁺.

Table 20 contains several other formation reactions of the 5-hexenyl radical, as well as the trapping reactions operating in each case.

The observation[/104a/] that the mercury-sensitized photoreaction of 1,5-hexadiene yields bicyclo[2.1.1]hexane and not the isomeric bicyclo[2.2.0]hexane is consistent with the behavior of the 5-hexenyl radical on cyclization. It is very important that it was possible to determine the cyclization rate constant (and its temperature dependence) for the 5-hexenyl radical (at $-40°C$: $k = 1.7 \times 10^3$ s^{-1} at $40°C$: $k \sim 10^5$ s^{-1}).[105] Using (*81*), one can now determine the absolute rate constants

127

128   **129**

TABLE 20 Formation and Trapping of the 5-Hexenyl Radical R^1

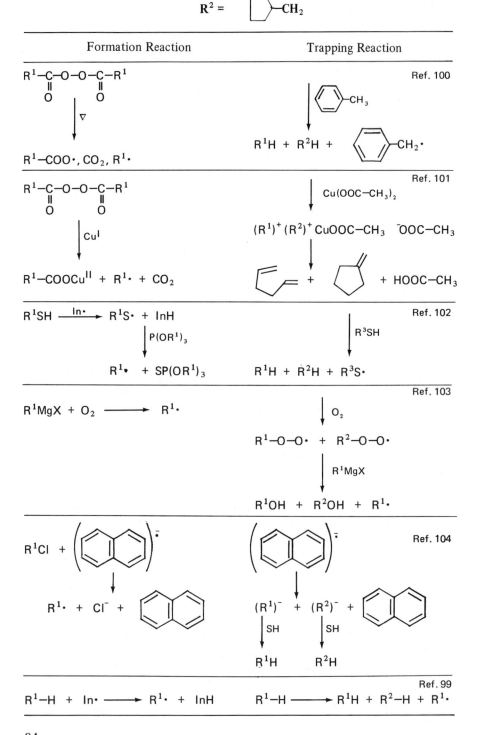

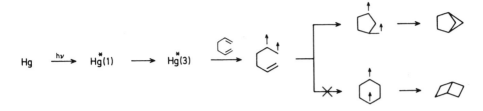

for the trapping reactions (e.g., ref. 101). These rate constants can be assumed to apply also to the corresponding reactions of other primary, sterically unhindered alkyl radicals. Of course, in a qualitative sense even mere cyclization of the 5-hexenyl radical can serve as a probe for alkyl radical reaction intermediates.[105a]

The linear dependence of the ratios of isomeric products (C and C′) on the concentration of reactant B (see Scheme 24) shows that the rearrangement to A′ proceeds as a discrete reaction and that C and C′ do not arise from the same delocalized and thus ambident precursor (e.g., 130).*

In this instance (83), the ratio of isomeric products, should be independent of the concentration of B, depending only on the electron distribution in 130:

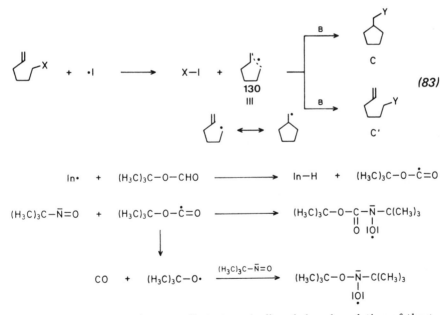

Scheme 32 Competition between "spin trapping" and decarbonylation of the *t*-butoxycarbonyl radical.

*Very rapid equilibration of A and A′ prior to reaction with B (which is equivalent operationally to 130) is also excluded.

Other situations that can be described by (81) are: competition between decomposition of a reactive free radical and its trapping by a diamagnetic stable molecule ("spin trapping," e.g., with nitroso compounds or nitrones; Scheme 32) to give a stable (easily measurable) radical.[106]

An essentially unanswered question in the chemistry of *organolithium compounds* is whether the reactive species are *monomers* or *oligomers* (e.g., hexamers in hydrocarbon solutions, tetramers in ether solutions). In competition reactions of the carbenoid *chloromethyllithium* [synthesized by *metal-halogen exchange* (Scheme 33) between butyllithium and bromochloromethane] with cyclohexene [to give

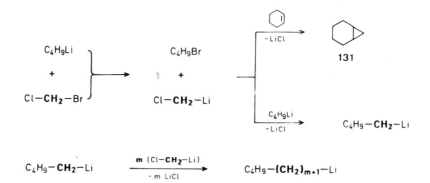

Scheme 33 Homologation of butyllithium and methylenation of cyclohexene (to norcarane) with chloromethyllithium.

norcarane (**131**)] and with butyllithium (to give its homologue) Huisgen and Burger[81] were able to show the involvement of two different, kinetically independent carbenoids.

The increase in concentration of cyclohexene affects only the *formation of norcarane* without a corresponding decrease in *homologation* yields. Treatment of 11 mmol of BuLi with 20 mmol of $BrClCH_2$ in 20 ml of hexane at various concentrations of cyclohexene gave the following results:

Cyclohexene (mmol)	Methylenation (% **131**)	Homologation (%)
0	0	59
4	15	60
21	34	58
41	39	53

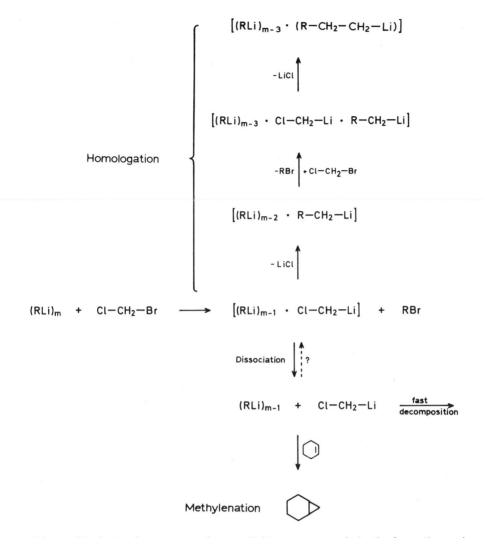

Scheme 34   Role of aggregates of organolithium compounds in the formation and subsequent competition reactions (homologation of RLi, methylenation of cyclohexene, decomposition) of chloromethyllithium.

The authors interpret these results in the following way. Chloromethyllithium is formed by halogen-metal exchange between chlorobromomethane and the butyllithium aggregate. It then becomes part of the aggregate and reacts by a Wurtz-type reaction to give homologues (Scheme 34).

Such aggregated (stabilized) chloromethyllithium does not react with cyclohexene. *Free Cl—CH$_2$—Li* generated by dissociation from the aggregate can undergo competitive *trapping* (methylenation of cyclohexene) and *decomposition* (to unknown products). The amount of trapping can be increased by increasing the concentration of cyclohexene.

In *solvolyses* performed in the presence of a nucleophile (Nu) dissolved in a large excess of solvent (*84*), [solv] is constant and the ratio [R—solv$^+$]:[R—Nu$^+$] should again show a dependence on [Nu] as expressed by (*81*).

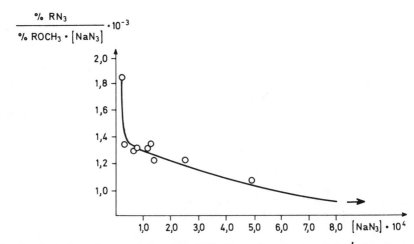

Thus, the quotient

$$\frac{[R-Nu^{\oplus}]}{[R-Solv^{\oplus}][Nu]} = \frac{k_2}{k_1[Solv]} \qquad (85)$$

should have the same value for all concentrations of Nu. Taking into account the salt effect of sodium azide on k_1, a situation of this sort is encountered in the S_N2 solvolysis of 2-propyl tosylate in aqueous alcohol in the presence of varying concentrations of sodium azide.[107]

The relationship between the ratio of rate constants of competing nucleophilic substitutions expressed by (*85*) and the ratio in which the

Fig. 8 Deviation from relationship (*85*) in methanolysis of 4,4′-dimethoxydiphenylmethyl-2,4,6-trimethyl benzoate [From C. D. Ritchie, *J. Amer. Chem. Soc.* **93**, 7324 (1971) (ref. 108).]

corresponding products are formed is often used as a criterion *for the identity of rate-determining and product-determining steps* in solvo-lyses [see schematic representation in *(84)*]. Deviations from *(85)* show that the two steps are different. The rate is then controlled by the nature of RX, whereas the properties of the intermediate ion pair(s) control the ratio in which the final products are formed (see also Section 5.2.6.3).

The deviation observed in the methanolysis of 4,4′-dimethoxydi-phenylmethyl-2,4,6-trimethyl benzoate (in the presence of sodium azide) (Fig. 8) was interpreted in the following way. At high concen-trations of sodium azide trapping occurs at the ion-pair stage with a low ratio $(k_2/k_1)_1$, whereas at low concentration of sodium azide the reacting species are the more selective mesomeric carbocations[108] $((k_2/k_1)_2$ higher) (Scheme 35).

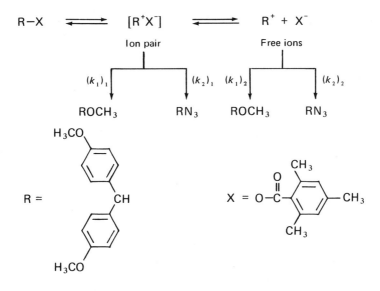

Scheme 35 Steps in heterolysis and product formation of substrates reacting via carbocations.

2.3 PRODUCT RATIOS IN CONSECUTIVE REACTIONS

2.3.1 Examples

The BF$_3$-catalyzed reaction of dicyclopropyl ketone with excess diazomethane[109] (for mechanism see p. 59) is typical of reactions where the product formed in the first step reacts in the same way at the same rate as the starting material. Scheme 36 shows reaction paths and amounts (%) of products obtained after work-up.

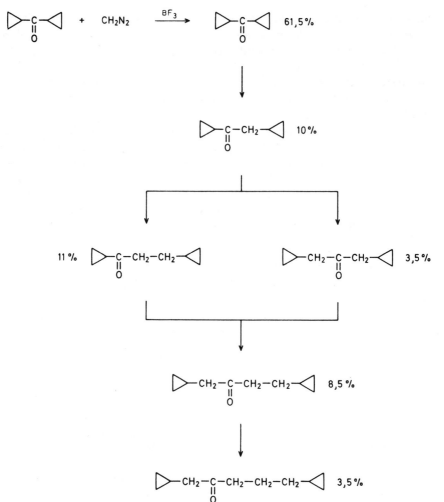

Scheme 36 Formation and amount (%) of ketones found in the reaction of dicyclopropyl ketone with diazomethane.

Other examples are *alkylations of ketones* and *primary and secondary amines*, the *Friedel-Crafts reaction of aromatic hydrocarbons*, some *oxidations of primary alcohols*, and many *isomerizations* (Scheme 10). The primary transformation of *the reactive groups* in these reactions is such that the repetition of the first step becomes possible. Similar situations arise in transformations of polyfunctional compounds.

The main problem in preparative work with such substances is to achieve a *specific partial transformation*. This requires the knowledge of rate constants for the individual steps and optimization of reaction times and concentrations. The kinetic problem is the determination of the rate constants for the various steps from the time dependence of concentrations. Exact solutions for the sets of differential equations are often impossible. We consider here only the simplest situations. A detailed treatment of the subject can be found in Bamford and Tipper.[110]

2.3.2 First-Order Consecutive Reactions

$$A \xrightarrow{k_1} B \xrightarrow{k_2} C \tag{86}$$

$$\frac{d[A]}{dt} = -k_1[A] \tag{87}$$

$$\frac{d[B]}{dt} = k_1[A] - k_2[B] \tag{88}$$

$$\frac{d[C]}{dt} = k_2[B] \tag{89}$$

$$[A] = [A]_o \cdot e^{-k_1 t} \tag{90}$$

$$\frac{d[B]}{dt} = k_1[A]_o \cdot e^{-k_1 t} - k_2[B] \tag{91a}$$

After rearrangement and substitution of $[B] = u \cdot v$, one can write

$$\frac{d[B]}{dt} + k_2[B] = k_1[A]_o \cdot e^{-k_1 t} \tag{91b}$$

$$u'v + v'u + k_2 \cdot uv = k_1[A]_o \cdot e^{-k_1 t} \tag{92}$$

$$u(v' + k_2 v) + u'v = k_1[A]_o \cdot e^{-k_1 t} \tag{93}$$

For $v' + k_2 v = 0$

$$u'v = k_1[A]_o \cdot e^{-k_1 t} \quad \text{and} \tag{94}$$

$$v = v_o \cdot e^{-k_2 t} \tag{95}$$

Combining (94) and (95) gives

$$u' = \frac{k_1[A]_o}{v_o} e^{(k_2 - k_1)t} \tag{96}$$

$$u = \frac{k_1[A]_o}{v_o(k_2 - k_1)} (e^{(k_2 - k_1)t} - 1) \tag{97}$$

$$[B] = \frac{k_1[A]_o}{k_2 - k_1} (e^{-k_1 t} - e^{-k_2 t}) \tag{98}$$

$$[C] = [A]_0 - [A] - [B] = [A]_0 \left[1 + \frac{1}{k_1 - k_2} (k_2 e^{-k_1 t} - k_1 e^{-k_2 t}) \right] \qquad (99)$$

For the case $k_1 \approx k_2$ the exponential in (97) is written as the series, which is cut off after the second term. This yields

$$[B]_{k_1 \approx k_2} = k_1 [A]_0 \cdot t \cdot e^{-k_2 t} \qquad (98a)$$

and

$$[C]_{k_1 \approx k_2} = [A]_0 [1 - e^{-k_1 t} (1 + k_1 t)] \qquad (99a)$$

Figure 9 illustrates the time dependence of concentrations of A, B, and C for a given combination of initial concentrations and rate constants.

A good example from practice is the aluminum bromide catalyzed isomerization (100) (Fig. 10, see also Scheme 10).

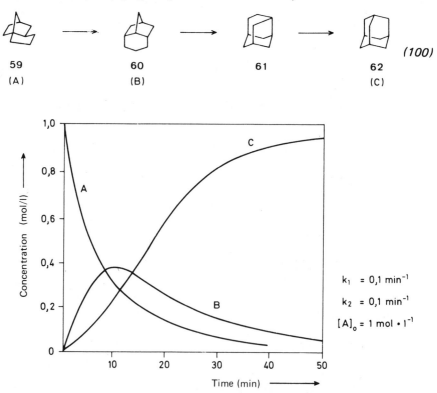

Fig. 9 Concentration-time dependence for a first-order consecutive reaction. (From K. Schwetlick, *Kinetische Methoden zur Untersuchung von Reaktionsmechanismen*, VEB Deutscher Verlag der Wissenschaften, Berlin, 1971.)

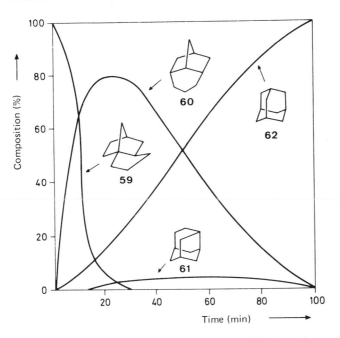

Fig. 10 Time dependence of relative concentrations of hydrocarbons present in (*100*). [From E. M. Engler et al., *J. Amer. Chem. Soc.* **95**, 5769 (1973) (ref. 53).]

Since protoadamantane (**61**) reacts much faster than it is formed, it cannot accumulate in large amounts (see p. 96). The maximum concentration of the intermediate B [see (*86*)], and the time when it is achieved, can be found by differentiation of the concentration-time function for B:

$$[B]_{max} = [A]_0 \left(\frac{k_2}{k_1}\right)^{\frac{k_2/k_1}{1-k_2/k_1}} \tag{101}$$

$$t_{max} = \frac{1}{k_1-k_2} \ln \frac{k_1}{k_2} \tag{102}$$

Figures 11 and 12 show the effect of k_1 and k_2/k_1 on the maximum for [B] ([A]$_0$ = 1); the larger k_1 and k_2/k_1, the faster the maximum is reached.

Application[111]

7-Phenylcycloheptatriene (**132**) undergoes a thermal 1,5-hydrogen shift to give 3-phenylcycloheptatriene (**133**), which in turn undergoes 1,5-hydrogen shifts setting itself in equilibrium with 1-phenyl-(**134**) and 2-phenyl-(**135**)-cycloheptatrienes [(*103*)].

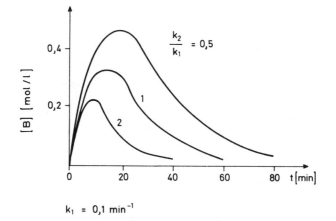

$$k_1 = 0,1 \text{ min}^{-1}$$

Fig. 11 Influence of k_2/k_1 on the value and on the time of the maximum concentration of intermediate B (*86*). (From K. Schwetlick, *Kinetische Methoden zur Untersuchung von Reaktionsmechanismen*, VEB Deutscher Verlag der Wissenschaften, Berlin, 1971.)

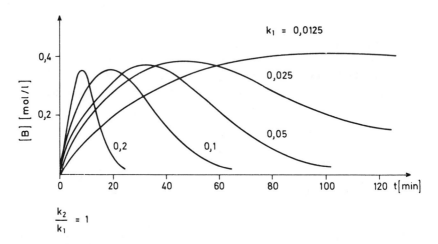

Fig. 12 Influence of the rate constant of the first step in (*86*) on the time at which the maximum for [B] is reached. (From K. Schwetlick, *Kinetische Methoden zur Untersuchung von Reaktionsmechanismen*, VEB Deutscher Verlag der Wissenschaften, Berlin, 1971.)

94

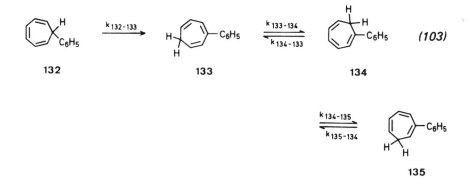

$k_{132\text{-}133}$ at 116°C is 2.03 × 10⁻⁵ s⁻¹; it can be estimated that $k_{133\text{-}134}$ at 116°C is 4.5 × 10⁻⁷ s⁻¹; in addition, the ratio $k_{133\text{-}134}$/ $k_{134\text{-}133}$ must be 2.7. Since the difference between $k_{132\text{-}133}$ and $k_{134\text{-}133}$ is large, the rearrangement of **134** to **133** can be neglected and the equations derived above can be applied to the system **132** → **133** → **134**:

$$t_{max} = 53 \text{ h}, \quad [133]_{max} = 93\%; \quad [132] = 2\%; \quad [134] = 5\%$$

After 70 h the reaction mixture consisted of 0.2% **132**, 91.4% **133**, and 8.4% **134**.

Similarly, substituting [in (87)] A = P, and $k_1 = k_r + k_p$, [in (88)] B = S, $k_1 = k_r$, $k_2 = k_s$, one can obtain the concentration-time relationships for the acetolysis of tosylate P and its isomer S, which is formed by partial rearrangement of P during acetolysis (Scheme 37).[112a] (This behavior is known as *"internal return,"* since the formation of S takes place in the *intimate ion* pair F and is not influenced by addition of external OTos⁻).

Using the expression for [P] [modified (90)] and [S] [modified (98)] it is possible to calculate the *"instantaneous rate constant"* k (which increases with time) [(104)] as the sum of k_p and k_s each multiplied by the corresponding fraction of P and S in the tosylate mixtures [(105)].

$$\frac{dx}{dt} = k(a - x) \tag{104}$$

where a = initial concentration of P
$\quad$ x = concentration of transformed tosylate (P + S) at time t

$$k = k_p \frac{[P]}{[P] + [S]} + k_s \frac{[S]}{[P] + [S]} \tag{105}$$

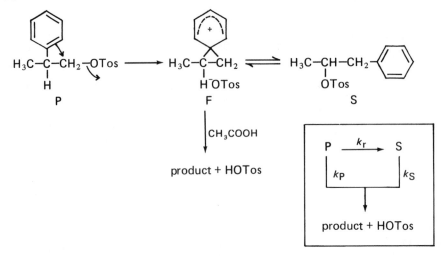

Scheme 37   Kinetic scheme for the acetolysis of 2-phenylpropyl tosylate (P) proceeding with "internal return" and partial isomerization to S.

Using the values of k, k_p (obtained by extrapolating k to zero time), and k_S (determined from the acetolysis of pure S; $k_S \approx 6k_p$), it was found that $k_r \approx 3k_p$.

2.3.3   Steady States

Many of the reactions described in preceding sections involved reactive intermediates. However, since in these instances k_2 is much larger than k_1 (86), the reactive intermediate is present in a very small steady concentration; it is consumed practically as fast as it is formed.

$$\frac{d[B]}{dt} = k_1[A] - k_2[B] = 0 \qquad\qquad (106)$$

The *steady state concentration* of the intermediate product is:

$$[B] = \frac{k_1}{k_2}[A] \qquad\qquad (107)$$

The rate of formation of the final product (unlike the case of reactions with relatively stable intermediates) *is equal to the rate at which A is consumed*:

$$\frac{d[C]}{dt} = k_2[B] = k_1[A] = -\frac{d[A]}{dt} \qquad\qquad (108)$$

The rate of the overall reaction (A → C) (86) is thus equal to the rate of its slowest step (A → B).

In reactions with accumulating intermediate products ($k_1 \simeq k_2$) the rate of formation of C is influenced by the rate constants of *both* steps [as can be seen by substituting (98) in (89)]. It must be emphasized that in a multistep reaction the reaction with the highest *rate constant* does not necessarily have the highest *rate*. If this reaction is preceded by a very unfavorable equilibrium [see (111) and (112)], it may become the slowest, even the rate-determining step.[112b]

The steady state method permits the kinetic analysis of many important systems whose exact mathematical treatment would be very difficult.

Examples

1. S_N 1-hydrolysis proceeding via free dissociated ions [(109)]:

$$RX \underset{k_{-1}}{\overset{k_1}{\rightleftharpoons}} R^{\oplus} + X^{\ominus} \xrightarrow{k_2[H_2O]} ROH \tag{109}$$

where $k_1 \ll k_{-1} + k_2$. The rate of formation of the final product is given by:

$$\frac{d[ROH]}{dt} = \frac{k_1[RX]}{1 + \dfrac{k_{-1} \cdot [X^{\ominus}]}{k_2[H_2O]}} \tag{110}$$

At the beginning of the reaction (when [X⁻] is very small) the hydrolysis is first order. As the reaction proceeds, deviations are observed because of an increase in [X⁻].

2. Reactions of C—H active compounds with electrophiles under the influence of bases [(111)]:

$$C\text{-}H + B^{\ominus} \underset{k_2}{\overset{k_1}{\rightleftharpoons}} C^{\ominus} + B\text{-}H$$

$$C^{\ominus} + E^{\oplus} \xrightarrow{k_3} C\text{-}E \tag{111}$$

The rate of formation of the final product is:

$$\frac{d[C\text{-}E]}{dt} = k_3 \cdot k_1 \frac{[C\text{-}H][B^-]}{k_2[B\text{-}H] + k_3[E^+]} \cdot [E^+]$$

When $k_3[E^+] \ll k_2[B\text{-}H]$, this becomes:

$$\frac{d[C{-}E]}{dt} = k_3 \cdot K \frac{[C{-}H][B^-]}{[B{-}H]} [E^+] \qquad (112)$$

The equilibrium of the first step is always restored. In catalytic reactions (e.g., aldol condensations, $E^+ = R_2CO$):

$$\frac{[B^-]}{[B{-}H]} = const$$

The scope of the steady state method can be determined by comparing the complete equation for [B] (98) with the one for [B] as steady state [the latter is obtained by substituting (90) in (107)]. The two expressions are identical when $k_2 \gg k_1$ *and* $t \gg 1/k_2$ (i.e., after an *induction period*). Figure 10 shows the buildup of the steady state concentration of the reactive intermediate protoadamantane **61**.

2.3.4 Second-Order Irreversible Consecutive Reactions

$$A + B \xrightarrow{\ k_1\ } C \qquad C + B \xrightarrow{\ k_2\ } D \qquad (113)$$

The new reagent B requires an additional differential equation,

$$\frac{d[A]}{dt} = -k_1[A][B] \qquad (114)$$

$$\frac{d[B]}{dt} = -k_1[A][B] - k_2[C][B] \qquad (115)$$

$$\frac{d[C]}{dt} = k_1[A][B] - k_2[C][B] \qquad (116)$$

This set of differential equations is required to describe, for example, a reaction as simple as exhaustive methylation of dimethylamine [(117)]:

$$2\ (H_3C)_2NH \ + \ CH_3I \ \xrightarrow{\ k_1\ } \ (H_3C)_3N \ + \ (H_3C)_2\overset{\oplus}{N}H_2 I^{\ominus}$$
$$\quad\ \ A \qquad\qquad\quad B \qquad\qquad\qquad\ C$$

$$\qquad\qquad\qquad\qquad\qquad\qquad\qquad\qquad\qquad\qquad (117)$$

$$(H_3C)_3N \ + \ CH_3I \ \xrightarrow{\ k_2\ } \ (H_3C)_4\overset{\oplus}{N} I^{\ominus}$$
$$\quad\ C \qquad\quad\ B \qquad\qquad\qquad D$$

Here one has to substitute $2k_1$ for k_1 in (114), since after slow methylation a second mole of $(CH_3)_2NH$ is consumed very fast on neutralization.

The solution of the set of equations (114)-(116) is possible only by introducing a new parameter $\theta = \int_0^t [B]\ dt$.

Substituting $dt = d\theta/[B]$, equations (114), (115), and (116) become

$$\frac{d[A]}{d\theta} = -k_1[A] \tag{118}$$

$$\frac{d[B]}{d\theta} = -k_1[A] - k_2[C] \tag{119}$$

$$\frac{d[C]}{d\theta} = k_1[A] - k_2[C] \tag{120}$$

Equations (118) and (120) correspond to those for first-order consecutive reactions (see above) and for $[C]_0 = 0$ the solutions are:

$$[A] = [A]_0\, e^{-k_1\theta} \tag{121}$$

$$[C] = \frac{k_1}{k_2 - k_1}\, [A]_0\, (e^{-k_1\theta} - e^{-k_2\theta}) \tag{122}$$

Substitution of (121) and (122) in (119) and integration gives:

$$\frac{[B]_0 - [B]}{[A]_0} = f_{k_2,\theta}\, (k_1) = 2 - 2e^{-k_1\theta} - k_1(e^{-k_2\theta} - e^{-k_1\theta})/(k_1 - k_2) \tag{123}$$

The relationship between θ and t is obtained by plotting $[B]$ versus t and graphic integration (Fig. 13).

The time dependence of $[CH_3I]$ (= $[B] \rightarrow \theta$) in the exhaustive methylation of dimethylamine could be determined by titration of unreacted amine $(CH_3)_2NH + (CH_3)_3N$.[113]

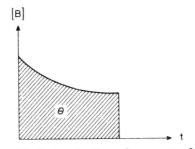

Fig. 13   Determination of parameter θ:

$$\theta = \int_0^t [B]\ dt$$

In the computation one must use the "modified"* equation (123). The value of k_2 is determined independently; k_1 is determined by plotting $f_{k_2,\theta}$ (k_1) versus k_1 (for a certain value of θ) and choosing the value of k_1 for which $f_{k_2,\theta}$ (k_1) = ([B]$_0$ - [B])/[A]$_0$.

The kinetic analysis of higher order consecutive reactions is also possible if one of the reagents is used in large excess. In this instance the reaction becomes a system of *irreversible first-order consecutive reactions*. Then k_1 for methylation of dimethylamine can be obtained by extrapolation of rate constants measured for various stages of the transformation to zero time (Fig. 14).

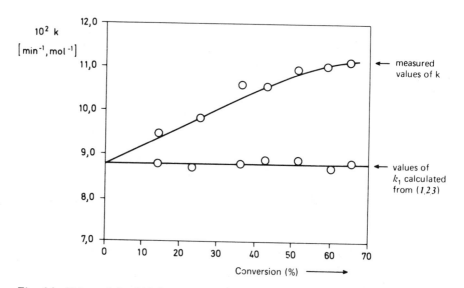

Fig. 14   Value of k_1 (117) determined by calculation and extrapolation of k to zero time. [From K. Okamoto, S. Fukui, and H. Shingu, *Bull. Chem. Soc. J.* **40**, 1920 (1967) (ref. 113).]

*Obtained by substition of $2k_1$ for k_1 in (118).

RATES

CHAPTER 3

Rates of Reaction

3.1 EMPIRICAL AND SEMIEMPIRICAL EQUATIONS CORRELATING STRUCTURAL CHANGES IN SUBSTRATES AND THEIR EFFECTS ON RATE CONSTANTS

Having considered what products may be formed in a given reaction, the next question to ask is at what *rate* these products are formed. Absolute values of rate constants for synthetically important reactions—involving larger molecules in solution—cannot be calculated by *ab initio* methods (i.e., on the basis of quantum mechanical considerations alone, without recourse to experimental data). For most of these reactions even empirical data give only *qualitative estimates* with respect to reactivity (see Section 5.2).

This chapter deals with situations that allow us to *quantitatively predict relative values* of rate constants from *empirical* or *semiempirical* relationships. Such relationships are important not only from a practical point of view—for example, storage of data, planning transformations of a given compound (or similar ones) for which they apply, but also for the classification of substituents, reagents, and solvents, the analysis of reaction mechanisms, and the identification of molecular interactions.

3.1.1 Aromatic Compounds; The Hammett Equation; Correlation of Reactivity and Quantum Mechanical and Inductive Parameters

The effect relative to unsubstituted substrates of substituents X in the *meta-* and *para* position on rate constants (k) and equilibrium constants (K) of side chain reactions is expressed by the well-known empirical Hammett equation (124)[114]

$$\log \frac{k_X}{k_H} = \rho \sigma_X \qquad (124)$$

TABLE 21 Values of Substituent Constants for Commonly Encountered Substituents (from ref. 114b and from sources mentioned in notes)[114c]

	σ_m	σ_p	σ_p^{+}[1]	σ_p^{-}[2]	σ_I[3]	σ_R[4]	F[5]	M[6]	$\sigma^*_{CH_2X}$[7]	E_s[8]
H	0	0	0	0	0	0	0	0	0	0
$N(CH_3)_2$	-0.21	-0.83	-1.5		0.06	-0.94				
NH_2	-0.16	-0.66	-1.3		0.12	-0.78	-0.28	-3.64		
OCH_3	0.12	-0.27	-0.78		0.27	-0.63	0.20	-2.58	0.52	
OC_2H_5	0.10	-0.24					0.17	-2.28		
CH_3	-0.07	-0.17	-0.31		-0.04	-0.07	-0.12	-0.77	-0.10	-0.07
C_2H_5	-0.07	-0.15	-0.30			-0.07	-0.12	-0.64	-0.12	-0.36
$i\text{-}C_3H_7$	-0.07	-0.15	-0.28			-0.08				-0.93
$cyclo\text{-}C_3H_5$	-0.07	-0.21	-0.45[a]						+0.011[b]	
$t\text{-}C_4H_9$	-0.10	-0.20	-0.26			-0.09			-0.17	-1.74
C_6H_5	0.06	-0.01	-0.18		0.10	-0.10			0.215[175]	-0.38
F	0.34	0.06	-0.07		0.50	-0.59	0.58	-1.61	1.10	-0.24
Cl	0.37	0.23	0.11		0.46	-0.35	0.65	-0.70	1.05	-0.24
Br	0.39	0.23	0.15		0.44	-0.34	0.68	-0.77	1.00	-0.27
I	0.35	0.28	0.14		0.39	-0.23	0.61	-0.84	0.85	-0.37

CO$_2$C$_2$H$_5$	0.37	0.45		0.64	0.30	0.03	0.64	0.91	
COCH$_3$	0.38	0.50		0.84	0.28	0.05	0.65	1.24	0.60
CN	0.56	0.66		0.88	0.56	0	0.97	1.23	1.30
NO$_2$	0.71	0.78		1.24	0.65	0	1.23	1.14	
$\overset{+}{N}$(CH$_3$)$_3$	0.88	0.82	0.41[c]	0.77	0.91[c]	0	1.52	0.39	1.90
CF$_3$	0.43	0.54		0.65	0.45	0	0.74	1.12	
SO$_2$CH$_3$	0.60	0.72		0.98	0.59	0	0.74	1.12	

[a] Y. Kusuyama and Y. Ikeda, *Bull. Chem. Soc. J.* **46**, 204 (1973).

[b] Y. E. Rhodes and L. Vargas, *J. Org. Chem.*, **38**, 4077 (1973).

[c] Ref. 114a.

NOTES

1. Substituent constant for cases of resonance between an electron-deficient reaction center and the substituent, see p. 108.
2. Substituent constant for cases of resonance between an electron-rich reaction center and the substituent, p. 108.
3. Substituent constant reflecting the inductive effect only.[123]
4. Substituent constant reflecting resonance effects.
5. Field effect parameter for the semiempirical derivation of substituent constants, see p. 117.[136]
6. Mesomeric effect parameter for the semiempirical derivation of substituent constants, see p. 117.[136]
7. Substituent constant for aliphatic compounds, see pp. 140, 142, and literature.[114a]
8. Substituent constant reflecting steric effects, see p. 143 and literature.[114a]

where ρ = reaction constant (susceptibility factor)
σ = substituent constant
k_H = rate constant of the unsubstituted substrate
k_X = rate constant of the substituted substrate

The basic assumptions are that *meta* and *para* substituents do not exert
any *steric effects* and that the ability to increase or decrease electron
density at neighboring atoms depends *exclusively* on the nature of X,
when attached to the *meta* or *para* positions. Since neither the nature
of the side chain (or of other substituents present in the molecule) nor
the nature of the reaction partner or solvent (when X is uncharged and
aprotic)[115a] has any effect, one can express the *polar character of X*
by a universal *"substituent constant"* σ for all side chain reactions
obeying the Hammett equation (Table 21, columns 2, 3).*

For the same reason the effect of double substitution may be obtained by summation of the σ constants (*125c*):

$$\lg \frac{k_{X,H}}{k_{H,H}} = \rho \cdot \sigma_X \quad (125a) \qquad \lg \frac{k_{X,Y}}{k_{X,H}} = \rho \cdot \sigma_Y \quad (125b) \qquad \lg \frac{k_{X,Y}}{k_{H,H}} = \rho \, (\sigma_X + \sigma_Y) \quad (125c)$$

where $k_{H,H}$ is the rate constant of the unsubstituted substrate and
$k_{X,H}$, $k_{X,Y}$ are the rate constants for the same reaction with

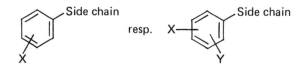

The empirical character of the Hammett equation resides essentially in
the *reaction constant* (susceptibility factor) ρ. This parameter shows to
what *extent* a reaction is affected by a polar substitutent effect and
is indicative of changes in charge at the reaction center as the reaction
proceeds.†

This constant must be computed for each reaction by determining
rate constants (or equilibrium constants) for a *reaction series*; that is,
for a series of compounds with different substituents whose σ values
vary from one extreme to the other (e.g., p-OCH$_3$, H, p-NO$_2$) (Table
22).[115b]

---

*Electron-donor substituents have a negative sign, electron acceptors have a posi-
tive sign.
†Generation of negative charge (reduction of positive charge) results in positive
values for ρ. Generation of positive charge (reduction of negative charge) gives
negative values for ρ. Low absolute values for ρ indicate small changes in charge,
high absolute values of ρ indicate significant charge changes (see Section 4.2.5.2).

TABLE 22 Reaction Constants for Rates of Some Side Chain Reactions and Aromatic Substitutions[116]

Reaction	Solvent	Temperature (°C)	ρ
$XC_6H_4COOC_2H_5 + OH^-$	C_2H_5OH, 85%	25	2.55
$XC_6H_4NH_2 + C_6H_5COCl$	C_6H_6	25	-3.21
$XC_6H_4CH_2CH_2J + C_2H_5O^-$	t-C_4H_9OH	30	2.07
$(XC_6H_4CO_2)_2 \longrightarrow 2XC_6H_4CO_2\cdot$	$C_6H_5COCH_3$	80	-0.201
$XC_6H_4CH_3 + Cl\cdot \longrightarrow XC_6H_4CH_2\cdot$	C_6H_6	80	$-1,5$
$XC_6H_5 + Cl_2$	CH_3COOH	25	-10.0^{117}
$XC_6H_5 + Hg(CH_3COO)_2$	CH_3COOH	25	-4.0^{117}
$XC_6H_4CH_3 + LiNHC_6H_{11}$	Toluene	50	4.0^{118}
$XC_6H_4F + CH_3O^-$	CH_3OH	0	7.55
$X-\langle\!\!\!\!\bigcirc\!\!\!\!\rangle-Cl + HN\langle\!\!\bigcirc$ NO_2	C_6H_6	45	3.80^{119}
$XC_6H_4Br + C_6H_5Li$	Ether	25	4.0^{120}

3.1.1.1 *Various Substituent Constants and Modified Hammett Equations*

Ortho substituents display steric effects in addition to the *polar* effect. These steric effects vary from one reaction to the next and as a result no universal substituent constants can be derived for these substituents (see however p. 121). σ_{meta} Constants are caused essentially exclusively by the inductive effect [(*126*), see also p. 128]:

$$\sigma_m \simeq \sigma_{I(nductive)}{}^{121} \qquad (126)$$

σ_p represents the sum of inductive and resonance effects. The difference between σ_{para} and σ_{meta} is a measure of the *resonance effect* of a given substituent. The inductive effect is assumed to have the same intensity for *meta* and *para* positions [(*127*)].

$$\sigma_p - \sigma_m = \sigma_p - \sigma_I \simeq \sigma_{R(esonance)}{}^{121} \qquad (127)$$

If conjugation involves an electron-deficient (*128*) or electron-rich (*129*) center in the side chain, the resonance effect of electron-donating substituents (e.g., OR, NR_2, phenyl, CH_3) or electron-accepting substituents (e.g., NO_2, CN, RCO) is particularly strong and one must use σ^+ or σ^- constants. Their numerical values, in particular

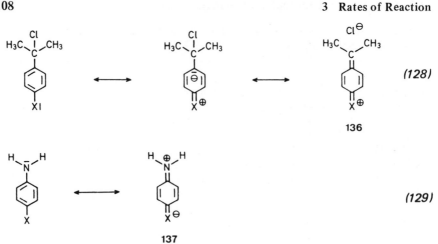

(128)

136

(129)

137

for the *para* position, are higher than those of the corresponding ordinary σ constants (see Table 21).

Solvolysis of ring-substituted phenyldimethylcarbinyl chlorides is best described by using σ⁺ because of the large contribution of structure **136** [(*128*)] to the transition state for solvolysis.

Because structure **137** has a large contribution to the ground state of the corresponding *p*-X-aniline, its reactions must be described by means of σ⁻. Quinoid structures **136** [(*128*)] and **137** [(*129*)] for side chain derivatives are very similar to the semiquinoid structures **138** [(*130*)], **139** [(*131*)], which contribute to the intermediates of aro-

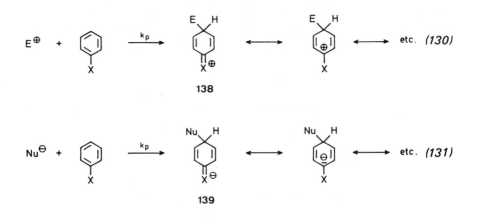

etc. (130)

138

etc. (131)

139

matic substitution. Consequently, relative rates of electrophilic and nucleophilic aromatic substitutions in *meta* and *para* positions with respect to X can also be described by means of constants σ⁺ and σ⁻. These constants are closely related (*132*) to the partial rate factors (see p. 54):

$$\left(\lg \frac{k_{p-X}}{k_{p-H}}\right) = \varrho \cdot \sigma^+_{p-X} = \lg f^X_p \qquad (132)$$

Certain reactions where resonance effects operate to a lesser or (rarely) larger extent than that corresponding to σ^+ or (σ^-) (but stronger than that for σ) can best be described by a linear combination of both substitutent constants, as shown in the *four-parameter* equation (133) (Yukawa-Tsuno equation) and in (134) (Ehrenson et al.).[123,/123 a/,114c]

Yukawa-Tsuno Equation

$$\log \frac{k_X}{k_H} = \rho[\sigma_X + r(\sigma^+_X - \sigma_X)] \qquad (133)$$

Frequently: $0 < r < 1$

Four-parameter equation of Ehrenson et al.

$$\log \left(\frac{k_X}{k_H}\right)_i = \sigma_{I,X}\rho^i_I + \sigma_{R,X}\,\rho^i_R \qquad (134)$$

where i = *meta* resp. *para*

In such cases the characteristics of a reaction are expressed by two reaction constants, ρ (for "normal" polar effects) and r (for resonance effects), resp. ρ^i_I (in combination with the universal inductive substituent constant σ_I) and ρ^i_R (in combination with the appropriate π-delocalization constant σ_R). (For higher precision Ehrenson et al. use different reaction constants for *meta* and *para* substitution and four different values of σ_R depending on the reaction type). Some reactions obeying the Yukawa-Tsuno equation are given in Table 23.

Apart from their ability to reproduce more accurately reaction rates for a known series of reactions, the significance of the four-parameter

TABLE 23 Reactions Obeying the Yukawa-Tsuno Equation[122]

Reaction	Solvent	Temperature (°C)	ρ	r
$XC_6H_4C(CH_3)_2Cl + H_2O$	Acetone-water, 1:9	25	−4.54	1.00
$XC_6H_5 + HOBr/HClO_4$	Dioxane-water, 1:1	25	−5.28	1.15
$XC_6H_4COCHN_2 + H^+$	Acetic acid	40	−0.82	0.56
$XC_6H_4C(CH_3)=NOH + H^+$	Sulfuric acid	51	−1.98	0.43
Beckmann rearrangement				
Semicarbazone formation				
from XC_6H_4CHO	Ethanol-water, 75:25	25	1.35	0.40

equations is questionable. Higher precision is always a result of using a
larger number of empirical parameters. For new reactions, however, the
difficulty of choosing adequate parameters and appropriate values when
their numbers keep increasing soon becomes overwhelming! In parti-
cular, in reactions where $r \simeq 1$ and $r \simeq 0$ (resp. $\rho_I^i \simeq \rho_R^i$) the occurrence
of r (resp. of two different ρ's) may be a result of experimental *errors*
or statistical *artifacts* in the analytical treatment of data. Even in the
intermediate region the appearance of two reaction constants and two
substituent constants does not necessarily mean that one is dealing with
a certain "mixture" of inductive and mesomeric substituent effects in a
single process. It may be a consequence of two different *parallel* or
consecutive processes each with its own ρ and σ. For instance, in reactions where the rate-determining step involves a positively charged
species in equilibrium [(*135*)] :

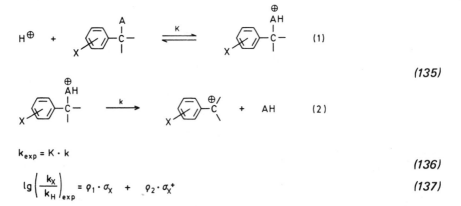

$$(135)$$

$$k_{exp} = K \cdot k$$

$$(136)$$

$$\lg \left(\frac{k_x}{k_H} \right)_{exp} = \rho_1 \cdot \sigma_x + \rho_2 \cdot \sigma_x^+$$

$$(137)$$

In such cases the simple Hammett equation gives good correlations
for substituents with $\sigma^+ \simeq \sigma$ ($\rho = \rho_1 + \rho_2$). However, for substituents
with $\sigma^+ > \sigma$ there are deviations from such a simple correlation (both
with σ and σ^+). Mechanistic differences within a reaction series can also
lead to deviations from the simple dual-parameter correlation.

A clear-cut example of changing resonance effects can be found in
the electrophilic substitutions of biphenyls.[117] The reactions chosen are
those for which precise values of rate constants and ρ's are available in
the benzene series (biphenyl belongs to these series as a phenyl-substi-
tuted derivative). Plotting

$$\log \frac{k_{4\text{-biphenyl}}}{k_{\text{benzene}}/6} = \log f \tag{138a}$$

for various reactions versus their corresponding ρ values (Fig. 15) does
not give the straight line defined by (*138b*), as expected for a simple
case, but a curve. [Since in Fig. 15 selectivities of a series of reagents

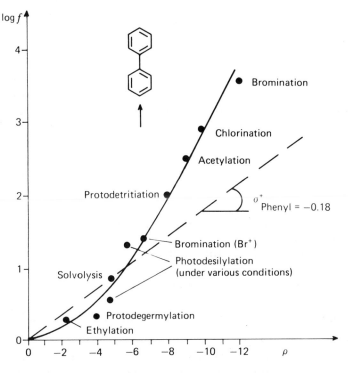

Fig. 15 Extended selectivity relationship for biphenyl: $\log f = \log(k_{4\text{-biphenyl}}/k_{\text{benzene}}/6)$. [From L. M. Stock and H. C. Brown, *Adv. Phys. Org. Chem.* **1**, 35 (1963) (ref. 117).]

for the pair biphenyl(4-position)-benzene ($\log f$) are compared with selectivities for a series of "normal" benzene derivatives (ρ), such diagrams were termed "extended selectivity relationships.]

Application of (*138b*) to the point corresponding to bromination (Fig. 15) gives (*138c*)

$$\lg f = \rho \cdot \sigma^{*}_{p\text{-Phenyl}} \tag{138b}$$

$$(\sigma^{+}_{p}\text{-phenyl})_{\text{effective}} = -0.29 \tag{138c}$$

This result can be reconciled with the normal value for $\sigma^{+}_{p\text{-phenyl}}$ (-0.18) only if in the Yukawa-Tsuno equation for this reaction one chooses for r the value 1.6 [(*138d*)].

$$\sigma^{+}_{\text{effective}} = \sigma + r(\sigma^{+} - \sigma)$$
$$-0.29 = -0.01 + r(-0.17) \tag{138d}$$
$$r = 1.6$$

Correspondingly, for ethylation the value of r turns out to be 0.8.

The occurrence of different resonance contributions in the electrophilic substitutions of biphenyl in the *para* position, as expressed by the curvature in Fig. 15, and of different values of r may be understood if one assumes that coplanarity of the two phenyl rings (e.g., in bromination) is developed much further than in ethylation. Thus in bromination resonance can manifest itself much more strongly. For a quantitative correlation of r and the degree of coplanarity of adjacent π systems, see ref. 117a.

The corresponding diagram for reactions of fluorene is a straight line (Fig. 16). The resonance effects for each reaction are equally strong in this planar derivative of biphenyl.

Actually, correspondence between absolute values of ρ and r (e.g., as in bromination and ethylation of biphenyl) should be expected in other cases as well, because when reactivity is strongly dependent on changes in charge, the influence of the resonance effect should also be larger. Consequently, one may suspect that combinations of high ρ and low r values in the Yukawa-Tsuno correlation may be due to some artifacts.

3.1.1.2 *Heterocycles and Polynuclear Aromatic Compounds*[114c]

The number of possible isomers in *heterocycles* and *polynuclear aromatic compounds* is substantially higher than in benzene derivatives and their reactivity is quite different. This is why quantitative correlations of reactivity data are relatively scarce in these systems. One can

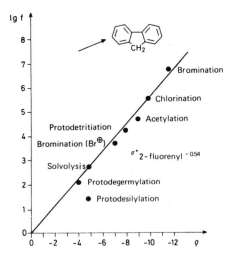

Fig. 16 Extended selectivity relationship for fluorene: $\log f = \log (k_{2\text{-fluorene}}/k_{\text{benzene}}/6)$. [From L. M. Stock and H. C. Brown, *Adv. Phys. Org. Chem.* **1**, 35 (1963) (ref. 117).]

describe their reactivity in terms of the Hammett equation as applied to benzene derivatives if one considers them to be *substituted benzenes*. Replacement of one or more CH= groups or one or more −CH=CH− groups by a heteroatom or an annulated ring is expressed by the so called $\sigma_{replacement}$ or $\sigma^+_{replacement}$ constants.* For reactions of a β-side chain in naphthalene (using ρ values from the benzene series) one can write

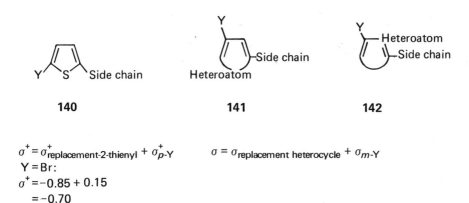

$$\log \frac{k}{k} = \rho \cdot \sigma_{replacement(2\text{-naphthyl})} \qquad (139)$$

Assuming that substituent constants are additive (see above), one can account for the influence of additional substituents on the basic benzene ring by adding σ_p (σ^+_p) or σ_m for that substituent (**140, 141**).

Computation of substituent constants for side chain reactions (or electrophilic ring substitution) in substituted heterocycles:

140 **141** **142**

$\sigma^+ = \sigma^+_{replacement\text{-}2\text{-thienyl}} + \sigma^+_{p\text{-}Y}$ $\sigma = \sigma_{replacement\ heterocycle} + \sigma_{m\text{-}Y}$

Y = Br:

$\sigma^+ = -0.85 + 0.15$

$= -0.70$

Exp: $\sigma^+ = -0.72$

To treat structure type **142** one must use the Yukawa-Tsuno equation (with $\sigma_{p\text{-}Y}$ or $\sigma^+_{p\text{-}Y}$, although Y is formally in the *meta* position), since the heteroatom is particularly well suited to transmit the resonance effect.[124] Some $\sigma_{replacement}$ constants are given in Tables 24 and 25. (However, since experimental data are scarce, many of them should be used for orientation only.)

*Normally, the subscript *"replacement"* is not applied. It is used here to avoid confusion with "normal" σ's expressing the substituent effect of the respective polycycle or heterocycle when attached to a benzene ring.

TABLE 24 $\sigma_{replacement}$ Constants for Side Chain
Reactions of Naphthalene and Heterocycles: Arrows
Show Position of Chain[114c]

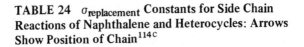

$\sigma = 0,042 \quad (\sigma^+ = -0,145 \, , \ \sigma^- = 0,166)^a$

$0,96^b$

$0,65^b$

$0,75^b$

$(\sigma^{*1} = 0,65)^d$

$0,33^c \quad (\sigma^{*1} = 1,08)^d$

$0^c \quad (\sigma^{*1} = 0,65)^d$

$0^c \quad (\sigma^{*1} = 0,93)^d$

[a] Ref. 125.   See pages 140 and 142.
[b] Ref. 126.
[c] Ref. 127.
[d] Ref. 128.

Using ρ values for corresponding reactions of benzene derivatives
and the substituent constants above one can at least approximately
describe:

1   Reactions of acids.[126-128]
2   Substitution reactions.

$$R{-}\overset{|}{\underset{|}{C}}{-}X \rightarrow R{-}\overset{|}{\underset{|}{C}}{-}Y^{129-132}$$

$$R{-}CH_2{-}CH_2{-}X \rightarrow R{-}CH_2{-}CH_2{-}Y^{127}$$

3   Electrophilic substitutions $R{-}H \rightarrow R{-}X.^{130}$

More accurate results may be obtained by writing, for each ring system, *its own* correlation equations. For example, in Hammett equations
for side chain reactions and ring substitutions of substituted thiophenes
k_H is no longer the rate constant of the unsubstituted benzene (derivative) for the identical reaction [as in (*139*)], but rather that of the
unsubstituted thiophene (derivative); $\sigma_{replacement}$ constants drop out,
and the usual $\sigma(\sigma^+, \sigma^-)$ constants are used in equations like (*140*).

TABLE 25 $\sigma_{replacement}^{+}$ Constants: Arrows Show Position of Electrophilic Ring Substitution or of Side Chains Able to Interact with Ring by Resonance Effect[114c]

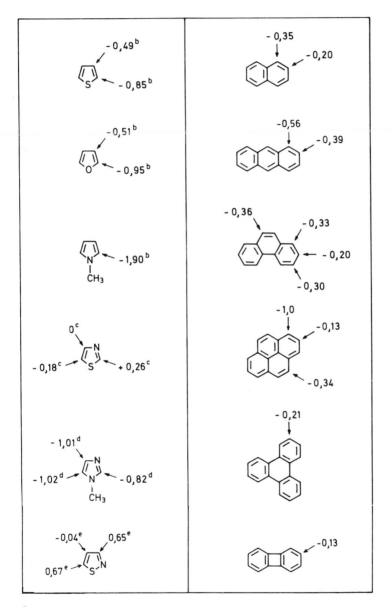

[a] Values for polynuclear aromatics from ref. 129.
[b] Ref. 130.
[c] Ref. 131.
[d] Ref. 132.
[e] Ref. 132a.

Values of ρ in such equations are different from those for benzene derivatives (Table 26).

TABLE 26 Electrophilic Substitution of Thiophene and Benzene Derivatives[133]

Electrophilic Substitution	ρ-Thiophenes	ρ-Benzenes
Bromination	−10.0	−12.1
Chlorination	− 7.8	−10.0
Protodetritiation	− 7.2	− 8.2
Acetylation	− 5.6	− 9.1
Mercuration	− 5.3	− 4.0

One no longer has complete additivity of effects of the heteroatom and the substituent, because in this instance ρ values in (141), derived with respect to benzene, and those in (140), derived with respect to the unsubstituted heterocycle, should be equal:

$$\log \frac{k_{het-X}}{k_{het-H}} = \rho_{het} \cdot \sigma_X \qquad (140)$$

$$\log \frac{k_{het-X}}{k_{benzene-H}} = \rho_{benzene} (\sigma_{replacement} + \sigma_X) \qquad (141)$$

Combining (140) and (141) and rearranging one can write

$$\log k_{het-X} = \rho_{het} \cdot \sigma_X + \log k_{het-H} = \rho_{benzene} (\sigma_{replacement} + \sigma_X) +$$
$$\log k_{benzene-H} \qquad (142)$$

and taking into consideration the definition of $\sigma_{replacement}$ [see (139)] :

$$\rho_{het} = \rho_{benzene} \qquad (143)$$

For the small number of reactions of naphthalene derivatives investigated so far the four-parameter equation (134) gives a reasonable correlation.[134]

This equation also describes (in some cases quite well, in others only approximately) reactivities of side chains and double (and triple) bonds in olefins and acetylenes (and even ring reactivities of some cyclopropanes); all these compounds possess a rigid polarizable molecular structure (as do aromatic compounds), where steric substituent effects play only a subordinate role and inductive and resonance effects of X seem to act in a way similar to that in aromatic compounds.

Reactions of substrates

for which correlation equations are available[135] are the following:

1 Reactions of acids (dissociation, esterification, reaction with di-phenyldiazomethane, saponification of esters).
2 Solvolyses and nucleophilic substitutions of halides and reactive esters.
3 Electrophilic additions (halogenation, oxymercuration, hydrobora-tion, formation of complexes with silver ions).
4 Nucleophilic additions of amines.
5 Additions of radicals.
6 Diels-Alder and 1,3-dipolar additions.
7 H/D exchange in acetylenes.

3.1.1.3 Semiempirical Derivation of Substituent Constants According to Dewar and Grisdale

Dewar and Grisdale[136] proposed (144) for the computation of σ con-stants for the general situation **A**;

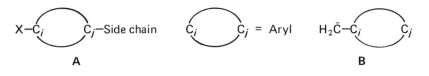

Dewar-Grisdale equation

$$\sigma_{ij} = \frac{F}{r_{ij}} + Mq_{ij} \qquad (144)$$

where F is a measure of the field effect of substituent X, M is its reso-nance effect (both are empirical values) (see below), r_{ij} is the distance between ring carbon atoms i and j, measured in units of the C–C distance in benzene. (The assumption of structures made of regular hexagons with C–C bond lengths as in benzene gives satisfactory results for polynuclear aromatics.) q_{ij} is the fraction of an elementary charge induced by a CH_2^- group at C_i (as a model for substitutent X) on C_j. It is equal to the square of coefficient a (see below) representing the contribution of the $2p$ orbital of C_j to the nonbonding molecular orbital (NBMO) of the arylcarbinyl system **B**.

TABLE 27 Values for r_{ij} and q_{ij}

	i	j	r_{ij}	q_{ij}
Benzene	1	3	$\sqrt{3}$	0
	1	4	2	1/7
Naphthalene	3	1	$\sqrt{3}$	0
	4	1	2	1/5
	5	1	$\sqrt{7}$	1/20
	6	1	3	0
	7	1	$\sqrt{7}$	1/20
Biphenyl	3'	4	$\sqrt{21}$	0
	4'	4	5	1/31

Values for r_{ij} and q_{ij} for benzene, naphthalene, and biphenyl are given in Table 27.

Directions for the calculation of NBMO Coefficients for Nonlinear, Alternant Hydrocarbons[137]*

1. In formulas representing the hydrocarbons carbon atoms are starred in such a way that starred positions alternate with nonstarred ones and the starred set is the more numerous one (see p. 119).

Only orbitals of the "active" starred atoms contribute to the NBMO.

2. The sum of the coefficients a of "active" carbon atoms adjacent to a nonstarred position must be zero.

3. Normalization. The sum of the squares of all coefficients must add up to unity (in our example $7a^2 = 1$, $a = 1/\sqrt{7}$). The NBMO wave function of the benzyl system is

$$\psi_{\text{NBMO,benzyl}} = -\frac{1}{\sqrt{7}}\,X_2 + \frac{1}{\sqrt{7}}\,X_4 - \frac{1}{\sqrt{7}}\,X_6 + \frac{2}{\sqrt{7}}\,X_7 \quad (145)$$

where X_i = wave function of atom i

*If in formulas representing *alternant conjugated hydrocarbons* (which may never contain odd-membered rings) carbon atoms are alternatively starred and unstarred, it can be seen that each starred atom has only unstarred neighbors, and vice versa. In *nonalternant hydrocarbons* (e.g., fulvene or azulene) the same starring procedure leads to formulas where a starred (unstarred) carbon atom has a neighbor of the same type. The terms *odd* and *even* indicate the number of carbon atoms in the conjugated system. An odd alternant hydrocarbon must exist as an ion or radical. The significance of the concept of alternant and nonalternant conjugated systems for molecular orbital theory is described, for example, in M. J. S. Dewar and R. C. Dougherty, *The PMO Theory of Organic Chemistry*, Plenum Press, New York and London, 1975.

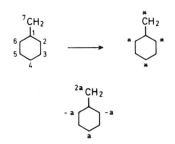

Values of F and M for individual substitutents are found by substituting their σ_m or σ_p and the corresponding values r_{ij} and q_{ij} for benzene in (144). They are given, along with other substituent constants in Table 21.

The values of F and M and (144) enable us to calculate constants for the various possibilities of orientation of substituents and side chain in polynuclear systems. Calculated σ values for sterically unhindered positions in naphthalene derivatives and in biphenyl derivatives with side chains at C-1 or C-4 are given in Tables 28 and 29 (see also Fig. 17).

The Dewar-Grisdale equation (144) applied to five-membered heterocyclic systems gives good agreement between σ_{25} and σ_{24} for the heterocyclic systems and σ_p and σ_m for the benzene system (see above).[138]

Calculation of σ^+ (σ^-) involves a similar procedure. In (144) q_{ij} is the charge generated at position j bearing substituent X, on addition of an

TABLE 28 Calculated σ Values for Sterically Unhindered Positions in Naphthalene Derivatives

Side chain

Substituent	σ_{31}	σ_{41}	σ_{51}	σ_{61}	σ_{71}
NO_2	0.71	0.84	0.52	0.41	0.53
CN	0.56	0.73	0.43	0.32	0.43
Br	0.39	0.19	0.22	0.23	0.21
Cl	—	0.19	0.21	0.23	—
CH_3	—	−0.21	−0.07	−0.04	−0.08
OCH_3	—	−0.42	−0.05	0.07	−0.07
HO	0.12	−0.57	−0.09	0.07	−0.08
H_2N	—	−0.87	−0.29	—	—

TABLE 29 σ Values for Sterically Unhindered Positions in Biphenyl Derivatives

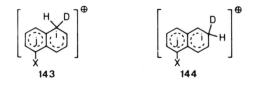

Substituent	$\sigma_{3'4}$		$\sigma_{4'4}$	
	Experimental	Calculated	Experimental	Calculated
NO_2	0.23	0.27	0.30	0.29
Br	0.12	0.15	0.13	0.15
Cl			0.13	0.15
CH_3			-0.02	-0.05
OCH_3			0.07	-0.04
HO			-0.19	-0.07
H_2N			-0.25	-0.19

electrophile (nucleophile) to atom i (see, e.g., **143, 144**). The charge is again equal to the square of the corresponding NBMO coefficients of the odd alternant hydrocarbon formed upon addition (**143, 144**, $X = H$).

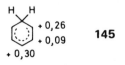

After standardization by means of σ^+ constants for benzene one can derive σ^+ constants for naphthalene[139]; they correlate reasonably well with the rates of protodeuteration of naphthalene derivatives via **143** and **144**.

Based on nuclear magnetic resonance (nmr) measurements, charge distribution in protonated benzene may be represented by **145**.

The ratio of charges in *ortho* and *para* positions of **145** is mirrored by the ratio of logarithms of partial rate factors (*148*) in protodetritiation of substituted benzenes [(*146*), (*147*)][140]:

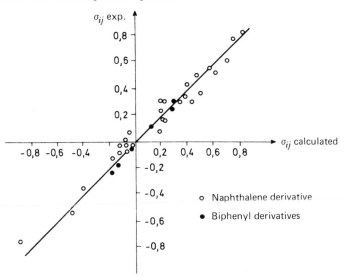

Fig. 17 Comparison of experimental and calculated σ constants. [From M. J. S. Dewar and P. J. Grisdale, *J. Amer. Chem. Soc.* **84**, 3548 (1962) (ref. 136).]

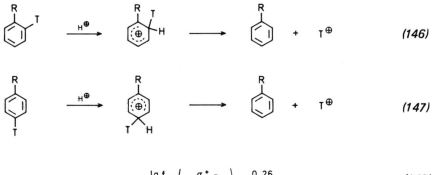

$$\frac{\lg f_o}{\lg f_p}\left(= \frac{\sigma^+_{o-R}}{\sigma^+_{p-R}} \right) = \frac{0,26}{0,30} \tag{148}$$

This result may be understood in terms of the Dewar-Grisdale equation if one assumes that steric and inductive effects play *no* role in proto-detritiation, so that:

$$\frac{\sigma^+_{o-R}}{\sigma^+_{p-R}} = \frac{q_{ortho}}{q_{para}} \tag{149}$$

Then σ^+_O constants valid for these special circumstances may be derived from σ^+_p constants via *(148)*:

$$\sigma^+_{o-OCH_3}: \ -0{,}67;$$
$$\sigma^+_{o-SCH_3}: \ -0{,}515;$$
$$\sigma^+_{o-CH_3}: \ -0{,}27;$$
$$\sigma^+_{o-C_6H_5}: \ -0{,}155.$$

$\sigma^+_{replacement}$ constants of several polynuclear aromatics (p. 115) can be correlated with calculated charges on the exocyclic CH_2 group of the corresponding $Ar-CH_2^+$ system.[129]

$$\triangleright\underset{R}{\overset{H}{\underset{|}{\overset{|}{C}}}}-X \quad \longrightarrow \quad \triangleright-\overset{H}{\underset{R}{\overset{\oplus}{C}}} \; X^{\ominus} \quad \left(\frac{k_{R=C_6H_5}}{k_{R=H}}\right) = 10^{8{,}4} \qquad (150a)$$

$$\underset{R}{\triangleright}-CH_2X \quad \longrightarrow \quad \underset{R}{\triangleright}-CH_2^{\oplus} \; X^{\ominus} \quad \left(\frac{k_{R=C_6H_5}}{k_{R=H}}\right) \approx 2{,}2 \qquad (150b)$$

The substantial difference between stabilizing effects of a phenyl group in 1- and 3-positions of a cyclopropylcarbinyl cation [see ratio of solvolysis rate constants (150a) and (150b)] may be attributed in part to the relative values of NBMO coefficients a in this system[141] (Fig. 18).

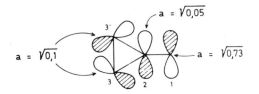

$$a = \sqrt{0{,}05}$$
$$a = \sqrt{0{,}1} \qquad\qquad a = \sqrt{0{,}73}$$

Fig. 18 NBMO coefficients a for the cyclopropylcarbinyl system. [From C. F. Wilcox, L. M. Loew, and R. Hoffmann, *J. Amer. Chem. Soc.* **95**, 8193 (1973) (ref. 141).]

3.1.1.4   Reactivity Indices N_t of Aromatic Systems

NBMO coefficients for odd alternant hydrocarbons play a role in empirical correlation equations for equilibria or rate constants of addition reactions of aromatic hydrocarbons. The change in π-electron energy (ΔE_π) on going from an even alternant hydrocarbon to an odd hydrocarbon, that is, the addition product (151)

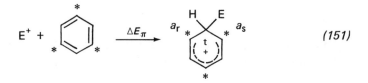

$$E^+ \; + \quad \xrightarrow{\;\;\Delta E_\pi\;\;} \qquad\qquad (151)$$

is given by[137]

$$\Delta E_\pi = -2\beta(a_r + a_s) = -\beta N_t \qquad (152a)$$

$$N_t = 2(a_r + a_s) \qquad (152b)$$

where β is the C—C resonance integral, a_r and a_s are the *NBMO coefficients* of the *sp²*-hybridized carbon atoms r and s in the addition product next to carbon atom t, where additions took place (these coefficients are calculated according to rules given on p. 118). N_t is the *reactivity index* at position t (Fig. 19). *The lower the reactivity index, the more reactive is this position.*

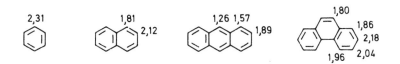

Fig. 19 Reactivity indices N_t.

Linear correlations of N_t (or similar parameters)/[142]/ and log K or log f_t (f_t = partial rate factor for position t) exist for:

1 *Protonation* equilibria (assuming that protonation occurs at the most reactive position t) and *Diels-Alder reactions* with maleic anhydride.[144]
2 Rates of *electrophilic substitution*,[145] *free radical substitution*,[146] *hydroxylation with OsO₄*,[147] and *ozonolysis*[147] (Figs. 20 and 21).

Since the transformation shown in (*151*) is only *partially* achieved in the *transition states* of this reaction, only part of ΔE_π is operative, and so empirical values of β (Table 30) which plays here the role of the reaction constant are *lower* than the real value (-84 kJ/mol). Logarithms of *rates of formation* of arylmethyl cations, arylmethyl anions, and arylmethyl radicals from arylmethane derivatives (*154*) may be similarly correlated with the change in π-electron energy (ΔE_π).

$$R^{(+, \cdot, -)} + \text{aryl–CH}_2X \xrightarrow{\quad \Delta E_\pi \quad} \text{aryl–CH}_2{}^{(+, \cdot, -)} + RX \qquad (154)$$

where R^+ = polar solvent X = halogen[150]

$R^\cdot = \cdot CCl_3$ X = H[151]

$R^- = {}^-NHC_6H_{11}$ X = H[152]

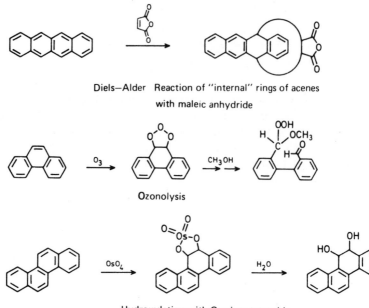

Diels–Alder Reaction of "internal" rings of acenes
with maleic anhydride

Ozonolysis

Hydroxylation with Osmium tetroxide

Fig. 20 Some addition reactions of polycyclic aromatic compounds where log k (or log f_t) can be correlated with N_t.

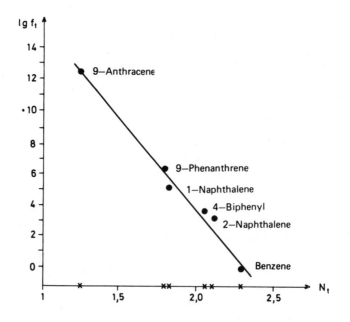

Fig. 21 Bromination of aromatic hydrocarbons[148]. log $f_t = -67(N_t - N_{C_6H_6})/2.3RT$.[153]

124

TABLE 30 β Values for Aromatic Substitution
Reactions[149] Obtained by Plotting log f_t versus
$(N_t - N_{C_6H_6})/2.3RT$

Reaction	β (kJ/mol)
Phenylation (radicalic)	$-14, -13$
Methylation (radicalic)	-23
Trichloromethylation (radicalic)	-42
Nitration (electrophilic)	-20
Chlorination (electrophilic)	-61
Bromination (electrophilic)	-65

A simpler (less accurate) treatment is possible by means of equation

$$\Delta E_\pi = 2\beta(1 - a_{or}) \qquad (155)$$

using NBMO coefficients obtained as described on page 118. Here a_{or} is the NBMO coefficient of the exocyclic CH_2 group of the arylcarbinyl species formed in these reactions.

The basis for all these correlations is the following: within a series of aryl derivatives undergoing the same reaction (reaction series), it is assumed that all factors determining reactivity are roughly equal for all compounds except the changes in π-electron energy so that the latter alone are essentially responsible for differences between individual compounds.

3.1.1.5 The Inductive Parameter $1/r_{ij}$

Logarithms of rate constants of reactions where a *negative* charge is produced in the rate-determining step (e.g., lithiumcyclohexyl amide catalyzed exchange of deuterium attached to aromatic carbon atoms whose rate-determining step is shown below)[153]

and saponification of unhindered methylaryl acetates[154] may be satisfactorily represented as linear functions of parameter Σ_j $(1/r_{ij})$ [Figs. 22 and 23; see also (144)]. The terms r_{ij} are distances in angstrom units between the reacting carbon atom (i) and all the other carbon atoms (j) of the aromatic system (their nuclei attract the negative charge generated in the reaction) (see Fig. 24). The large discrepancies noticed

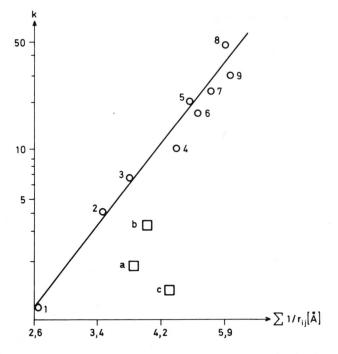

Fig. 22 Correlation of relative D exchange rates of aromatic hydrocarbons with lithium cyclohexylamide: 1, benzene (k = 1, definition); 2, 2-naphthalene; 3, 1-naphthalene; 4, 1-anthracene; 5, 2-pyrene; 6, 9-phenanthrene; 7, 1-pyrene; 8, 9-anthracene; 9, 4-pyrene; a, 4-biphenyl; b, 3-biphenyl; c, 2-biphenyl. [From A. Streitwieser, Jr., and R. G. Lawler, *J. Amer. Chem. Soc.* **87**, 5388 (1970) (ref. 153).]

for biphenyl in the first case and for methyl esters of the 1-naphthyl acetate type in the second case are probably caused by lack of co-planarity of the two phenyl rings (see p. 112) and inhibition of OH⁻ attack by the *peri*-hydrogen atoms.

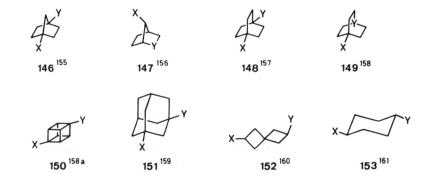

146 [155]   147 [156]   148 [157]   149 [158]

150 [158a]   151 [159]   152 [160]   153 [161]

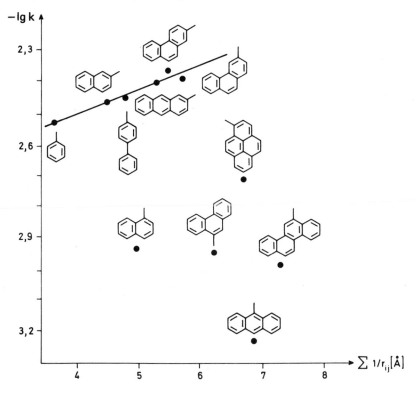

Fig. 23 Correlation of hydrolysis rates ($1 \cdot mol^{-1}\ s^{-1}$) of methyl aryl acetates (– = CH$_2$–COOCH$_3$). [From N. Acton and E. Berliner, *J. Amer. Chem. Soc.* **86**, 3312 (1964) (ref. 154).]

Fig. 24 Distances that must be considered in reactions at the 2-position of pyrene.

3.1.2 Rigid Alicyclic Compounds; σ_I Correlations; Bridgehead Reactivities; Correlation of Reactivities and Spectroscopic Properties

Compounds **146-153** (X = substituent, Y = reaction center) correspond in a sense to aromatic compounds; steric effects and resonance effects between substituent and reaction center are out of the question; the rigidity of polycyclic compounds and the strong favoring of the di-

equatorial conformation in *trans*-1,4-disubstituted cyclohexanes gua-
rantee that within a reaction series the distance between X and Y will
remain constant. Consequently, for these and similar compounds, σ_m
constants are again a good measure for the *polar effect*, the only one
active here.

Even more appropriate are σ_I constants (Table 21) defined by means
of the dissociation constants K of 4-substituted bicyclo[2.2.2]octane-1-
carboxylic acids (148, Y = COOH)[157] (*156a*):

$$\log \frac{K}{K_0} = 1.65\,\sigma_I \quad \text{(ethanol-water, 1:1, 25°C)} \qquad (156a)$$

They reflect exclusively the polar effect of X, whereas σ_m reflects to a
small extent its influence on Y as well, via indirect resonance effects
[(*156b*)] **154** ↔ **155**. These correlations hold for common *reactions
of acids* (dissociation, esterification, rates of saponification of
esters),[155,157-161] also rates of S_N1 *reactions* (Y = Hal, OTos, C—Hal,
C—OTos)[162], S_N2 *reactions* (e.g., X—R—CH$_2$OTos + SR⁻),[163] and
hydrogen abstraction by radicals (X—R—H + ·CCl$_3$ → XR· + HCCl$_3$).[163]

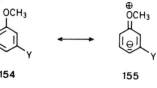

154   **155**

$$\sigma_m = \sigma_I + 0{,}3\,\sigma_R \qquad\qquad (156b)$$

3.1.2.1 Bridgehead Reactivities

Logarithms of rate constants of reactions at bridgehead atoms in many
alicyclic systems (e.g., *157a-157d*) may be correlated with strain differences between R—H and R⁺,[164] calculated by means of potential functions for bond lengths, bond angles, torsion, and van der Waals interactions./[163a]/

R—X	⟶	R⊕ +	X⊖		(157a)
R—N=N—R	⟶	R—N=N· +	·R		(157b)
R—CO$_3$—*t*-C$_4$H$_9$	⟶	R· +	CO$_2$ +	·O—*t*-C$_4$H$_9$	(157c)
R—H + ·X	⟶	R· +	HX		(157d)

R =

RH   R$^\oplus$

The basis for this correlation is the special steric situation at the reaction center. Direction of attack by reagents, steric hindrance, stereochemistry of solvation, and electronic interactions with the rest of the molecule as well as entropy effects are the same for all compounds, so that the only variable is the *difference in strain*.

This method enables the correct prediction of regioselectivity in cationic bridgehead bromination at C-6 in protoadamantane[165] [(*158*)]:

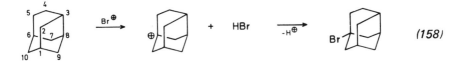

(*158*)

The calculation predicts that ionization at the C-6 position will result in a smaller increase in the strain of the carbon skeleton than ionization at the other positions (C-1, C-3, and C-8).

Calculated differences in strain energy between ground states and models of transition states have also been correlated with relative reaction rates in the following cases: solvolysis of 2-alkyl-2-adamantyl esters,[165a], chromic acid oxidation of secondary alcohols,[165b] reduction of ketones by sodium borohydride,[165c] and trifluoroethanolysis of cycloalkyl tosylates.[165d]

3.1.2.2 The Foote-Schleyer Equation

Rates of S_N1 acetolysis of polycyclic and bicyclic secondary tosylates whose structure prevents S_N2 reactions with solvent as well as stabilization of the carbocation by neighboring group participation (see p. 30 and 207) are relatively uncomplicated and thus amenable to correlation. The prototype for these compounds is 2-adamantyl tosylate **156**. The axial hydrogen atoms shown inhibit the S_N2 reaction (Fig. 25) and the bridgehead nature of the CH groups surrounding C-2 makes hyperconjugation impossible.

Only *conformational* factors are responsible for reactivity; the most influential are bond angles. The larger the deviation of the C–C$^+$–C angle in the cation from the ideal 120° value (toward smaller values) prompted by restrictions imposed by the molecular framework, the slower the ionization. A measure of these deviations are the C=O vibrational frequencies of ketones corresponding to the carbocations. (Car-

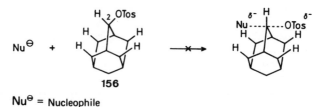

$Nu^{\ominus}$ = Nucleophile

Fig. 25 Steric hindrance in the $S_N 2$ reaction of 2-adamantyl tosylate.

bonyl compounds are often chosen as models for carbocations, since their polar resonance structure is that of an oxidocarbocation). A smaller $>C=O$ bond angle corresponds to a higher C=O frequency.[166] The correlation between rate of ionization k (s⁻¹) and carbonyl frequency $\nu_{C=O}$ (cm⁻¹) is given by:

$$\log k \;=\; -a\nu_{C=O} + C \tag{159}$$

The value of the constant C is obtained from the rate of acetolysis of cyclohexyl tosylate at 25°C:

$$\lg k_{rel} = \lg k - \lg k_{\langle\rangle - OTos} = a \, (1715 - \nu_{C=O}) \tag{160}$$

The slope was determined by plotting the empirical values of $\log k_{rel}$ (160) for some bridge tosylates versus the carbonyl frequencies of the corresponding ketones [Fig. 26 and (161)].[167]

$$\log k_{rel} \;=\; \frac{1}{8}\,(1715 - \nu_{C=O}) \tag{161}$$

For higher precision one must take into consideration additional contributions from torsion effects,[163a] van der Waals interactions, and polar effects of functional groups[168]:

Foote-Schleyer equation

$$\log k_{rel} \;=\; \frac{1}{8}\,(1715 - \nu_{C=O}) + 1.32\sum_{i}\,(1 + \cos 3\phi_i) + \text{inductive term} + \frac{GS - TS}{5.6} \tag{162}$$

where ϕ_i = smallest torsion angle between the HCOTos group and its two neighboring bridgehead CH bonds (i)

    GS – TS = difference in van der Waals energies (kJ/mol) between ground state and transition state

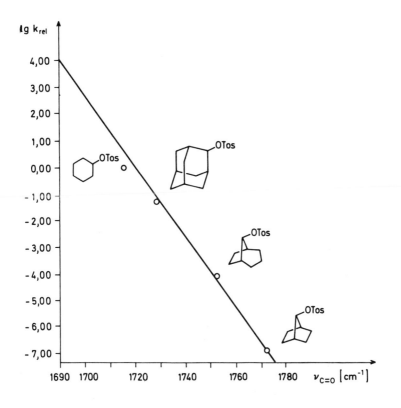

Fig. 26 Determination of the bond angle dependence of relative acetolysis rates of secondary tosylates; k_{rel} ⬡—OTos = 1

The second term of the complete equation (*162*) is derived from the potential function for internal rotation in ethyl chloride and accounts for the fact that smaller values ($\phi_i < 60°$) of the angle of torsion between the HCOTos group and its two neighboring bridgehead CH groups (*i*) result in higher ground state energies. This is shown in Fig. 27 for the eclipsed rotamer **157** of ethyl chloride chosen as model. Transition from **157** to the ethyl cation **158**, which according to calculations[168a] must have the same energy in any conformation, requires less energy (and thus proceeds faster) than transition from the staggered rotamer **159**. This rotamer, just like the chair form of cyclohexyl toxylate, completely lacks torsional effects.

The inductive term due to *polar effects* of aryl groups or double bonds in the β-position is taken[168] to be -0.9;[168e] the estimated value for cyclopropane rings in β-position is -0.5; inductive effects of other

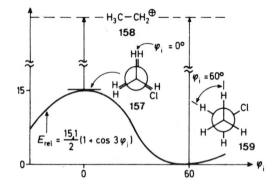

$$E_{rel} = \frac{15.1}{2}(1 + \cos 3\varphi_i)$$

Fig. 27 Concerning the second term of the Foote-Schleyer equation. Torsion energies and ionization of rotamers of ethyl chloride.

substituents can be estimated from $\sigma\rho$ or $\sigma^*\rho^*$ correlations (p. 142). [If one uses $\sigma^*_{CH_2X}$ (X = phenyl, ethyl, see Table 21), then ρ^* for these acetolysis reactions estimated from the ratio log $(k_{\beta\text{-phenyl}}/k_{\beta\text{-ethyl}}) \simeq -0.9$ turns out to be -3. These ρ^* values can then be used to estimate rate effects of other substituents by means of their $\sigma^*_{CH_2X}$].

The last term takes into account the possibility that van der Waals interactions in the ground state (GS) and in the transition state (TS) may contribute to differences in energy between those two states and thus increase or decrease the reaction rate as compared to the normal case (solid line in Fig. 28).

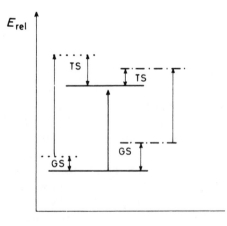

Fig. 28 Contributions (TS) of unfavorable van der Waals interactions to energies of transition states for acetolysis [relative to the normal case (solid lines)] can be larger (dots) or smaller (dots and dashes) than the corresponding contributions (GS) to the ground state energy of the reacting tosylates.

The energy contribution to GS can be estimated by inspection of models of the given molecules and their comparison with model compounds for which steric energies are known (p. 39). For instance, to evaluate steric interactions at the reaction site of aliphatic compounds listed in Table 31, one chooses as the basic model the corresponding cyclohexane derivative (generally the axial cyclohexyl tosylate with a conformational energy of 2.5 kJ/mol); corrections for specific situations must then be applied. (Van der Waals interactions in *endo*-2-norbornyltosylate are actually much stronger than in the axial cyclohexyl tosylate.[168b] A direction for the estimation of van der Waals energies can be found in the literature).[168c]

To estimate the energy contribution of TS is practically impossible, since this would require precise knowledge of the changes in position of the corresponding atoms in moving toward the transition state and of potential functions for bonding relationships in the transition state. Unequivocal interpretations are not possible, as can be seen for instance in the ionization of *endo*-2-norbornyl chloride (*endo*-2-chloro-bicyclo-[2.2.1]-heptane). If one starts from ground state **160**, Scheme 38, the

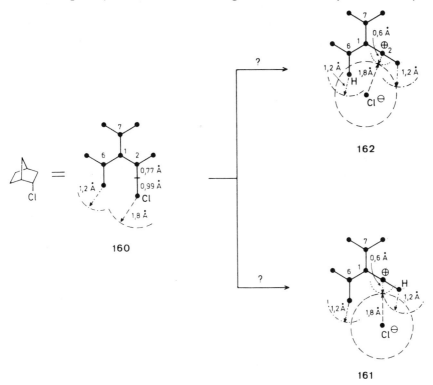

Scheme 38 Possible modes of bond breaking in the ionization of *endo*-2-norbornyl chloride. [From: H. C. Brown, I. Rothberg, P. v R. Schleyer, M. M. Donaldson, and J. J. Harper, *Proc. Nat. Acad. Sci. (USA)* **56**, 1653 (1966) (ref. 170).]

TABLE 31 Relative Rate Constants for Acetolysis of Tosylates[168]

Tosylate	Ketone ν_{CO} (cm^{-1})	ϕ_i (degrees)	GS – TS (kJ/mol)	log k_{rel} Calculated[a]	Experimental
7-Norbornyl	1773	60.60	1.7	−7.0	−7.00
endo-8-Bicyclo[3.2.1]octyl	1750	60.60	2.5	−4.2	−4.11
endo-2-Benznorbornenyl	1756	0.40	1.7	−2.4	−2.22
endo-2-Norbornenyl	1745	0.40	1.7	−1.0	−1.48
2-Adamantyl	1727	60.60	2.5	−1.1	−1.18
Cyclohexyl	1716	60.60	0.00	−0.1	0.00
Cyclotetradecyl	1714	60.60?	0.00	+0.1	+0.08
Isopropyl	1718	60.60	0.00	−0.4	+0.15
endo-2-Norbornyl	1751	0.40	5.5	−0.2	+0.18
Cyclopentadecyl	1715	60.60?	0.00	0.0	+0.42
cis-4-t-Butylcyclohexyl	1716	60.60	2.5	+0.4	+0.42
Cyclododecyl	1713	60.60?	0.00	+0.3	+0.50
2-Butyl	1721	60.60	2.5	−0.3	+0.53
3,3-Dimethyl-2-butyl	1710	60.60	5.0	+1.5	+0.62
Cyclotridecyl	1713	60.60?	0.00	+0.3	+0.66
3-Methyl-2-butyl	1718	60.60	5.9	+0.6	+0.93
Cyclopentyl	1740	0.20?	0.00	+1.5	+1.51
Cycloheptyl	1705	45.45	0.00	+2.0	+1.78
trans-2-t-Butylcyclohexyl	1700	60.60	2.5	+2.2	+2.20
cis-2-t-butylcyclohexyl	1700	60.60	5.0	+2.6	+2.61

1,4-α-5,8-β-Dimethanoperhydro-

9-anthracyl	1696	60.60	2.5	+2.8	+2.67
Cyclopropyl	1815	0.00	0.00	−7.2	−5.32
9-Bicyclo[3.3.1]nonyl	1726	60.60	2.5	−1.0	+0.48
exo-Trimethylene-norborn-exo-2-yl	1751	0.40	1.3	−0.6	+0.84
cis-3-Bicyclo[3.1.0]hexyl	1739	0.40	0.00	−0.2	+1.14
axial-2-Bicyclo[3.2.1]octyl	1717	50.60	2.5	+0.4	+1.62
2-Bicyclo[2.2.2]octyl	1731	0.60	1.7	+0.9	+1.85
equat-2-Bicyclo[3.2.1]octyl	1717	50.60	0.8	+0.1	+0.47
trans-3-Bicyclo[3.1.0]hexyl	1739	0.40	2.5	+0.2	+0.17
syn-8-Bicyclo[3.2.1]oct-2-enyl	1758	60.60	4.6	−5.5	−5.54
Cyclooctyl	1703	40.40?	0.00?	+2.8	+2.76
Cyclononyl	1703	40.40?	0.00?	+2.8	+2.70
Cyclodecyl	1704	40.40?	0.00?	+2.7	+2.98
Cycloundecyl	1709	40.40?	0.00?	+2.1	+2.05

[a] By means of the Foote-Schleyer equation.

C—Cl bond can be stretched in the direction of the original bond (161), in which case TS $\simeq$ 0 is a fair approximation. If, however, the leaving group moves in a direction that is at all times perpendicular to the front side of the carbocation being formed (so as to maximize the overlap between the leaving group and the $2p$ orbital at C-2), one must take into consideration considerable steric hindrance by H-6$_{endo}$ and GS – TS < 0 (p. 32).

All data in Table 31 were obtained under the assumption TS = 0.[170] This is often a valid assumption ("evidently because leaving groups are generally able to find a propitious avenue for departure[99])[168] and has been proved to be correct for solvolyses of axial cyclohexane derivatives.[168d]

Experimental acetolysis rates for tosylates 163[169] and 164-166[170] are significantly lower than values calculated for TS = 0. Obviously, in this instance GS – TS < 0. (In addition to the possibility of some physical barrier inhibiting the departure of the leaving group, one must take into consideration steric hindrance of solvation by alkyl groups; see, e.g., p. 254 and ref. 173a).

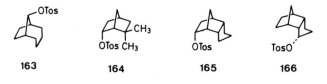

163   164   165   166

[It is surprising that the Foote-Schleyer equation correctly predicts reactivities of simple secondary cycloalkyl tosylates, even though the additional effect of nucleophilic solvent participation plays an important role in their solvolyses (see p. 201, 203, and 212)].[170a]

Relative epimerization rates of the following esters [(*163*) and Table 32] show that C=O stretching frequencies of the corresponding ketones correlate quite well with *rates of formation* of the *mesomeric* (ideally sp^2-hybridized) *carbanions*.[171]

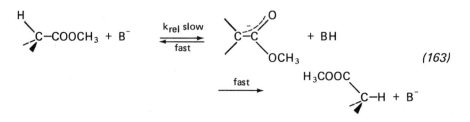

(*163*)

TABLE 32 Relative Epimerization Rates of Esters and C=O Stretching Frequencies of the Corresponding Ketones

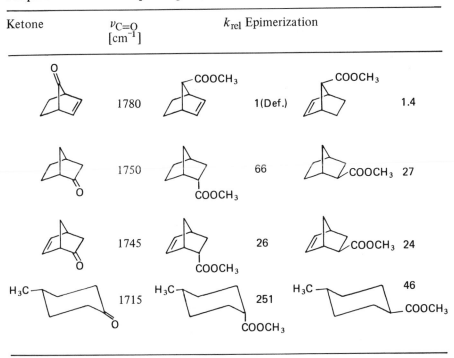

Ketone	$\nu_{C=O}$ [cm^{-1}]		k_{rel} Epimerization		
(bicyclic ketone, alkene)	1780	(ester, alkene)	1(Def.)	(ester)	1.4
(bicyclic ketone)	1750	(ester, COOCH$_3$)	66	(ester, COOCH$_3$)	27
(bicyclic ketone, alkene)	1745	(ester, alkene, COOCH$_3$)	26	(ester, alkene, COOCH$_3$)	24
(H$_3$C cyclohexanone)	1715	(H$_3$C cyclohexane, COOCH$_3$)	251	(H$_3$C cyclohexane, COOCH$_3$)	46

3.1.2.3 CH Acidities and s-Character

We should mention here another correlation between spectroscopic data and reactivities investigated mainly in alicyclic compounds. Logarithms of rate constants for metallation of 3,3-dimethylcyclopropene and of bridgehead CH bonds in bicyclobutanes with an additional alkyl bridge by methyllithium (*164*) may be approximately correlated with the $^{13}C-^{1}H$ coupling constants (*J*) for these bonds[172] (Table 33).[172a]

The same is true for rate constants of tritium exchange between cycloalkanes and N$-^{3}$H$-$ cyclohexylamine, catalyzed by cesium cyclohexylamide[173] (Table 34 and Fig. 29).

$$ \text{H}_3\text{C}-\text{(bicyclobutane)}-\text{CH}_3 \;+\; \text{CH}_3\text{Li} \;\xrightarrow{\;k\;}\; \text{H}_3\text{C}-\text{(bicyclobutane-Li)}-\text{CH}_3 \;+\; \text{CH}_4 $$

(*164*)

TABLE 33 Dependence of the Rate of Metallation of C–H by Methyllithium on Their $^{13}C-^{1}H$ Coupling Constants

Substrate	$J_{^{13}C-H}$ [Hz]	k/k_o	$\lg k/k_o$
H₃C, CH₃ cyclopropene with H, H	$221 \pm 0,5$	2 500	3,4
H₃C—◇—CH₃	212 ± 2	65	1,8
⬠—CH₃	206 ± 1	12	1,1
⬡	200 ± 1	$1 (k_0)$	0,0

$$\lg \frac{k}{k_o} \approx 0,16\, J_{^{13}C-H} - 32$$

These correlations (which, depending on the chosen model compound and the base used differ mainly in their constant terms) are based on the fact that $J^{13}C-H$ is a *measure of the s-character* of the considered C–H bond and that the contribution of the unshared electron pair to the energy of the carbanion is the smaller the higher the

TABLE 34 Relative Rate Constants in the Isotope Exchange Reaction

R–H + HN̄—⬡ $\xrightarrow{\text{slow, } k_{rel}}$ R⁻ + H₂N—⬡

R⁻ + THN—⬡ $\xrightarrow{\text{fast}}$ R–T + HN̄—⬡

and $J^{13}C-H$ of cycloalkanes R–H

R–H	k/k_0	$J^{13}C-H$ [Hz]
Cyclopropane	$(7.0 \pm 0.9) \times 10^4$	161
Cyclobutane	28 ± 10	134
Cyclopentane	5.72 ± 0.27	128
Cyclohexane	$1.00\ (k_0)$	123
Cycloheptane	0.76 ± 0.09	123
Cyclooctane	0.64 ± 0.06	122

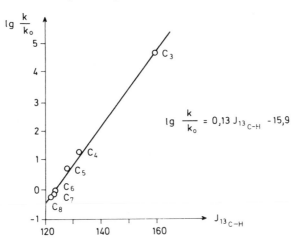

Fig. 29 Graphic representation of data from Table 34. [From A. Streitwieser, Jr., R. A. Caldwell, and W. Young, *J. Amer. Chem. Soc.* **91**, 529 (1969) (ref. 173).]

s-character of "its" orbital. Within a series of compounds the ratio between the s-character of the C—H bond of the hydrocarbon and the s-character of the carbanion formed from this hydrocarbon must remain constant and other factors should not have any influence on anion formation. (The simplest assumption is that the hydrocarbon and the carbanion have the same degree of hybridization, since their molecular skeletons should resist conformational changes caused by possible changes in hybridization on carbanion formation.)

3.1.3 Flexible Compounds; Separation of Steric and Polar Effects; The Taft Equations

Basic conditions for correlations involving substituent constants presented so far:

Absence of steric substituent effects,
Constant distance between substitutent and reaction center within a given reaction series

are no longer present in freely rotating aliphatic compounds. Their capacity to minimize substituent effects by conformational changes becomes evident if one compares the relative dissociation constants of 4-bromobicyclo[2.2.2]octane-1-carboxylic acid **167** and ω-bromo-valeric acid (**168**).
Such consequences of the flexibility of aliphatic chains can be taken into account if—starting with the reaction center—one considers the *whole* group bearing the substituent as substituent.

(H) Br—⟨benzene ring⟩—COOH

167

$$\lg \frac{K_{Br}}{K_H} = 0,49$$

(H) Br—CH_2—CH_2—CH_2—CH_2—COOH

168

$$\lg \frac{K_{Br}}{K_H} = 0,09$$

One would then describe the reactivity of ω-bromovaleric acid not by means of σ_I-Br, but by means of σ^*-$(CH_2)_4$-Br. Taft computed the σ^* constants for the *polar effect* of substituents X directly attached to the reaction center by means of rate constants k for acid- and base-catalyzed hydrolysis of esters X—COOR.[114,175]

$$\log \left(\frac{k_X}{k_{CH_3}}\right)_{\text{basic}} - \log \left(\frac{k_X}{k_{CH_3}}\right)_{\text{acidic}} = 2.48\ \sigma_X^* \qquad (165)$$

$$\sigma_{CH_3}^* = 0 \qquad (166)$$

The factor 2.48 was chosen to obtain for σ^* constants absolute values comparable to those of aromatic σ constants. The essence of Taft's analysis relies on the assumption that *steric* and *polar* effects of X act *independently*. The subsequent procedure is based on two postulates:

First Taft Postulate

Both *polar* and *steric* effects of X are operative in *base-catalyzed* hydrolysis; on the other hand, rates of *acid-catalyzed hydrolysis* are affected only by the *steric effect*.

The low values of ρ in the acid hydrolysis of esters of benzoic acid ($\rho = +0.14$) attest to the fact that polar effects have very little influence on this reaction. The reason is that the rate-determining step (*168*) consists of attack of water on the protonated ester **170**, formed in pre-equilibrium [(*167*)]:

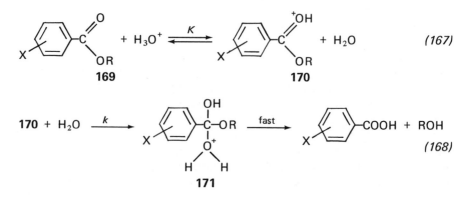

$$(167)$$

$$(168)$$

For the rate-determining formation of **171** one can write

$$\frac{d[171]}{dt} = k\,K\,[169]\,[H_3\overset{\oplus}{O}] = k_{exp}\,[169]\,[H_3\overset{\oplus}{O}] \qquad (169)$$

Application of the Hammett equation to k_{exp} gives:

$$\lg\left(\frac{k_X}{k_H}\right)_{exp} = \lg\frac{k_X}{k_H} + \lg\frac{K_X}{K_H} = (\rho_{(167)} + \rho_{(168)})\,\sigma_x = \rho_{exp}\cdot\sigma_x \qquad (170)$$

Opposite signs for $\rho_{(167)}$ and $\rho_{(168)}$ [an electron-withdrawing substituent X favors reaction (*168*), but disfavors reaction (*167*)] are responsible for the low value of ρ_{exp}.

Second Taft Postulate
(Steric effect of X – steric effect of CH_3)$_{base}$ – (steric effect of X – steric effect of CH_3)$_{acid}$ = 0   *(171)*

Application of both postulates to (*165*), where the log k terms represent the sum of the corresponding polar (P) and steric (S) effects $[(P_X + S_X)_{base} - (P_{CH_3} + S_{CH_3})_{base} - (P_X + S_X)_{acid} - (P_{CH_3} + S_{CH_3})_{acid} = 2.48\,\sigma_X^* \;(165a)]$ gives:

$$\text{polar effect of X – polar effect of } CH_3 = 2.48\sigma^* \qquad (172)$$

The assumption that differences in steric effects of two substituents in the acid hydrolysis of esters are the same as differences in base hydrolysis (independent of the nature of substituents) makes sense, since aside from solvation effects, the structures of the tetrahedral intermediates in both reactions (**172, 173**) differ only by two protons and thus by only a small steric effect, which remains constant in various systems.

(In general one must anticipate that even *differences* in steric effects will depend on the nature of the reacting systems. For instance, replacement of H by CH_3 in the 2-adamantyl system (**174**, R = H or CH_3, respectively) results in an *increase* in acetolysis rate, whereas in the *endo*-2-norbornyl system (**175**, R = H or CH_3, respectively) the rate decreases![173a])

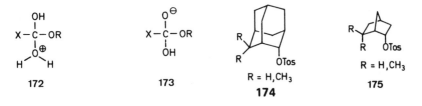

Neither structure 172 nor 173 allows for resonance between X and the reaction center. Since the initial states are the same in both reactions (with potential resonance interactions between X and COOR), resonance contributions to both terms of equation (165) must be the same and must cancel out.

The hypothesis that σ^* constants reflect only the *inductive effect of* X is consistent with the observation that σ_X^* is proportional to $\sigma_{I,X}$ and that insertion of a CH_2 group between X and the rest of the molecule diminishes the σ^* values of *all* substituents by the same factor.

$$\sigma_X^* = 6,23\, \sigma_{I,x} \tag{173}$$

$$\sigma_{CH_2x}^* = 0,36\, \sigma_X^* \tag{174}$$

Charton described correlations that allow the computation of inductive substituent constants from dissociation constants of substituted acetic acids.[174] $\sigma_{CH_2-X}^*$ constants can be found in Table 21.

1 First Taft equation

Reactions where *steric* and *conjugative effects* do not play any role may be described by (175) and (176):

$$\log \frac{k_X}{k_{CH_3}} = \rho^* \cdot \sigma_X^* \tag{175}$$

and for several substituents

$$\log \frac{k_X}{k_{CH_3}} = \rho^* \sum \sigma_X^* \tag{176}$$

where k = rate constant or equilibrium constant

Examples[175] are various acid-base dissociation equilibria (carboxylic acids, ammonium and phosphonium ions, thiols) and also rate constants of:

(a) Acid-catalyzed hydrolysis of acetals and epoxides.[176]
(b) Esterification of carboxylic acids with diphenyl diazomethane.
(c) S_N1 and S_N2 reactions[177a] (see, however, p. 144).
(d) Additions to olefins [e.g., Br_2, Hg^{2+} (oxymercuration), $\cdot CCl_3$].
(e) Oxidations with chromic acid[177b] and abstraction of α-protons from esters.[177c]

According to Taft's first postulate the relationship between structure and rates of acid hydrolysis of esters is exclusively a consequence of *steric factors*, which thus can be determined quantitatively by means of this reaction (*177*):

$$\log\left(\frac{k_X}{k_{CH_3}}\right)_{\text{acid hydrol. of esters}} = E_S \qquad (177)$$

where E_S = steric parameter; for values of E_S, see Table 21.

Charton[178] showed that the E_S parameter is a linear function of van der Waals radii r_X, and

$$E_{S,\ CH_2X} = -0.412\,r_x + 0.445 \qquad (178)$$
$$E_{S,\ CHX_2} = -2.88\,r_x + 3.49 \qquad (179)$$
$$E_{S,\ CX_3} = -3.49\,r_x + 4.14 \qquad (180)$$

[Values of ΔG for equilibria between axial and equatorial conformers of monosubstituted cyclohexanes (p. 39) do not correlate with van der Waals radii of substituents.[179] This shows that steric effects in organic molecules depend to a great extent on the specific character of a given system. In the cyclohexane case one is dealing with 1,3-interactions between C–X and axial C–H groups. The C–H bond is exceedingly short (1.09 Å); as the van der Waals radius of X becomes larger, so does the C–X bond length, and large substituents "grow" somehow "over" the C–H groups; therefore the increase in size of X does not result in stronger interaction with C–H groups. In interactions of C–X^1 with C–C or C–X^2 this situation obtains to a much lesser degree.]

2   Second Taft equation

When reaction rates are affected exclusively by *steric effects* of the type observed in reactions used to define E_S (additions to sp^2-hybridized carbon atoms) (*181*) (with the susceptibility constant δ) should be valid.

$$\log\frac{k_X}{k_{CH_3}} = \delta E_S \qquad (181)$$

Examples[175] are acid-catalyzed transesterification, catalytic hydrogenation of olefins, hydroboration with "disiamylborane" (bis[3-methylbutyl-(2)]-borane),

$$\underset{\displaystyle (H_3C-\overset{\displaystyle CH_3}{\underset{\displaystyle |}{CH}}-\overset{\displaystyle CH_3}{\underset{\displaystyle |}{CH}})_2 BH}{}$$

and quaternization of heterocycles [(*182*)] [180]:

 (*182*)

Compounds in which free rotation of R is impossible (e.g., **176**) no longer obey the linear E_s correlation.

176

3   In a third group of reactions polar *and* steric effects act independently; they can be described by a combination of equations mentioned above (*183*):

$$\lg \frac{k_X}{k_{CH_3}} = \rho^* \sigma_X^* + \delta E_{s,X}$$ (*183*)

Examples are (within the limits of applicability of E_s) polar carbonyl additions (e.g., base-catalyzed transesterification and reduction with BH_4^-)[181] and additions to olefins[182] when substrates bear sterically demanding substituents (branched alkyl groups).

4   Reactions where polar and steric effects do *not* seem to operate *independently* or where steric interactions differ in character from those in the reaction used to define E_s cannot be described by means of Taft's σ^*-E_s correlations. Examples are S_N2 reactions of alkyl halides.[183] Their rates may be described by (*184*)

$$\lg \left(\frac{k_{RX}}{k_{C_2H_5X}} \right) = r \; \alpha_R$$ (*184*)

where α_R is the "alkyl reactivity constant" of the alkyl group R and r is the reaction constant. The α_R values in Table 35 show very clearly the enhanced reactivity of methyl halides and the reduced reactivity of branched isomers.

TABLE 35   Alkyl Reactivity
Constants α_R for S_N2 Reactions[183]

R	α_R
CH_3	+ 1,308
C_2H_5	0,000
C_3H_7	− 0,359
C_4H_9	− 0,371
i–C_3H_7	− 1,615
i–C_4H_9	− 1,387
Neopentyl	− 4,875

3.2   EMPIRICAL CORRELATION EQUATIONS FOR CHANGES IN REAGENTS*

Three types of reaction are most important for organic compounds:

1   Redox reactions (e.g., Birch reduction or oxidation with metal ions).
2   Pericyclic reactions (Section 5.2.4.1).
3   Lewis acid-base reactions.

Extensive studies concerning structure-reactivity relationships exist only for type 3[184] and then only with respect to variations of the base (or nucleophile, see below).

3.2.1   General Aspects: Relative Nucleophilicity Orders; the HSAB (Hard and Soft Acids and Bases) Principle

Nomenclature
To describe *equilibrium properties* it is common to use the concepts of *basicity* and *acidity*. The corresponding kinetic terms are *nucleophilicity* and *electrophilicity*.

A first insight into the role of some important factors may be gained from the thermodynamic analysis of the reaction of a Lewis base (Nu⁻) and a Lewis acid (M^+) (*185*):

$$Nu^- + M^+ \xrightarrow{\ K_{MNu}\ } MNu \qquad (185)$$

The free energy of reaction and the equilibrium constant K_{MNu} are described by (*186*) and Scheme 39.

*"Usually ... the interest of the chemist is focused on only one of the starting materials and on the end products arising from this compound. The other starting materials are then called *reagents*."[183a]

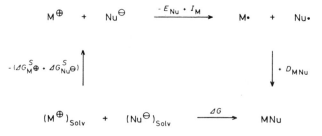

Scheme 39 Thermodynamic cycle for the reaction of a solvated Lewis acid $(M^+)_{solv}$ and a solvated Lewis base $(Nu^-)_{solv}$ and the derivation of (186). In the individual steps energies absorbed have negative sign; energies released have positive sign.

$$\Delta G = -RT \ln K_{MNu} = D_{MNu} + I_M - E_{Nu} - \Delta G^S_{M\oplus} - \Delta G^S_{Nu\ominus} + T\sum \Delta S \quad (186)$$

where D = dissociation energy
I = ionization potential
E = electron affinity
ΔG^S = free energy of solvation of ions
ΔS = entropy of reaction

For a series of bases Nu^- reacting with a common acid M^+ (if one assumes that entropy terms are identical and steric effects are neglected), one can write (187):

$$\Delta \Delta G = -RT \Delta \ln K_{MNu} = \Delta D_{MNu} - \Delta(E_{Nu} + \Delta G^S_{Nu\ominus}) \quad (187)$$

Table 36 gives values of the terms in (187) for the reactions of halide ions with the proton and the Hg^{II} ion, respectively.

TABLE 36 Free Energy Differences for Reactions of Halide Ions with the Proton and the Hg^{II} Cation (in water)[184a]

Nu	$E_{Nu}+\Delta G^S_{Nu\ominus}$ [kJ/mol]	D_{HNu} [kJ/mol]	$D_{HgNu\oplus}$ [kJ/mol]	$\Delta\Delta G =$ $\Delta G_{HNu}-\Delta G_{HF}$ [kJ/mol]	$\Delta\Delta G =$ $\Delta G_{HgNu\oplus}-\Delta G_{HgF\oplus}$ [kJ/mol]
F	−857	−561	−419	0	0
Cl	−761	−427	−339	38	−16
Br	−714	−365	−302	53	−26
I	−653	−297	−251	60	−36

In the hydride series the decrease in HNu bonding energy in going from one halogen to the next higher one is larger than the decrease in absolute value of the term $(E_{NU} + \Delta G_{Nu^-}^S)$. $\Delta\Delta G$ becomes positive, which corresponds to a decrease in stability of HNu, that is, basicity toward the proton decreases in the order $F^- > Cl^- > Br^- > I^-$. On the other hand, in the same series basicity toward the Hg^{II} ion increases, since decrease in bonding energy is now offset by the decrease in the $(E_{Nu} + \Delta G_{Nu^-}^S)$ term; $\Delta\Delta G$ becomes negative (and $HgNu^+$ becomes more stable). These results show that the natures of the acid and the solvent have a profound influence on the order of relative basicities. Since the *dependence of the rate constant k on structure often corresponds to that of the respective equilibrium constant (K)* [(*188*)] (see Section 5.2.2.2.),

$$\lg k_{Nu^\ominus \cdot M^\oplus} \sim \lg K_{MNu} \qquad (188)$$

the same may be expected for the orders of relative nucleophilicities. Rate constants of a large number of *proton transfer* reactions may be correlated with *pK* values of the conjugate acids of the participating nucleophiles (Brönsted relationships, p. 167); consequently, *relative reactivities* of a series of nucleophiles with respect to *carbon* should correspond to their *basicities with respect to carbon*. Table 37 shows that there are various basicity scales even for different hybridization states and bonding states of carbon.

Whereas in the case of the *acetyl* group one can still see a certain parallelism between K_B and K_C, for the *methyl* group K_C values are completely different from K_B values. One must point out in particular the different orders of the $O-CH_3$, $S-CH_3$, and CN groups on the two basicity scales. Also, in contrast to the methyl series, in the phenyl series $O-CH_3$ is almost as basic as $S-CH_3$ with respect to carbon (resonance between OCH_3 and the aryl residue is larger than for SCH_3).

A similar picture emerges if gas phase enthalpies of reaction are compared (Table 37a).[236e] Proceeding from a heteroatom nucleophile to nucleophilic carbon proton affinities (ΔH_{H^+}) are seen to decrease, whereas methyl cation affinities ($\Delta H_{CH_3^+}$) increase (e.g., $OH \rightarrow MeC{\equiv}C$, $OCH_3 \rightarrow C_6H_5CH_2$, and $F \rightarrow CN$).

Orders of *relative nucleophilicities toward carbon* expected on the basis of the proportionality of log k and log K (*188*) and the data in Table 37, and actually observed in practice (see Section 3.2.3.2) were reproduced by Klopman[186] by calculating for models of the transition state electrostatic interactions and interactions ΔE between *occupied* molecular orbitals of nucleophile R and *empty* molecular orbitals of electrophile S (see Fig. 30).

TABLE 37 Carbon Basicities (K_C) and Proton Basicities (K_B) of Some Nucleophiles[185]

$$HOR + Nu^- \underset{\longleftarrow}{\overset{K_C}{\longrightarrow}} RNu + OH^- \text{ (in water)}$$

$$HOH + Nu^- \underset{\longleftarrow}{\overset{K_B}{\longrightarrow}} HNu + OH^-$$

	Nu	K_B	K_C
$R = CH_3$	O—⬡	$9{,}0 \cdot 10^{-7}$	$1{,}3 \cdot 10^{-5}$
	SH	$1{,}0 \cdot 10^{-9}$	$8 \cdot 10^{-1}$
	S—⬡	$3{,}0 \cdot 10^{-10}$	2
	$O-CH_3$	$1{,}1 \cdot 10^{-1}$	$1{,}2 \cdot 10$
	$S-CH_3$	$1{,}9 \cdot 10^{-6}$	$4 \cdot 10^4$
	CN	$1{,}5 \cdot 10^{-7}$	$1 \cdot 10^9$
$R = C_6H_5$	$O-CO-CH_3$	$2{,}6 \cdot 10^{-12}$	$1{,}1 \cdot 10^{-15}$
	$S-CH_3$	$1{,}9 \cdot 10^{-6}$	$1{,}1$
	$O-CH_3$	$1{,}1 \cdot 10^{-1}$	$1{,}5$
$R = CO-CH_3$	O—⬡—Cl	$2{,}4 \cdot 10^{-7}$	$7{,}2 \cdot 10^{-11}$
	O—⬡	$9{,}0 \cdot 10^{-7}$	$3{,}9 \cdot 10^{-10}$
	O—⬡—OCH_3	$1{,}5 \cdot 10^{-6}$	$2{,}4 \cdot 10^{-9}$
	$S-CH_2-CH_2-NH-CO-CH_3$	$2{,}7 \cdot 10^{-7}$	$1{,}6 \cdot 10^{-8}$
	$OPO_3^{2\ominus}$	$4{,}5 \cdot 10^{-5}$	$6{,}3 \cdot 10^{-8}$
	$O-CH_2-CF_3$	$2{,}1 \cdot 10^{-4}$	$5{,}5 \cdot 10^{-6}$
	$O-CH_2-CH_2-Cl$	$1{,}8 \cdot 10^{-2}$	$1{,}7 \cdot 10^{-2}$
	$O-C_2H_5$	$9 \cdot 10^{-1}$	$6{,}1$

In the fundamental equation (*189*) of Klopman's "general multi-electron perturbation theory" ΔE_{total} is the total energy change that

$$\Delta E_{\text{total}} = \underbrace{-q_r q_s T_{rs} + \Delta\text{solv}(1)}_{\text{Electrostatic term}} + \underbrace{\sum_j \sum_k \frac{2\,C_{rj}^2 C_{sk}^2 \beta_{rs}^2}{\alpha_j - \alpha_k}}_{\text{Covalent term}} \quad (189)$$

results on partial bond formation (in the transition state) between atoms r and s of the donor molecule R and the acceptor molecule S;

TABLE 37a Gas Phase Reaction
Enthalpies (kJ/mol)

$$Nu^- + H^+ \longrightarrow NuH - \Delta H_{H^+}$$

$$Nu^- + CH_3^+ \longrightarrow NuCH_3 - \Delta H_{CH_3^+}$$

Nu	$-\Delta H_{H^+}$	$-\Delta H_{CH_3^+}$
OH	1635	1154
$CH_3C\equiv C$	1588	1187
OCH_3	1586	1130
$C_6H_5CH_2$	1586	1169
CF_3	1571	1155
F	1554	1077
CN	1469	1093
C_6H_5O	1464	1002
CH_3CO_2	1458	994

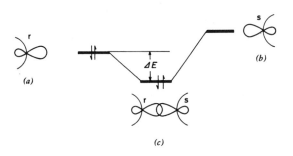

Fig. 30 Energy gain (ΔE) on interaction (in the transition state for nucleophilic attack) of an occupied molecular orbital of a nucleophile (through the atomic orbital of its atom r) and an empty molecular orbital of an electrophile (through the atomic orbital of its atom s). (a) Part of an occupied molecular orbital of nucleophile R with the atomic orbital of atom r. (b) Part of an empty molecular orbital of electrophile S with the atomic orbital of atom s. (c) Transition state of the nucleophilic attack of atom r of nucleophile R at atom s of electrophile S.

T_{rs} is the Coulombic interaction between r and s, which have charges q_r and q_s; Δsolv(1) is a partial desolvation energy that must be expended for r to approach s; C_{rj} (C_{sk}) are the coefficients of the atomic orbital r(s) in various occupied (unoccupied) molecular orbitals of R (S); β_{rs} is the resonance integral for interaction of r and s; α_j (α_k) is the energy of the molecular orbitals of R (S).

For model situations, where

$$q_r = -1 \qquad C_r^2 = C_j^2 \equiv 1 \qquad T_{sr} = \frac{e^2}{\varepsilon R_{r-H}} \qquad \Delta \text{solv}(1) = 0$$
$$q_s = +1 \qquad \varepsilon = 80$$
$$\beta_{rs} = \sqrt{\beta_{rr}\,\beta_{ss}}$$

ϵ (= 80) = dielectric constant of water
e = elementary charge
R_{r-H} = bond length in molecule R–H
ΔE_{total} was calculated for the nucleophiles given in Table 38 and three different types of acceptor orbitals:

$$(\alpha_k = -7 \text{ eV}, \alpha_k = -5 \text{ eV}; \alpha_k = +1 \text{ eV}) \qquad (\text{Table 39})$$

TABLE 38   Properties of Some Nucleophiles

	$I^{\ominus}$	$Br^{\ominus}$	$Cl^{\ominus}$	$F^{\ominus}$	$SH^{\ominus}$	$CN^{\ominus}$	$OH^{\ominus}$
R_{r-H} [Å]	1,60	1,41	1,27	0,92	1,34	1,06	0,96
$-2\beta_{rs}$ [eV]	3,16	3,60	4,10	4,48	4,0	3,8	4,50
$-\alpha_j$ [eV]	8,31	9,22	9,94	12,18	8,59	8,75	10,45

TABLE 39   Relative Nucleophilicities Toward Centers of Different Electrophilicities (α_k)

$\alpha_k = -7\,\text{eV}$	$-\Delta E_{Total}$ [eV]	$SH^{\ominus}$ >	$I^{\ominus}$ >	$CN^{\ominus}$ >	$Br^{\ominus}$ >	$Cl^{\ominus}$ >	$OH^{\ominus}$ >	$F^{\ominus}$
		2,64	2,52	2,30	1,75	1,54	1,49	1,06

$\alpha_k = -5\,\text{eV}$	$-\Delta E_{Total}$ [eV]	$SH^{\ominus}$ >	$CN^{\ominus}$ >	$I^{\ominus}$ >	$OH^{\ominus}$ >	$Br^{\ominus}$ >	$Cl^{\ominus}$ >	$F^{\ominus}$
		1,25	1,17	1,07	1,01	0,98	0,97	0,82

$\alpha_k = +1\,\text{eV}$	$-\Delta E_{Total}$ [eV]	$OH^{\ominus}$ >	$CN^{\ominus}$ >	$SH^{\ominus}$ >	$F^{\ominus}$ >	$Cl^{\ominus}$ >	$Br^{\ominus}$ >	$I^{\ominus}$
		0,58	0,56	0,55	0,53	0,52	0,48	0,45

When the acceptor orbital has high electronegativity ($\alpha_k = -7$ eV, $|\alpha_j - \alpha_k|$ is small) the contribution of the covalent term is dominant; one speaks about an "orbital controlled order of relative nucleophilicities"; this situation pertains in nucleophilic substitutions at peroxidic oxygen [(190)].[186a]

$$Nu| \longrightarrow \overset{R^1}{O} - \overset{R^2}{O} \longrightarrow NuOR^1 + IOR^2 \qquad (190)$$

where $Nu = R_3C^-, NR_3, SR_2, RI;$
$R^1, R^2 = H$, alkyl, acyl (no charges drawn)

For more electropositive acceptors (increase in α_k), the importance of the covalent term decreases, whereas that of the electrostatic term (in the model) remains the same. For $\alpha_k = +1$ eV one has a "charge-controlled" order, reflecting relative affinities toward the proton. In the intermediate region ($\alpha_k = -5$ eV) the order roughly corresponds to the order found with respect to saturated carbon (see p. 182).

Support for Klopman's method can be found in Metzger's study on the kinetics of methylation of the thiocarbonyl group in heterocycles[187] [(191)]:

$$\overset{X}{\underset{Y}{\bigcirc}} C=S \ + \ CH_3J \ \xrightarrow{\text{Acetone}} \ \overset{X}{\underset{Y}{\bigcirc}} \overset{\oplus}{C}-SCH_3 \ + \ J^{\ominus} \qquad (191)$$

Calculations showed no systematic relationship between charge on the sulfur atom of the thiocarbonyl group and the observed reactivity. On the other hand C_{rj} of the sulfur atom turns out to be the same for all compounds considered (0.976 ± 0.010). Ionization potentials determined by photoelectron spectroscopy indicated the low energy (α_j) of the highest occupied molecular orbitals. This situation, typical for orbital control, can be further simplified: since the various heterocycles are similar and since the electrophile and the solvent are constand, both β_{rs} and the electrostatic term will be the same within a given series. Klopman's equation (189) then becomes

$$\Delta E = a + \frac{b}{\alpha_j - \alpha_k} \qquad (192)$$

where a and b are constants valid for the entire series. If one chooses for α_j the negative value of the ionization potential I of the unshared electron pairs on sulfur, and $a = 21.34$, $b = 52.68$, $\alpha_k = -5$, one can reproduce the observed relationship between reactivity ($-\log k$) and ionization potential I (Fig. 31, Table 40). Since ΔE is proportional to $-\log k$ (see Section 4.1) values of a and b were chosen such that (192) gives directly $-\log k$; that is, in this particular case $\Delta E = -\log k$. Klopman's results provide a theoretical background for the *principle of "hard" and "soft" acids and bases* (HSAB principle) enunciated some time ago mainly by Pearson.[188]

TABLE 40 Ionization Potentials (I), Experimental Rate Constants (k), and Rate Constants (ΔE) Calculated from Equation (192) for Reaction (191) of Heterocyclic Thiocarbonyl Compounds

No.	Nucleophiles	I (eV)	$-\log k$, Experimental	ΔE, Calculated (eV)
1		7.78	2.50	2.40
2		7.94	3.36	3.43
3		7.95	3.48	3.49
4		7.98	3.57	3.67
5		7.98	3.84	3.67
6		8.02	3.80	3.91
7		8.03	3.85	3.96
8		8.04	3.97	4.02

TABLE 40 (Continued)

No.	Nucleophiles	I (eV)	$-\log k$, Experimental	ΔE, Calculated (eV)
9	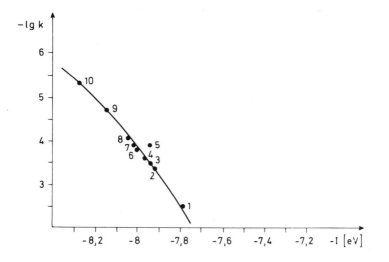	8.16	4.77	4.68
10		8.30	5.40	5.39

This purely empirical principle states that *hard acids* coordinate best with *hard bases*, and *soft acids* with *soft bases*. The (hard + hard) interactions in the sense of this principle correspond to Klopman's charge-controlled interactions; consequently, *hard acids* are species of *low electron affinity*, with *unoccupied orbitals* of high energy; their *small ionic radius* and *high charge* enhance the importance of the electrostatic term and the necessary loss of *strong solvation* makes covalent inter-

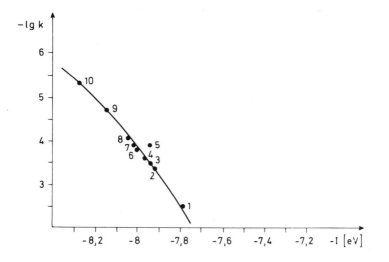

Fig. 31 Relationship between rate constants (k) for reaction (*191*) and ionization potential (I) of the free electron pairs of the sulfur atom in the heterocycles (the numbering corresponds to that in Table 40). [From M. Arbelot, J. Metzger, M. Chanon, C. Guimon, and G. Pfister-Guillonzo, *J. Amer. Chem. Soc.* **96**, 6217 (1974) (ref. 187).]

actions more difficult. Examples are H^+, Li^+, Na^+, Mg^{2+}, and Al^{3+}. In the carbon series hard centers are $R-\overset{+}{C}=O$ and $\overset{+}{C}\equiv N$. The same considerations are valid for the *ionic radius, charge* and *solvation* of *hard bases*; their *high electron affinity* is a consequence of the *very low energy* of *their highest occupied orbitals*. Examples are: H_2O, OH^-, ROH, RO^-, F^-, Cl^-, $SO_4{}^{2-}$, $NO_3{}^-$, $PO_4{}^{3-}$, $CO_3{}^{2-}$, $ClO_4{}^-$, NH_3, RNH_2, and N_2H_4.

Soft-soft interactions correspond to Klopman's orbital-controlled processes. *Soft bases* have *occupied orbitals* of high energy, relatively *low electron affinities*, and *large ionic radii* (R_2S, RS^-, I^-, SCN^-, $S_2O_3{}^{2-}$, Br^-, R_3P, CN^-, CO, RNC, C_2H_4, C_6H_6, H^-, R^-). *Soft acids* have *low-lying unoccupied orbitals* and relatively *high electron affinities*, along with *large ionic radii* (Hg^{2+}, Ag^+, Cd^{2+}, Cu^+, Pt^{2+}). In the carbon series, CH_3^+ is a soft acid.

Hard Acids

$$H^+, Li^+, Na^+, K^+, Mg^{2+}, Al^{3+}, R-\overset{+}{C}=O, \overset{+}{C}\equiv N$$

Species with low electron affinity.

Unoccupied orbitals are of high energy.

Small ionic radius.

High charge.

Strong solvation.

Hard Bases

$$H_2O, HO^-, ROH, RO^-, F^-, Cl^-, SO_4^{2-}, NO_3^-, PO_4^{3-}, CO_3^{2-}, ClO_4^-,$$
$$NH_3, RNH_2, N_2H_4$$

Species with high electron affinity.

Occupied orbitals are of low energy.

Small ionic radius.

High charge.

Strong solvation.

Soft Acids

$$Hg^{2+}, Ag^+, Cd^{2+}, Cu^+, Pt^{2+}, CH_3^+$$

Species with high electron affinity.

Unoccupied orbitals are of low energy.

Large ionic radius.

Soft Bases

R_2S, RS^-, I^-, SCN^-, $S_2O_3^{2-}$, Br^-, R_3P, CN^-, CO, RCN, C_2H_4, C_6H_6, H^-, R^-

Species with low electron affinity.

Occupied orbitals are of high energy.

Large ionic radius.

Combinations of reaction partners with similar interaction capabilities are more favorable than combinations in which one partner prefers Coulombic interactions, whereas the other is capable only of orbital-controlled interactions. Consequently, if one assumes that RS^- is softer than HO^-, it follows from the position of the equilibrium in (193a) that the hypothetical methyl cation is softer than the proton. (Data in Table 37 give $K = K_{C(SCH_3)}/K_{B(SCH_3)} = 2 \times 10^{10}$).

If the CH_3 group is replaced by $COCH_3$ [(193b)] K for RS^- = $^-SCH_2CH_2NHCO-CH_3$ is only 6×10^{-2}. The upper side now represents the (hard + hard) + (soft + soft) combination and the hypothetical

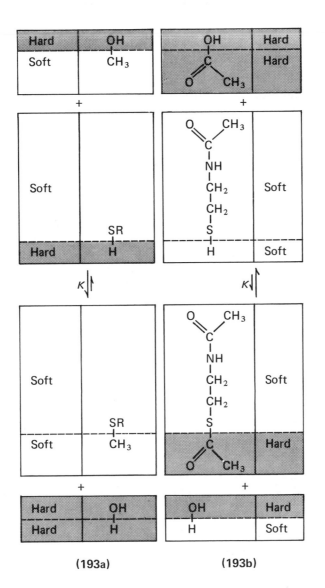

(193a)   (193b)

COCH$_3$ cation is consequently harder than the proton. Other *hypothetical* Lewis acid (base)-fragments can be put on this scale in the same way.

$$\text{Hard} \quad R\overset{+}{C}O > H^+ > C_6H_5^+ > t\text{-}C_4H_9{}^+ > (CH_3)_2CH^+ > {}^+C_2H_5 >$$

$$^+CH_3 > CH_2 > {}^-CH_3 > -C{=}C{-}O^- > CN^- > H^- \quad \text{Soft} \qquad (193)$$

The order shows that substitution by methyl makes C$^+$ harder (compare the relative positions of $^+CH_3$ and t-C$_4$H$_9^+$).[188a] Replacement of hydrogen by the more electronegative carbon increases the positive charge of the central carbon atom. (Molecular orbital calculations also show that the customary belief that methyl groups are electron donors and thus are able to stabilize carbocations, e.g., by hyperconjugation, is not quite correct and that methyl groups actually increase the positive charge of neighboring positive carbon atoms.[189] In the gas phase alkyl groups increase the acidity of amines and alcohols.[190])

Predictably sp^2-hybridization increases hardness. On the other hand, methylene and carbon radicals are relatively soft, since they lack positive charge. For the same reason *arynes* are soft electrophiles. Electrophilic *silicon* is hard.[188a]

3.2.2 Ambident Substrates, Symbiosis, and Leaving Group Effects

The existence of many log k – log K correlations makes it possible to use the HSAB principle to predict and interpret relative reaction rates. Of two competitive reactions, the one having according to the HSAB principle the larger K should also have the higher rate constant. One of the most fruitful applications deals with *regioselective reactions of ambident substrates*. Some examples follow.

3.2.2.1 *Reactions of Enolates and Phenolates (Scheme 40)*

In general *alkylation* of enolates takes place predominantly at carbon, whereas *acylation* occurs at the harder alkoxide oxygen atom. Alkylation reactions of enolates and phenolates are subject to strong *solvent effects*.[191] For instance, the amount of enol ether present in the reaction product of the sodium enolate of propiophenone and butyl bromide is 5% in ether, tetrahydrofuran and diglyme, 23% in dimethyl sulfoxide, and 40% in hexamethylphosphoric triamide.[191c]

The ratio of *O*-alkylation product to *C*-alkylation product is larger for free anions (which exist, e.g., in dipolar aprotic solvents; see p. 187) than for ion pairs or higher aggregates (present, e.g., in benzene or ether) or for strongly hydrogen-bonded anions in protic media. Association results in shielding of the oxygen atom. The same is true for N/C and N/O reactivities of imine[192] and oxime anions[193] (Scheme 41).

The reactivity of other ambident systems similar to enolates was described by Gompper and Wagner.[191c]

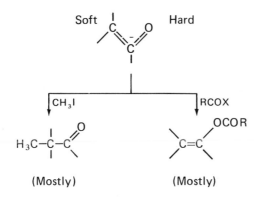

Scheme 40 Reactions of enolates and phenolates with hard and soft electrophiles.

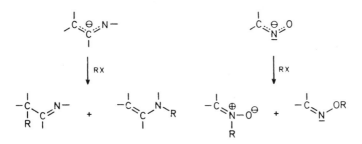

Scheme 41 Ambident behavior of imine and oximate anions.

3.2.2.2 S_N2 Versus Other Nucleophilic Reaction Modes

Most conspicuous are competing E2 reactions as exemplified in Scheme 42.

A good example is the increase in the amount of E2 reaction (at the expense of S_N2) with increasing pK of HNu (increasing hardness of the nucleophile) on treatment of 1,2-dichloroethane with bases[195] (Fig. 32).

Similar situations are found in reactions of nucleophiles with vinyl halides,[195a] [e.g., (194),[195b]] and with appropriate cyclopropane derivatives, [e.g. (194a)[195c]].

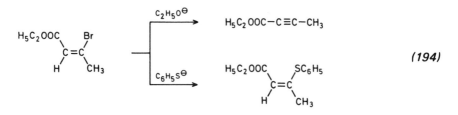

(194)

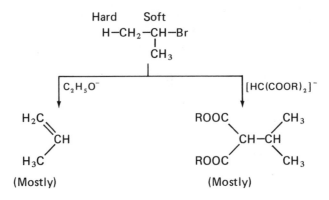

Scheme 42 Competition between E2 and $S_N 2$ reactions of isopropyl bromide.[194]

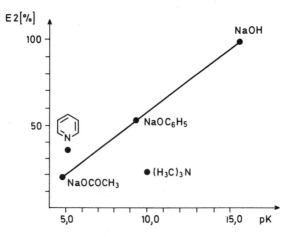

Fig. 32 Percent E2 reaction from 1,2-dichloroethane with different bases in water at 120°C. [From K. Okamoto, H. Matsuda, H. Kawasaki, and H. Shingu, *Bull. Chem. Soc. Jp.* **40**, 1917 (1967) (ref. 195).]

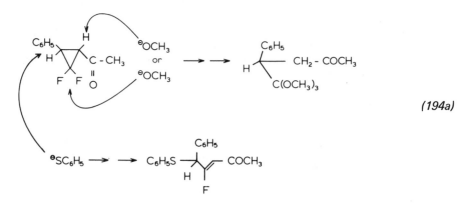

(194a)

The competition between C=O attack and S_N2 has been influenced by changing the leaving group, (*194b*)[195d] thereby making C-α of similar hardness to C=O.

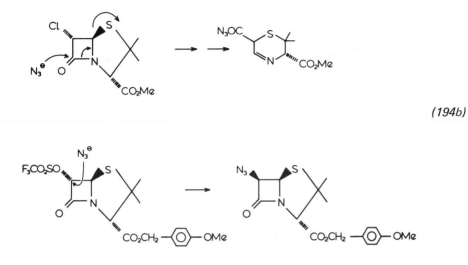

(194b)

3.2.2.3 Nitrite[196] and Cyanate Anions[197] (Schemes 43 and 44)

Here, also the influence of the gegenion (silver nitrite yields mainly nitrite esters, sodium nitrite mainly nitro compounds) may be understood in terms of the HSAB principle: the soft silver ion coordinates preferentially with the relatively soft nitrogen atom and shields it from the electrophile.

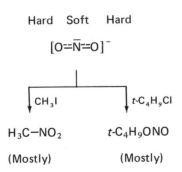

Scheme 43 Favored reaction paths of nitrite anion with soft and hard electrophiles.

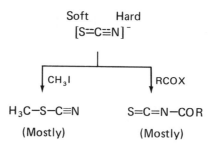

Scheme 44 Favored reaction paths of thiocyanate anion with soft and hard electrophiles.

3.2.2.4 *Aromatic Substitution*

Quantum mechanical calculations show[198] that in the ground state of toluene the largest negative charge resides on C-2, whereas the atomic orbital of C-4 has the largest contribution (coefficient) to the highest occupied molecular orbital. Consequently, C-2 may be considered the *harder* center and C-4 the *softer* center. Increasing hardness of the electrophilic reagent (as long as steric effects are unimportant or remain the same) should then lead to an increase in *ortho* isomer in the reaction mixture. The results on p. 14 show that this is indeed the case. Soft reagents (e.g., Br_2 or Hg^{2+}) attack the 4-position preferentially.

Charge effects observed in addition reactions of *homoaromatic* ions (*195*) can also be rationalized on the basis of the HSAB model. Nmr data for substituted bishomocyclopropenyl cations (177) show that C-2 and C-3 bear a substantially larger amount of positive charge than does C-7.[199] The hard base methoxide ion and other negatively charged

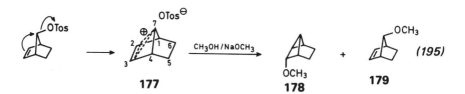

TABLE 41   Effect of Sodium Methoxide Concentration on the Methanolysis of *anti*-7-Norbornenyl Tosylate[200]

$NaOCH_3 [mol \cdot l^{-1}]$	% **178**	% **179**
$3,5 \cdot 10^{-8}$	0,3	99,7
0,2	6,4	93,6
0,88	20,3	79,7
1,93	34,7	65,3
3,93	51,5	48,5

nucleophiles (e.g., CN⁻ and H⁻) attack those positions preferentially, whereas the addition of CH_3OH where electrostatic factors are smaller takes place at C-7 (Table 41; see also p. 30).

Similar results were obtained[201] on decomposition of the *exo*-bicyclo[5.1.0]oct-8-yl diazonium ion [(*196*)], (proceeding via a disrotatory electrocyclic reaction, see Section 5.2.4.2).

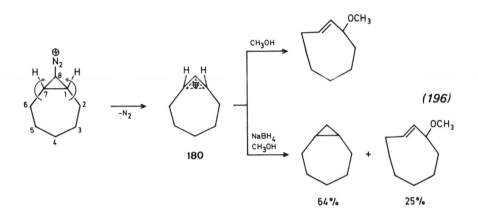

MO calculations show[202] that on slight disrotation of carbon atoms 1 and 7 in the direction shown by curved arrows a "half-opened" cation **180** may appear as intermediate and that in this ion positive charge on C-8 should be larger than on C-1 and C-7. Preferential formation of the cyclopropane derivative with the harder nucleophile H⁻ (relative to CH_3OH) may be interpreted in this way./[201a]/

Even though the HSAB concept can rationalize the results, it must be kept in mind that in detailed analyses of ambient behavior (especially in species having reaction centers that are not very different but may lead to very different reaction products), one must take into account other factors as well, for example, *geometries* of various transition states, *thermodynamics* of competing processes (i.e., relative stabilities of products; see Section 5.2.2.2), *steric effects*, the *position of transition states* on various reaction coordinates (see Section 5.2.2.1), or the *principle of least nuclear motion*/[203]/ (see Section 5.2.2).

3.2.2.5 *Addition of Nucleophiles to α,β-Unsaturated Carbonyl Compounds*

The HSAB model turned out to be a useful simple principle for ordering the multifaceted *additions of nucleophiles to α,β-unsaturated carbonyl compounds*, **181** (see also pp. 12-13).

181

1 Reactions with Organometallic Compounds $R^{\delta-}$–$Met^{\delta+}$ [204]

(a) *Variation of the metal.* The relatively polar organolithium compounds add preferentially to the hard 2-position; the more
covalent Grignard compounds form mostly 1,4-adducts.

(b) *Variation of R.* Vinyl- and phenylorganometallic compounds
have a larger tendency toward 1,2-additions than do the corresponding alkyl derivatives. Branching of the alkyl group increases the tendency toward 1,4-addition. In addition to steric
effects, the softening of R^- may also be responsible for this
behavior (electron-attracting alkyl substituents diminish the
hard base properties of the central C^- atom). In accord with
their softness, enolates and cyanide add to the 4-position
(*Michael addition*).

(c) *Variation of X.* When X varies from $X = H \rightarrow X = CH_3 \rightarrow X = OR$
$\rightarrow X = C_6H_5$, switching from predominantly 1,2-addition to
predominantly 1,4-addition within a series of $R^{\delta-}$–$Met^{\delta+}$
should take place progressively later with increasing hardness of
R^- because the hardness of the 2-position increases in the order
shown (Scheme 45).

Property of C-2 in 181

	Soft ⟶ Hard			
Property of R in $R^{\delta-}$–$Met^{\delta+}$	$X = H$	$X = CH_3$	$X = OR$	$X = C_6H_5$
Soft ↓ Hard	Reactant combinations leading predominantly to 1,2-additions, because the hardness of R is equal or greater than that of C-2		Combinations of reactants leading mainly to 1,4-additions because C-2 is supposedly harder than required by the softness of R	

Scheme 45 Schematic representation of the increase in 1,2-addition
within a series of compounds of structure **181** with increasing hardness
of the organic component R of the reaction partner $R^{\delta-}$–$Met^{\delta+}$.

2 Reduction with Complex Hydrides. Similar events take place in reductions with complex hydrides.[205] Since CH_3O^- is harder than H^-, $LiAlH(OCH_3)_3$ is harder (more ionic) than $LiAlH_4$ (see Section 3.2.2.6). Consequently, the latter reagent should lead predominantly to 1,4-reductions (Scheme 46).[205b]

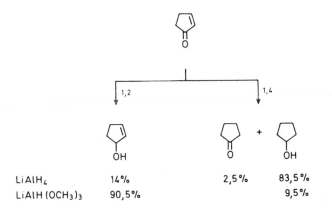

	1,2		1,4
	OH	O + OH	
LiAlH$_4$	14%	2,5%	83,5%
LiAlH(OCH$_3$)$_3$	90,5%		9,5%

Scheme 46 1,2- and 1,4-reduction of cyclopentene-3-one by complex hydrides. (Cyclopentanol is formed by further reduction of the initially formed enolate of cyclopentanone.)

Since the hardness of the reagent increases on successive replacement of the four hydrogens of $LiAlH_4$,

$$[LiAlH_4 \xrightarrow{R_2C=O} LiAlH_3OCHR_2 \xrightarrow{R_2C=O} LiAlH_2(OCHR_2)_2, \text{ etc.}]$$

smaller amounts of $LiAlH_4$ or *inverse addition* (addition of the hydride to the solution of the carbonyl compound (*197*)) allows an increase in the relative amount of the 1,2-reduction product.

The softer borohydrides yield more 1,4-reduction product than aluminum hydrides (*198*):

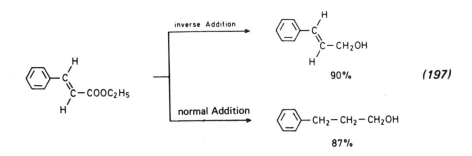

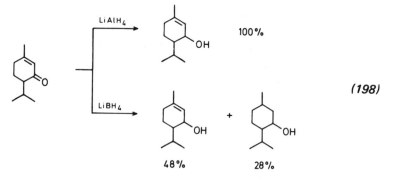

(198)

3 Deviations by Possible Changes of Mechanism.

Results of kinetically controlled additions of group IVA anions to cyclohexenones are given in Table 42*a*:

TABLE 42*a*   Additions of Group IVA Anions to Cyclohexenones[205c]

Lithium Compound	Solvent	Mode of Addition
MeLi	(THF–HMPA)	1,2
Me$_3$CLi	(THF–HMPA)	1,2 + 1,4
Me$_3$SiLi	(THF–HMPA)	1,4
Me$_3$SnLi	(THF)	1,2

The behavior of trimethylstannyllithium is opposite to predictions based on the HSBA principle. The trimethylstannyl anion is expected to be softer than the anions of silicium and carbon. Consequently, it should be *more* disposed to 1,4-addition. The discrepancy is explained by the assumption of different mechanisms for 1,2 and 1,4-additions. Direct nucleophilic additions of anions are hypothesized to be generally 1,2, whereas 1,4-additions are considered to be generally two-step *electron transfer* additions. *t*-Butyllithium and trimethylsilyllithium are known to be prone to electron transfer, since in these systems corresponding radicals are more stable than their anion counterparts. This does not apply to the corresponding tin system, where both the radical *and* the anion are stable, and to the simple methyl case, where the anion is more stable than the radical.

3.2.2.6 Symbiosis and Leaving Group Effects

The nature of a ligand affects the character of the central atom, and vice versa. B^{3+} becomes softer on coordination with H$^-$, which strives to achieve covalent bonding. This makes it easier to coordinate a second hydride ion and even easier a third and a fourth ligand of the same type. Similarly, a first ionic coordination of F^- enables further coordination of hard ligands. Mixed complexes (e.g., BHF_3^-) are unstable with respect to the soft BH_4^- and the hard BF_4^-./205a/

The tendency to *accumulate several ligands of the same type (symbiosis)* should account for the difference in effectiveness of various leaving groups in S_N2 reactions, since the transition state of these reactions can be viewed as an acid-base complex of nucleophile Nu and CR_3X:

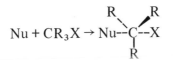

Data in Table 42 show that soft leaving groups are particularly effective when the nucleophile is also soft.[206]

TABLE 42   Dependence of $k_{CH_3OTos}:k_{CH_3I}$ on the Nature of Nucleophile Nu^- in

$$Nu^\ominus + CH_3X \xrightarrow{\ k_i\ } Nu - CH_3 + X^\ominus$$

X = OTos, i = CH₃OTos

X = I,   i = CH₃I

$\dfrac{k_{CH_3OTos}}{k_{CH_3I}}$	210	4.6	2.8	0.95	0.72	0.28	0.18	0.13	0.13
$Nu^\ominus$	CH_3OH	$CH_3O^\ominus$	$Cl^\ominus$	$N(C_2H_5)_3$	$Br^\ominus$	$SCN^\ominus$	$P(C_6H_5)_3$	$I^\ominus$	$C_6H_5S^\ominus$

The reversal of relative reactivities of methyl tosylate and methyl iodide on changing the nucleophile from methoxide to thiophenolate can also be viewed as a consequence of the increase in relative nucleophilicity of thiophenolate vis à vis methoxide, when replacing the tosylate by the iodide (see p. 147):

$$\left(\frac{k_{CH_3OTos}}{k_{CH_3I}}\right)_{CH_3O^\ominus} \left(\frac{k_{CH_3I}}{k_{CH_3OTos}}\right)_{C_6H_5S^\ominus} = \frac{4.6}{0,13} \qquad (199a)$$

Exchanging subscripts and rearranging terms gives (*199b*):

$$\left(\frac{k_{CH_3O^\ominus}}{k_{C_6H_5S^\ominus}}\right)_{CH_3OTos} \cdot \left(\frac{k_{C_6H_5S^\ominus}}{k_{CH_3O^\ominus}}\right)_{CH_3I} = \frac{4,6}{0,13} \qquad (199b)$$

Using experimental values

$$(k_{CH_3O^-})_{CH_3OTos} = 1.16 \times 10^{-3} \text{ l mol}^{-1} \text{ s}^{-1}$$

and

$$(k_{C_6H_5S^-})_{CH_3OTos} = 1.42 \times 10^{-1} \text{ l mol}^{-1} \text{ s}^{-1}$$

$$\left(\frac{k_{CH_3O^-}}{k_{C_6H_5S^-}}\right)_{CH_3OTos} = \frac{1}{122}$$

and from (*199b*):

$$\left(\frac{k_{CH_3O^-}}{k_{C_6H_5S^-}}\right)_{CH_3I} = \frac{1}{4320} \qquad\qquad (199c)$$

The often-quoted[207] ratio of relative reactivities of alkyl halides

$$RCl:RBr:RI = 1:50:100$$

is valid only for such hard nucleophiles as alkoxides, amines, and similar ones in protic media. For carbanions the ratio $k_{RI}:k_{RCl}$ can become much higher[208] (*200a-200d*):

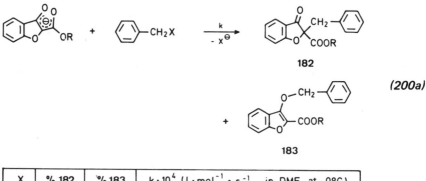

182

183

(200a)

X	% 182	% 183	$k \cdot 10^4$ (l·mol^{-1}·s^{-1} , in DMF at 0°C)
Cl	40	50	0,08
Br	64	29	13
I	72	18	200

$$\frac{k\left(\text{C}_6\text{H}_5\text{-CH}_2\text{I} \rightarrow 182\right)}{k\left(\text{C}_6\text{H}_5\text{-CH}_2\text{I} \rightarrow 182+183\right)} \cdot \frac{k\left(\text{C}_6\text{H}_5\text{-CH}_2\text{Cl} \rightarrow 182+183\right)}{k\left(\text{C}_6\text{H}_5\text{-CH}_2\text{Cl} \rightarrow 182\right)} = \frac{72}{90} \cdot \frac{90}{40} \qquad (200b)$$

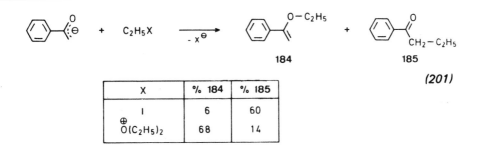

$$\frac{k \quad \langle = \rangle -CH_2Cl \longrightarrow 182 + 183}{k \quad \langle = \rangle -CH_2 I \longrightarrow 182 + 183} = \frac{0{,}08}{200} \qquad (200c)$$

$$\frac{k \quad \langle = \rangle -CH_2 I \longrightarrow 182}{k \quad \langle = \rangle -CH_2Cl \longrightarrow 182} = \frac{72 \cdot 200}{40 \cdot 0{,}08} = \frac{4500}{1} \qquad (200d)$$

The observation that enolates give more *O*-alkylation products with alkyl tosylates and oxonium salts than with alkyl iodides may also be interpreted as a *symbiotic effect*[191b] (*201*):

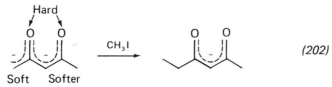

X	% 184	% 185
I	6	60
$\overset{\oplus}{O}(C_2H_5)_2$	68	14

(201)

3.2.2.7 Final Remarks

One must again emphasize the qualitative nature of the HSAB principle and Klopman's model calculations. Both treatments state that the best chances for a reaction to occur at a hard (soft) center are to use a hard (soft) reaction partner. Neither method, however, can predict *what* reagent has in practice the required hardness (softness) or even whether such a reagent exists. For instance, dianions of β-dicarbonyl compounds yield products resulting exclusively from attack at the terminal position even with the softest of the electrophiles used so far[59] [(*202*); see also p. 50]. In this context one should again mention the remarks on page 161.

(202)

3.2.3 Correlation Equations

3.2.3.1 Proton Transfer; Brönsted Relationships[208a]

For a series of similar acids (and within a not too wide range of p*K*'s) rates of protonation of substrate S by acid AH (*203*) obey very often the Brönsted equation (204); similarly, deprotonation by a series of bases A⁻ obeys (205):

$$A{-}H + IS \underset{k_{A^-}}{\overset{k_{AH}}{\rightleftharpoons}} IA^- + H{-}\overset{+}{S} \qquad\qquad (203)$$

Brönsted relationships

$$\log k_{AH} = \alpha \log K_{AH} + C_1 \qquad\qquad (204)$$

$$\log k_{A^-} = -\beta \log K_{AH} + C_2 \qquad\qquad (205)$$

$$\frac{k_{AH}}{k_{A^-}} = K = \frac{K_{AH}}{K^+_{S-H}} \qquad\qquad (206)$$

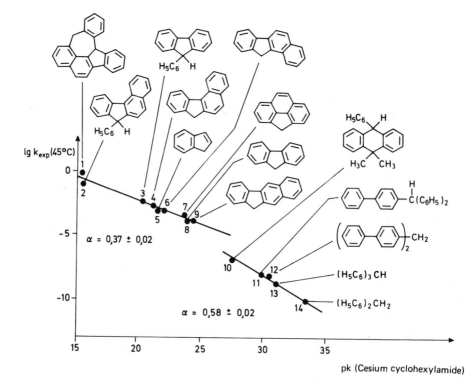

Fig. 33 Brönsted plot of the rate constants k_{exp} (1 mol⁻¹ s⁻¹) for the exchange reaction (detritiation)

$$AT \xrightarrow[k_{exp}]{CH_3OH/NaOCH_3} AH$$

of fluorenes and polyarylmethanes (AH).[209] [From A. Streitwieser, Jr., W. B. Hollyhead, G. Sonnichsen, A. H. Pudjaatmaka, C. J. Chang, and T. L. Kruger, *J. Amer. Chem. Soc.* **93**, 5096 (1971) (ref. 209).]

where C_1, α, C_2, and β are constants for a given series of similar acids AH; however, they change with substrate (S, H—S$^+$), solvent, and temperature. Substituting (204) and (205) in (206) and comparing coefficients (K^+_{S-H} = const) it is seen that in a given system the variation of AH, or A$^-$ in the forward and reverse directions must obey the relation $\alpha + \beta = 1$. In general values of α and β are positive and less than 1 (see following examples, Fig. 33 and Table 43).

The different values for α in Fig. 33 show that the rate-determining step (207)

$$\frac{CH_3OH}{CH_3O^-} + AT \underset{k_{-1} \text{ (fast)}}{\overset{k_1 \text{ (slow)}}{\rightleftharpoons}} \frac{CH_3OT}{CH_3OH} + A^- \xrightarrow{k_2 \text{ (fast)}} \frac{CH_3OT}{CH_3O^-} + AH \qquad (207)$$

is different for the two classes of compounds (possibly because of additional aromatic stabilization in cyclopentadiene derivatives). Because of "internal return," (k_{-1}), k_{exp} is smaller than the rate constant k_1 of the rate-determining step. The different extent of "internal return" in various compounds has only a small effect on the slope in the Brönsted plot[209] (see also Section 5.2.3.3).

Table 43, which gives β values for some elimination reactions in ethanol,[210] shows that variations in β values may be used as a diagnostic tool in the analysis of the reaction mechanism. Compounds such as t-butyl chloride, which undergo E1 elimination, show low values for β. Typical E2 substrates, such as 2-arylethyl bromides, show intermediate values and DDT, which reacts by the E1cB mechanism, shows high values for β.

TABLE 43 β Values for $A^- + {}_{{}_{\blacktriangleleft}}\overset{H}{\underset{X}{C-C}} \rightarrow AH + \overset{}{C=C} + X^-$

Compound	Temperature ($^\circ$C)	β
$t\text{-}C_4H_9Cl$	45	0.17
$t\text{-}C_4H_9\overset{+}{S}(CH_3)_2$	25	0.46
$(H_3C-CH_2-CH_2)_2CHBr$	60	0.39
$H_5C_6-CH_2-CH_2-Br$	60	0.56
$4\text{-}NO_2-C_6H_4-CH_2-CH_2-Br$	60	0.67
$(4\text{-}Cl\text{-}C_6H_4)_2CH-CCl_3$	45	0.88

Transgressions of the normal region ($0 < \alpha, \beta < 1$) were found for deprotonation of nitroalkanes[211] (208) (see Section 5.2.3.3 for the interpretation of this behavior).

$$R_2CH-NO_2 \quad + \quad OH^{\ominus} \quad \longrightarrow \quad R_2C=NO_2^{\ominus} \quad + \quad H_2O$$

$$R_2 = H,H; \; H,CH_3 \; ; \; CH_3,CH_3 : \; \alpha = -0,7 \tag{208}$$

$$R_2 = CH_3, CH_2 - \langle \text{⬡} \rangle_X \quad : \quad \alpha = \; 1,61$$

If in acid-base-catalyzed reactions proton transfer becomes rate determining ("general acid or base catalysis"), rate constants representing the contribution of various "general" acids or bases to the rate constant of the total reaction are given by (204) and (205).

[Such a situation was first encountered in the general base-catalyzed decomposition of nitramide to N_2O and water (209) by J. N. Brönsted and K. Y. Pedersen[212]; (204) and (205) therefore are known as *Brönsted's catalysis laws*.]

$$A^- + HN=\overset{+}{N}\overset{\displaystyle OH}{\underset{\displaystyle O^-}{\Big\langle}} \quad \xrightarrow{\text{slow}} \; A-H + N_2O + OH^-$$

$$\tag{209}$$

$$AH + OH^- \xrightarrow{\text{fast}} A^- + H_2O$$

An example of such a situation and of the method used in analysing general acid-base catalysis is the *detritiation of t-butylmalononitrile*.[213] In *aqueous* solution the base is not only the carboxylate ion (A^-) but also H_2O (and when $A^- = CH_3COO^-$ also OH^-). This can be seen from the fact that k_{exp} is different from zero even when the concentration of the acetate buffer is zero (Fig. 34; see also Fig. 46, below). The observed reaction must be catalyzed by H_2O and HO^-. The rate constant for water catalysis is given by $k_{exp(acid)}$ as found on working in acidic solutions.

The difference between the intercept on the ordinate for the acetate reaction and $k_{exp(acid)}$ is then $k_{OH^-}[OH^-]$. In a more basic buffer ($\cdot$) the intercept is higher because the contribution of OH^- is larger.

The characteristics of general catalysis are: linear relationship between rate constant k_{exp} and concentration of buffer containing the catalytically active acid or base and simultaneous action of all general acids or bases present; in base catalysis:

$$\text{rate} = k_{exp}[\text{substrate}] = \sum_i k_{A_i}[A_i^-] \cdot [\text{substrate}] \tag{212a}$$

$$k_{exp} = \sum_i k_{A_i}[A_i^-] \tag{212b}$$

Under these conditions (constant pH) rates of "specific" H^+ or OH^--catalyzed reactions (see below) remain practically unchanged.

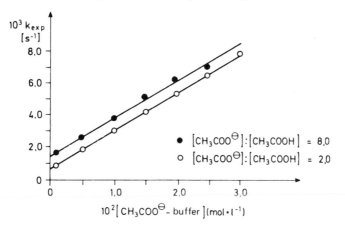

Fig. 34 Detritiation of t-butylmalononitrile in water.

Overall reaction:

$$t-C_4H_9-CT(CN)_2 \xrightarrow{k_{exp}} t-C_4H_9-CH(CN)_2 \qquad (210)$$

Discrete steps:

$$t-C_4H_9-CT(CN)_2 + CH_3COO^- \xrightarrow{k_{CH_3COO^-}} t-C_4H_9-C(CN)_2^- + CH_3COOT \quad (211a)$$

$$t-C_4H_9-CT(CN)_2 + H_2O \xrightarrow{k_{H_2O}} t-C_4H_9-C(CN)_2^- + H_2TO^+ \qquad (211b)$$

$$t-C_4H_9-CT(CN)_2 + OH^- \xrightarrow{k_{OH^-}} t-C_4H_9-C(CN)_2^- + THO \qquad (211c)$$

$$t-C_4H_9-C(CN)_2^- + H_2O \xrightarrow{fast} t-C_4H_9-CH(CN)_2 + OH^- \qquad (211d)$$

In base detritiation of t-butylmalononitrile, k_A values for different bases, obtainable from the slope of the respective k_{exp} – [A$^-$-buffer] plot are given in Table 44 and Fig. 35. (Deviations of k_{H_2O} and k_{OH^-} from Brönsted straight lines are often observed and are attributed to the nature of the species H_2O and OH^- in aqueous medium).

TABLE 44 Brönsted Relationship in Detritiation of
t-Butylmalononitrile

Base (A$^\ominus$)	pK (HA)	$k_{A^\ominus}$[l·mol^{-1}·s^{-1}]
H_2O	– 1,74	5,89 · 10^{-6}
$ClCH_2COO_2^\ominus$	2,85	3,25 · 10^{-3}
$HCO_2^\ominus$	3,75	2,52 · 10^{-2}
$CH_3-COO^\ominus$	4,76	0,231
$OH^\ominus$	15,75	2,48 · 10^5

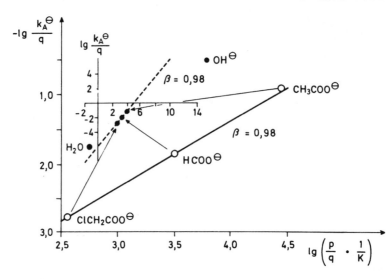

Fig. 35 Brönsted plot of statistically corrected data from Table 44. [From F. Hibbert, F. A. Long, and E. A. Walters, *J. Amer. Chem. Soc.* **93**, 2829 (1971) (ref. 213).]

Statistical Corrections

Rate and equilibrium constants used in the example in Scheme 9 and in the reaction above were statistically corrected. The probability (p) of a proton transfer from, for example, a molecule of water is twice that in dissociation of a COOH group, which has only one proton; on the other hand, in the reverse reaction the COO⁻ ion has two sites for the addition of the proton, whereas in OH⁻ there is only one (probability q). Consequently, for a general acid dissociation equilibrium (*213a*) the statistically corrected equilibrium constant K_{corr} can be expressed by (*213c*)[214]:

$$AH_p \underset{qk^-}{\overset{pk}{\rightleftharpoons}} AH_{p-1}^{q\left(\frac{1}{q}\ominus\right)} + H^{\oplus} \qquad (213a)$$

$$K_{exp} = \frac{pk}{qk^-} \qquad (213b)$$

$$K_{corr} = \frac{k}{k^-} = \frac{q}{p} K_{exp} \qquad (213c)$$

Examples of general acid-catalyzed reactions where rates are equal to the sum of the rates of individual protonation reactions of the substrate are: the *hydrolysis of vinyl ethers*[215], (*214a-214c*) and the related tritium exchange in 1,3,5-trimethoxybenzene-2-*t*[216] (*215*).

$$H_2C=CH-OC_2H_5 + HA \xrightarrow[\alpha=0.70]{slow} H_3C-CH\overset{\pm}{}OC_2H_5 + A^-$$ (214a)

$$H_3C-CH\overset{\pm}{}OC_2H_5 + H_2O \xrightarrow{fast} H_3C-\underset{\underset{OH}{|}}{CH}-OC_2H_5 + H^+$$ (214b)

$$H_3C-\underset{\underset{OH}{|}}{CH}-OC_2H_5 \xrightarrow{fast} H_3C-CHO + C_2H_5OH$$ (214c)

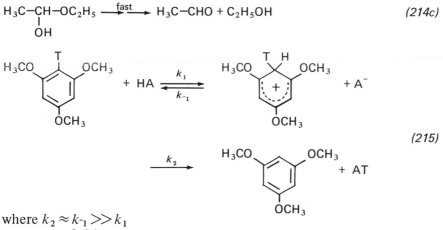

(215)

where $k_2 \approx k_{-1} \gg k_1$
$\alpha = 0.56$

The Brönsted plot for reaction (*215*) (Fig. 36) clearly shows that *optimal correlations are possible only within families of very similar types of compound*.

If all substrates are considered, $\alpha = 0.56$; for correlations of carboxylic acids only $\alpha = 0.59$; for correlation of bisulfate, bioxalate, and biphosphate $\alpha = 0.71$, for all substrates, omitting H_2O and H_3O^+, $\alpha = 0.69$ (Fig. 36). In particular, if α values are to be used to draw conclusions with respect to the structure of the transition state (see p. 326 and Section 5.2.3.3) one cannot ignore the variation of α with the nature of the acid chosen as catalyst.

The general acid catalysis in the *dehydration of the hydrate of acetaldehyde* (Fig. 37) can be associated with the rate-determining attack of conjugate base A^- following the very rapid equilibrium protonation of the substrate (216a, 216b):

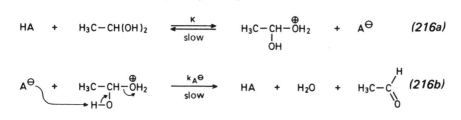

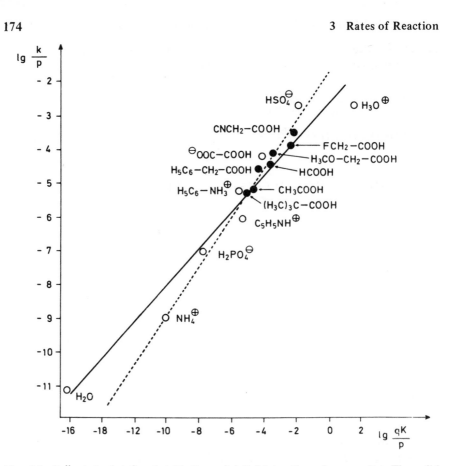

Fig. 36 Brönsted plot for detritiation of 1,3,5-trimethoxybenzene-2-*t*. The solid line results from considering all points ($\alpha = 0.56$). The dashed line results when points for H_2O and H_3O^+ are left out ($\alpha = 0.69$). [From A. J. Kresge, S. Slae, and D. W. Taylor, *J. Amer. Chem. Soc.* **92**, 6312 (1970) (ref. 216).]

If K is the ratio of acid dissociation constants for HA and $H_3CCH-\overset{+}{O}H_2$ (K_{HA} and $K_{H_3C-CH-\overset{+}{O}H_2}$), and if k_{A^-} is expressed by (205) and $\alpha + \beta = 1$, the rate of formation of acetaldehyde—and, the

$$\left(\frac{d[H_3C-CHO]}{dt}\right)_{HA} = k_{A\ominus}[A^\ominus]\left[H_3C-\underset{\underset{OH}{|}}{CH}-\overset{\oplus}{O}H_2\right] = k_{A\ominus}\cdot K\cdot[H_3C-CH(OH)_2][HA]$$

$$= C\cdot K_{HA}^{-\beta}\cdot\frac{K_{HA}}{K_{H_3C-CH-\overset{\oplus}{O}H_2}}[H_3C-CH(OH)_2][HA] \qquad (217a)$$

$$= C'\cdot K_{HA}^{\alpha}\cdot[H_3C-CH(OH)_2][HA]$$

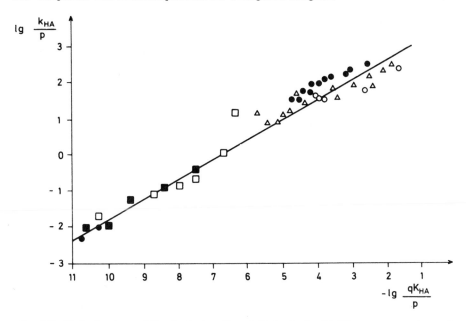

Fig. 37 Dehydration of the hydrate of acetaldehyde in 92.5% aqueous acetone at 25°C.[217] Triangles, aliphatic carboxylic acids; open circles, *ortho*-substituted aromatic carboxylic acids; solid circles; *meta*-substituted and *para*-substituted aromatic carboxylic acids; open squares, *ortho*-substituted phenols; solid squares, *meta*-substituted and *para*-substituted phenols. [From K. Schwetlick, *Kinetische Methoden zur Untersuchung von Reaktionsmechanismen*, VEB Deutscher Verlag der Wissenschaften, Berlin, 1971.)

rate constant k_{HA} for the action of acid HA—can be expressed by (*271a*) and (*271b*), respectively, where C and C' are constants.

$$\lg k_{HA} = \alpha \lg K_{HA} + \lg C' \qquad (217b)$$

In the corresponding acid-catalyzed hydrolysis of acetals a proton can no longer be abstracted by A⁻. The alternative is slow *unimolecular decomposition* of the protonated substrate (*218*):

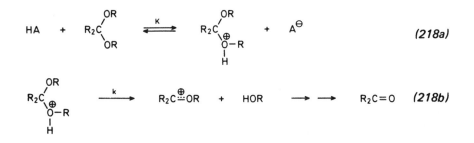

$$\frac{d[R_2CO]}{dt} = k \cdot K \cdot \frac{[HA][R_2C(OR)_2]}{[A^\ominus]} \qquad (218c)$$

In addition, under the reaction conditions the pair $HA-A^-$ is in acid-base equilibrium with water (K_{HA}) and:

$$K = \frac{K_{HA}}{K_{R_2C\underset{\underset{H}{|}}{\overset{OR}{\underset{\oplus}{\diagdown}}}OR}}$$

$$\frac{d[R_2CO]}{dt} = \frac{k}{K_{R_2C\underset{\underset{H}{|}}{\overset{OR}{\underset{\oplus}{\diagdown}}}O-R}} \cdot [R_2C(OR)_2][H_3O^\oplus] \qquad (219)$$

Catalytic reactions whose rates depend *exclusively on the concentration of the hydronium ion* are termed to be subject to *specific acid catalysis.*

    General acid catalysis in the hydrolysis of acetals is observed only when structural factors favor the formation of the alkoxycarbocation to such an extent that the protonated substrate is never fully formed; decomposition to the cation takes place even on partial bonding of the proton. Examples are **186** and **187** and orthoesters **188** (see also Section 3.3.1.2)[218]:

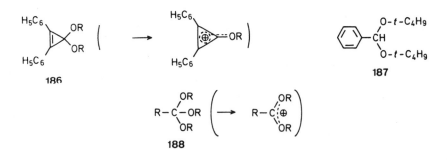

    The rapid formation of the carbocation from **186** and **188** is due to its special stability; in **187** the reason is that on ionization the unfavorable steric interactions between the two *t*-butyl groups are relieved.

    The importance of the Brönsted correlations is mainly derived from *mechanistic information* that can be obtained once the occurrence of general catalysis is established.[219a,219b] Values of α and β shed light on the properties of transition states (see above and Section 5.2.1). Of practical interest is the possibility of determining pK values of weak acids where equilibrium dissociation can no longer be measured directly.

Example

Determination of the "ion-pair pK value" of benzene.[220] The "ion-pair pK values" of polyfluorobenzenes (ArH) were determined by measuring the equilibrium constant K (in cyclohexylamine at 34°C) of the acid-base equilibrium

$$ArH + R^-Cs^+_{ion\ pair} \underset{}{\overset{K}{\rightleftarrows}} Ar^-Cs^+_{ion\ pair} + RH$$

$$K = \frac{K_{ArH}}{K_{RH}} \qquad\qquad (220)$$

TABLE 45 (A) "Ion-Pair pK Values" (in cyclohexylamine) of Fluorobenzenes and the Extrapolated Value for Benzene
(B) Logarithms of Partial Rate Factors for Exchange in *ortho-*, *meta-*, and *para-*Fluorobenzenes by NaOCH$_3$−CH$_3$OH or LiNHC$_6$H$_{11}$−C$_6$H$_{11}$NH$_2$

A		B		
ArH	pK (per Hydrogen)		NaOCH$_3$-CH$_3$OH	LiNHC$_6$H$_{11}$-C$_6$H$_{11}$NH$_2$
(pentafluorobenzene structure)	25.8	$\log f_o^F$	5.25	5.43
(tetrafluorobenzene structure)	31.5	$\log f_m^F$	2.07	1.95
(difluorobenzene structure)	35	$\log f_p^F$	1.13	1.03
(benzene structure)	43			

(RH = polyarylmethane of known K_{RH}). These pK values (see Table 45) are correlated with the logarithms of the relative rate constants for hydrogen isotope exchange at a single *ortho* position of ArH in $NaOCH_3-CH_3OH$ or $LiNHC_6H_{11}-C_6H_{11}NH_2$ and extrapolated to the pK of benzene [rate constants are obtained by summation of the corresponding values of log $f_{o,m,p}^F$ (Table 45), where f = partial rate factors, see p. 54]. The extrapolation gives $pK_{benzene}$ (per hydrogen) = 43 (Fig. 38). The extrapolation is based on the (assumed) Brönsted correlation of log K and log k, but since we are dealing with a series of closely related compounds, this seems to be perfectly valid.

pK values depend strongly on solvation energies. This means that "ion-pair pK values" determined in the nonpolar cyclohexylamine cannot be compared to common pK_a values determined in water. However, for work with "phenyl anion reagents," which almost always involves

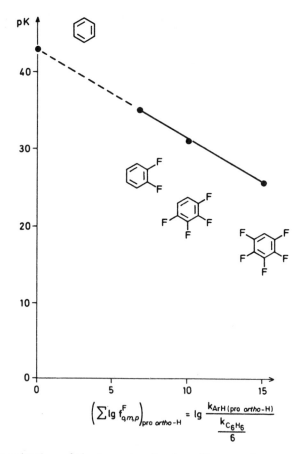

Fig. 38 Determination of the ion-pair pK value of benzene by means of rates of H isotope exchange of polyfluorobenzenes (ArH).

ion pairs in aprotic media, the pK values given in Table 45 are quantitatively significant.

In the case of cyclopentadiene it could be shown that the value of pK in water differs only slightly from the cesium ion-pair pK value in cyclohexylamine [15.6 (per hydrogen atom) and 16.25 (per hydrogen atom), respectively]. The same should be true for other hydrocarbons where extensive delocalization of negative charge in the carbanions minimizes specific ion-pair interactions.[220a]

3.2.3.2 Carbon Nucleophiles

Additions to the Carbonyl Group. Since the carbonyl group is polarized, one can expect that reactivity will be controlled by charge and that relative nucleophilicities with respect to this hard center will be approximately proportional to those toward the proton.

This is confirmed by reactions of several acetates with nucleophiles (Nu⁻) [(221) and Fig. 39.[221]].

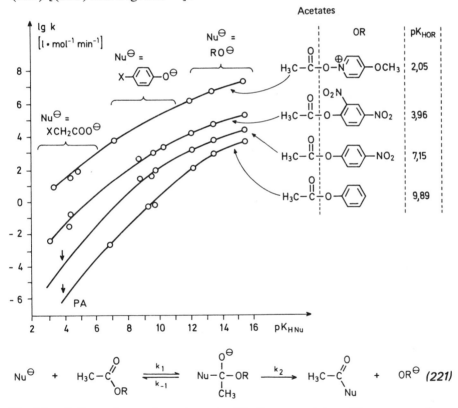

Fig. 39 pK Dependence of reaction rates (k) of some acetates with oxygen nucleophiles. [From W. P. Jencks and M. Gilchrist, J. Amer. Chem. Soc. 90, 2622 (1968) (ref. 221).]

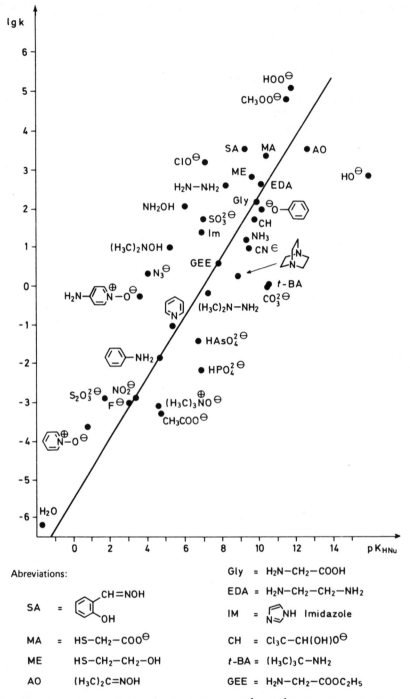

Abreviations:

SA = (benzene ring with CH=NOH and OH substituents)

MA = $HS-CH_2-COO^\ominus$

ME $HS-CH_2-CH_2-OH$

AO $(H_3C)_2C=NOH$

Gly = H_2N-CH_2-COOH

EDA = $H_2N-CH_2-CH_2-NH_2$

IM = (imidazole ring) NH Imidazole

CH = $Cl_3C-CH(OH)O^\ominus$

t-BA = $(H_3C)_3C-NH_2$

GEE = $H_2N-CH_2-COOC_2H_5$

Fig. 40 Dependence of rate constants k ($1\ mol^{-1}\ min^{-1}$) of nucleophiles (Nu^-) and p-nitrophenyl acetate on the pK of the conjugate acid (HNu). (For oximes, peroxides, and thiols: dependence on the pK of the compound shown.) [From W. P. Jencks and Y. Carriuolo, *J. Amer. Chem. Soc.* **82**, 1778 (1960) (ref. 222).]

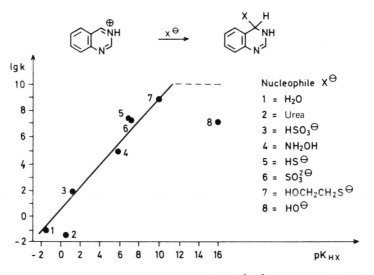

Fig. 41 pK Dependence of rate constants k (1 mol^{-1} s^{-1}) of nucleophiles (X$^-$) and the quinazolinuim cation. [From M. J. Cho and I. H. Pitman, *J. Amer. Chem. Soc.*, **96**, 1843 (1974) (ref. 223).]

Interpretations of these results, especially the nonlinear relationships, must take into account that reaction (*221*) proceeds in two steps.

The addition step may be expected to be rate determining only in the case of strongly basic nucleophiles ($pK_{HNu} > pK_{HOR}$). When $pK_{HNu} < pK_{HOR}$ the first step is *reversible* (the tendency of Nu$^-$ to leave the tetrahedral intermediate is greater than that of the OR$^-$ group) and the second step becomes *rate determining*. The stronger pK_{HNu} dependency of the reaction rate in this region ($\beta \simeq 1$) shows that pK_{HNu} has a greater effect on the *equilibrium* of *addition* than on its *rate*.

A rough pK dependence (β *ca.* 0.8) is observed even if one compares reactivities of nucleophiles belonging to different structural types; however, scattering due to various specific interactions becomes much larger[222] (Fig. 40). Linear pK dependence ($\beta = 0.88$) was also observed for rates of one-step addition of some nucleophiles to the chinazolinium cation[223] (Fig. 41).

S$_N$2-Reactions, Swain-Scott and Edwards Equations. Families of closely related nucleophiles, where both the attacking atom and its surroundings remain the same (e.g., series of ring-substituted phenolates, Fig. 42 and Table 46) show linear correlations between relative carbon nucleophilicity and pK even under S$_N$2 conditions. However, β values are lower than in additions to the carbonyl group, which means that in these soft-soft interactions the role of factors determining proton affinities is small.

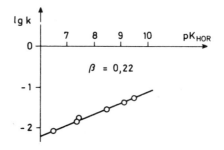

Fig. 42   pK Dependence of rate constants (1 mol^{-1} min^{-1}) of phenolates RO⁻ and 3-bromopropanol (see Table 46).[224] [From R. F. Hudson and G. Loveday, *J. Chem. Soc.*, **1962**, 1068 (ref. 224).]

TABLE 46   pK Dependence in Nucleophilic Substitution[225]

Substrate	Nucleophile	β [a]	Substrate	Nucleophile	β
$H_3C-O-SO_2-CH_3$	Pyridine	0,09	$C_3H_5)_2CHBr$ [b]	$R-C_6H_4O^\ominus$	0,27
$H_5C_2-O-SO_2-CH_3$	Pyridine	0,11	$C_8H_{17}Br$ [b]	$R-C_6H_4O^\ominus$	0,36
$H_3C-O-SO_3^\ominus$	$R-C_6H_4O^\ominus$	0,16	C_4H_9Br [b]	$R-C_6H_4O^\ominus$	0,37
$Br-(CH_2)_3-OH$	$R-C_6H_4O^\ominus$	0,22	$H_5C_6-(CH_2)_2-Br$ [b]	$R-C_6H_4O^\ominus$	0,35
$Cl-CH_2-COO^\ominus$	$R-COO^\ominus$	0,20	$H_5C_6-CH_2-Br$ [c]	$R-C_6H_4S^\ominus$	0,20
$Br-CH_2-COO^\ominus$	$R-COO^\ominus$	0,20	$H_2C=CH-CH_2-Br$ [d]	Pyridine	0,37
	$R-C_6H_4O^\ominus$	0,32			

[a] in H_2O [c] in CH_3OH
[b] in C_2H_5OH [d] in CH_3NO_2

The Swain-Scott equation *(222)*[226] allows larger variations in nucleophiles

Swain-Scott Equation

$$\log \frac{k}{k_0} = s \cdot n \qquad\qquad (222)$$

where k is the rate constant of the substrate with a given nucleophile, k_0 is the rate constant for the reaction of the same substrate with water, s is the susceptibility constant (reaction constant) of the substrate (by definition $s = 1$ for methyl bromide in aqueous solution) and n is the nucleophilicity parameter of a given nucleophile (by definition $n = 0$ for water) (Table 47). Several carbonyl compounds were investigated as well. Their s parameters have higher values than those of the sp^3-hybridized reaction centers.

TABLE 47 Swain-Scott Parameters

Nucleophile	n^a (H$_2$O, 0°C)	n^a (CH$_3$OH, 0°C)	Nucleophile	n^a (H$_2$O, 0°C)	n^a (CH$_3$OH, 0°C)
$S_2O_3^{2-}$	6.35	6.73	NO_3^-	2.1	0.4
SO_3^{2-}	5.67	6.92	HSO_4^-	2.1	
CN^-	5.13	4.98	ClO_2^-	2.1	
PO_4^{3-}	5.06		ClO_4^-	2.1	
I^-	4.93	5.70	H_2O	0.00	
SCN^-	4.80	5.07	Amines		
CO_3^{2-}	4.36		Pyrrolidine	5.67	
HO^-	4.23		Butylamine	5.13	
$C_6H_5O^-$	4.16	4.02	Morpholine	5.13	
Br^-	4.02	4.26	4-Methylaniline	4.52	
			4-Picoline	4.45	
N_3^-	3.92	4.24	Aniline	4.35	
$4\text{-}ClC_6H_4O^-$	3.86		Pyridine	4.27	
NO_2^-	3.71	3.94	Ammonia	4.16	
HPO_4^{2-}	3.52		Triethylamine	3.98	
$B_2O_7^-$	3.45		Pyridazine	3.98	
$4\text{-}NO_2C_6H_4O^-$	3.08		2-Picoline	3.64	
Cl^-	2.99	3.05	Imidazole	3.58	
CH_3COO^-	2.76	2.87	Pyrimidine	3.30	
			Pyrazole	2.81	
HCO_3^-	2.62	2.78	Chinoxaline	2.60	
HSO_3^-	2.3		2,4,6-Collidine	2.13	
$H_2PO_4^-$	2.2		2,6-Lutidine	2.04	
F^-	1.88	2.15	Urea	0.86	
SO_4^{2-}	<2.1		Acetamide	0.31	

Substrate	S (aqueous media)	S (benzene, 100°C)	Substrate	S (aqueous media)	S (benzene, 100°C)
CH_3Br^b	1	0.57[c]	CH_3NCS	0.99[c]	
CH_3I	1.27[a]		![] CHO (benzodioxole-CHO)	1.61[c,d]	
CH_3ClO_4	0.92[a]				
C_2H_5Br		0.78	Cl, NO$_2$, NO$_2$ (substituted benzene)	2.68[c,d]	
C_2H_5OTos	0.66[b]				
$H_5C_6\text{-}CH_2Cl$	0.87[b]				
$H_2C=CH\text{-}CH_2Br$		0.89	$(H_3C)_2COCl$	1.96[c,d]	
(epoxide)	1.19[c]		$\overset{+}{S}\text{-}CH_2CH_2Cl$ (thietanium)	0.95[b]	
(methyl epoxide) -CH$_3$	0.99[c]		C_6H_5COCl	1.43[b]	
$ClCH_2\text{-}COO^-$	1.08[c]		(β-propiolactone)	0.77[b]	
CO_2	1.40[c]				
$ClCOOC_2H_5$	2.37[c]				

[a] Ref. 227. [b] Ref. 226. [c] Ref. 228. [d] In alcohol.

183

The Swain-Scott equation allows more rigorous correlations when comparisons are limited to series of very similar nucleophiles[227]. Larger deviations noticed in reactions of the β-chloroethylthiiranium ion with OH⁻ led to an equation with four parameters (223),

Edwards equation[229]

$$\log \frac{k}{k_0} = \alpha E_n + \beta H \tag{223}$$

where k and k_0 are the same as in the Swain-Scott equation, and α and β are substrate parameters. Nucleophiles (Nu⁻) are characterized by E_n and H:

$$E_n = E^0_{Nu^-} + 2.60 \,(\text{Volt}) \tag{224}$$

$$H = pK_{HNu} + 1.74 \tag{225}$$

$E^0_{Nu^-}$ is the oxidation potential for the reaction

$$2Nu^- \xrightleftharpoons{\;E^0_{Nu^-}\;} Nu_2 + 2e^- \tag{226}$$

This equation allows the use of the same nucleophilicity parameters for reactions with very *different centers* (mainly from inorganic chemistry); unfortunately, only very few substrate parameters for attack on carbon are known for this equation. Relationships between α and β values and the structure of substrates are discussed by Davis.[230]

One can interpret the two terms in the Edwards equation as corresponding to the orbital-controlled and charge-controlled terms in the Klopman equation (189).

Zook and Miller[231] determined the Swain-Scott parameters for *C*- and *O*-alkylation of enolates in DMSO[231] (Table 48).

TABLE 48   Swain-Scott Parameters for the Alkylation of Enolates (DMSO, 30°C)

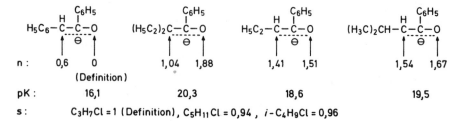

n :	0,6	0	1,04	1,88	1,41	1,51	1,54	1,67
		(Definition)						
pK :		16,1		20,3		18,6		19,5

s : $C_3H_7Cl = 1$ (Definition), $C_5H_{11}Cl = 0,94$, $i\text{-}C_4H_9Cl = 0,96$

Nucleophilicities (n) increase with pK values of the acetophenone derivatives; steric hindrance is probably responsible for the abnormally low value of n for the carbon center in diethylacetophenone.

The N_+ Parameter (Table 49). Ritchie et al.[232] found a very simple correlation for rates of reaction of *stable cations* **189-191**

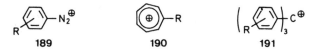

189   **190**   **191**

and of *reactive carbonyl compounds*

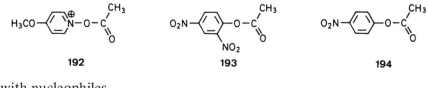

192   **193**   **194**

with nucleophiles

$$\lg k = N_+ + C \qquad\qquad (227)$$

where N_+ is a characteristic constant of the nucleophile, C is a constant of the substrate (crystal violet **191**, R = 4-N(CH$_3$)$_2$: $C = -5$; 4-nitrobenzenediazonium ion: $C = 1$). The absence of individual susceptibility factors means that two substrates have the same relative reactivity toward all nucleophiles: log $(k_1/k_2) = C_1 - C_2$. Recently correlation with N_+ has also been found for the rates of the following reactions:

TABLE 49 The N_+ Parameter[a]

Nucleophile	Solvent	N_+	Nucleophile	Solvent	N_+
C$_6$H$_5$S$^-$	DMSO	13.1	N$_3^-$	H$_2$O	5.4
C$_6$H$_5$S$^-$	CH$_3$OH	10.7	(CH$_2$NH$_2$)$_2$	H$_2$O	5.04
CN$^-$	DMF	9.4	H$_2$NCH$_2$COO$^-$	H$_2$O	4.95
CN$^-$	DMSO	8.6	C$_2$H$_5$NH$_2$	H$_2$O	4.88
N$_3^-$	CH$_3$OH	8.5	NH$_2$OH	H$_2$O	4.8
HOO$^-$	H$_2$O	8.4	OH$^-$	H$_2$O	4.5
SO$_3^{2-}$	H$_2$O	7.6	CN$^-$	H$_2$O	3.8
CH$_3$O$^-$	CH$_3$OH	7.5	C$_6$H$_5$SO$_2^-$	CH$_3$OH	3.8
ClO$^-$	H$_2$O	7.0	CH$_3$OH	CH$_3$OH	0.5
CN$^-$	CH$_3$OH	5.9	H$_2$O	H$_2$O	0
N$_2$H$_4$	H$_2$O	5.6			(Definition)

[a] A more extensive listing is found in ref. 232a.

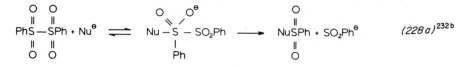

$(228a)^{232b}$

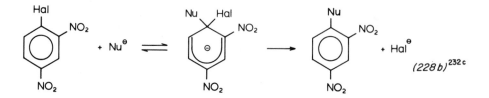

$(228b)^{232c}$

Correlations with nonunit slope have been observed too:

191 $(R = 4\text{-OCH}_3) + Nu^-$ $\log k = 0.81N_+$ $(228c)^{232d}$

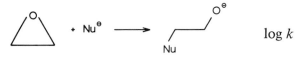

$\log k = 0.6N_+ - 6.01 \; (228d)^{232e}$

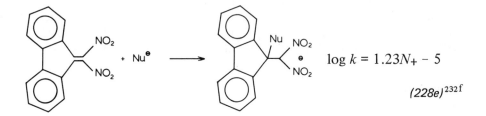

$\log k = 1.23N_+ - 5$

$(228e)^{232f}$

3.2.4 Medium Effects in Reactions of Nucleophiles

3.2.4.1 General Considerations

The fact that the susceptibility factor in the N_+ correlation has the same value of 1 for several substrates shows that in these instances relative reactivities of substrates toward given nucleophiles are determined only by the properties of the individual nucleophiles as expressed by N_+. Characteristics of substrates that may influence selectivity (bond strength in product, steric factors, solvation of the substrate, and Coulombic effects) seem to be the same with respect to all nucleophiles (and to affect only relative reactivities of various substrates). Since N_+ is not related to n (222) or to the equilibrium constants of reactions

$$R^{\oplus} + Nu^{\ominus} \underset{}{\overset{K_1}{\rightleftharpoons}} R-Nu$$

$$H^{\oplus} + Nu^{\ominus} \underset{}{\overset{K_2}{\rightleftharpoons}} H-Nu$$

it is assumed that it represents the energy required to detach a molecule of solvent from the solvent shell surrounding the nucleophile and thus make room for the electrophile (see, however, p. 367). The *larger* the value of N_+ the *easier* this process. The solvent dependence of N_+ is given in Table 50.

TABLE 50 Solvent Dependence of N_+

Nucleophile	$N_+(H_2O)$	$N_+(CH_3OH)$	$N_+(DMSO)$
$CN^{\ominus}$	3,8	5,9	8,6
$N_3^{\ominus}$	5,4	8,5	10,7
$C_6H_5S^{\ominus}$		10,7	13,1

Similar enhancements in reactivity of anionic nucleophiles (by factors of *ca.* 10,000) in going from *protic* media to *dipolar-aprotic* ones [e.g., dimethyl sulfoxide (DMSO)] are observed in other reactions as well (Table 51). They are caused by a strong activation of the anion as a result of loss of *hydrogen-bonding* and of *ion-dipole* interactions, both of which are stabilizing.[233a]

Hydrogen bonding is impossible in the dipolar aprotic dimethyl sulfoxide 195, in alkylated amides (e.g., dimethylacetamide 196, dimethylformamide 197, hexamethylphosphoric triamide 198, N-methylpyrrolidone 199), tetramethylenesulfone (200), acetonitrile, and acetone.

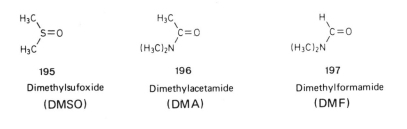

195	196	197
Dimethylsufoxide	Dimethylacetamide	Dimethylformamide
(DMSO)	(DMA)	(DMF)

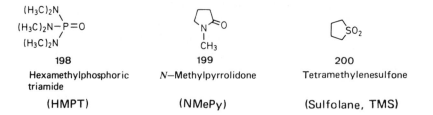

198	199	200
Hexamethylphosphoric triamide	N–Methylpyrrolidone	Tetramethylenesulfone
(HMPT)	(NMePy)	(Sulfolane, TMS)

TABLE 51   Applications of Dipolar Aprotic Solvents

Nucleophilic substitution at primary and secondary halides and similar compounds, nucleophilic aromatic substitution[a]

Enolates as nucleophiles[b]

Bicarbonate as nucleophile[c]

Synthesis of CH_2I_2 from CH_2Cl_2[d]

Nitriles from neopentyl-type halides[e]

Aryl nitriles from aryl halides and CuCN[f]

Elimination from $(H_5C_6)_3C-CH_2-CH_2-X$[g]

$(NBu_4)^+F^-$ as very strong base[h]

Cyclizing C=O condensations[i]

Michael additions[j]

Wittig reactions[k]

Olefin isomerizations[l]

Esterifications of sterically hindered carboxylic acids[m]

Decarboxylation of β-ketoesters[n]

Anionic oxidation of arylalkyl compounds[o]

[a] A. J. Parker, *Quart. Rev. (London)* **16**, 163 (1962) and ref. 245.

[b] Ref. 231; see also D. E. Butler and J. C. Pollatz, *J. Org. Chem.* **36**, 1308 (1971); R. J. Cregge, J. L. Herrmann, C. S. Lee, J. E. Richman, and R. H. Schlessinger, *Tetrahedron Lett.* **1973**, 2425.

[c] N. Bosworth and P. D. Magnus, *J. Chem. Soc., Perkin Trans 1*, **1973**, 2694.

[d] N. Altabev, R. D. Smith and N. S. I. Suratwala, *Chem. Ind. (London)* **1973**, 331.

[e] L. Friedman and H. Shechter, *J. Org. Chem.* **25**, 877 (1960).

[f] L. Friedman and H. Shechter, *J. Org. Chem.* **26**, 2522 (1961); M. S. Newman and H. Boden, *ibid.* **26**, 2525 (1961).

[g] R. O. C. Norman and C. B. Thomas, *J. Chem. Soc. C* **1967**, 1115.

[h] N. Ono, *Bull. Chem. Soc. J.* **44**, 1369 (1971).

[i] J. J. Bloomfield and P. V. Fennessey, *Tetrahedron Lett.* **1964**, 2273.

[j] B. R. Baker and P. M. Tanna, *J. Org. Chem.* **30**, 2857 (1965); see also J. E. Hofmann and A. Schriesheim, *Chem. Abstr.* **67**, 53290 (1967).

[k] R. Greenwald, M. Chaykovsky, and E. J. Corey, *J. Org. Chem.* **28**, 1128 (1963).

[l] Ref. 50c.

[m] J. E. Shaw, D. C. Kunerth, and J. J. Sherry, *Tetrahedron Lett.* **1973**, 689.

[n] P. Müller and B. Siegfried, *Tetrahedron Lett.* **1973**, 3565.

[o] J. E. Hoffmann, A. Schriesheim, and D. D. Rosenfeld, *J. Amer. Chem. Soc.* **87**, 2523 (1965).

Also, the positive pole essential for the stabilization of anions is buried in the interior of these molecules. (In addition, the central atoms of DMSO, HMPA, and TMS are relatively soft and therefore not suited to solvate hard anions.) Such solvent effects are strongest with small, hard ions (Cl⁻ suffers higher losses in hydrogen bond stabilization than I⁻; large, polarizable anions such as I⁻, or anionic S_N2 transition states are stabilized even in dipolar-aprotic solvents by ion-dipole interactions and dispersion forces)[233]:

Such solvent effects can even reverse the order of relative nucleophilicities in S_N2 reactions[234]:

In protic media $I^- > Br^- > Cl^- > F^-$
In dipolar-aprotic media $F^- > Cl^- > Br^- > I^-$

Leaving groups—receding nucleophiles—released into more or less favorable surroundings are also sensitive to solvent effects; for example, the loss of hydrogen bonding on going from aqueous alcohol to dimethylformamide affects chlorides most. The "normal" reactivity ratios RCl:RBr:RI = 1:50:100[207] change to about 1:500:3000.[235] (However, because of the activation of the nucleophile, even chlorides are substantially more reactive in dimethylformamide than in protic media.)

The strongly accelerating effect of dipolar aprotic solvents on reactions of anionic nucleophiles is very important for synthesis; many reactions can be performed at lower temperatures. Because of that, or because of shorter reaction times, side reactions can be suppressed and higher yields are secured. Even reactions that do not take place in normal (protic) solvents become possible. Some applications are given in Table 51.

"Naked Ions"

The ultimate in the separation of solvent effects from intrinsic factors influencing reactivity is reached when the same reaction series can be studied both in solution and in the gas phase. In the past decade a great number of such studies have been carried out. Of immediate concern in the present context are gas phase S_N2 reactions.[236a]

As is shown in Table 52 for some of these reactions—which are thought to involve the interconversion of loose ion-molecule complexes

TABLE 52 Absolute Rates in Various Media of $S_N 2$ Reactions at 25°C (log k in $M^{-1} s^{-1}$)[236a]

Reactants	H_2O	CH_3OH	DMF	Gas Phase $(X \cdots CH_3Y)^- \to (XCH_3 \cdots Y)^-$
$OH^- + CH_3F$	−6.2			10.2
$OH^- + CH_3Cl$	−5.2			12.0
$OH^- + CH_3Br$	−3.9			12.0
$F^- + CH_3Cl$	−7.8			11.7
$F^- + CH_3Br$	−6.5			11.6
$F^- + CH_3I$	−7.2	~−7.3		
$CH_3O^- + CH_3Br$				11.6
$CH_3^-O + CH_3I$		−3.6		
$Cl^- + CH_3Cl$				9.6
$Cl^- + CH_3Br$	−5.3	−5.2	−0.4	9.9
$Cl^- + CH_3I$	−5.5	−5.5	0.5	
$CN^- + CH_3Br$				10.3
$CN^- + CH_3I$	−3.2	−3.2	2.5	
$Br^- + CH_3Br$				<9.8
$Br^- + CH_3I$	−4.4	−4.1	0.1	

held together by ion-dipole and ion-induced dipole forces—they are faster by many orders of magnitude than their solution counterparts. In Table 53 relative rates and a similar measure of reactivity, relative reaction efficiencies (= fraction of collisions resulting in reaction) are compared (**A**) with enthalpies of reaction of various nucleophiles with methyl chloride and (**B**) with enthalpies of reaction of methoxide with methyl derivatives carrying various leaving groups.

TABLE 53 Reaction Enthalpies and Relative Reactivities in Gas Phase $S_N 2$ Reactions[236a]

A. $Nu^- + CH_3Cl \to NuCH_3 + Cl^- - \Delta H$

Nu:	H	>	NH_2	>	OH	>	CH_3O	>	CH_3S	~	F	~	CN	>	Cl
$-\Delta H$(kJ/mol):	361		210		199		167		130		119		114		0

Nu:	OH	~	NH_2	>	F	~	H	~	CH_3O	>	CH_3S	>	Cl	>	CN
Rel. Eff.:	210		210		120		120		83		15		1		<0.07

B. $CH_3O^- + CH_3Y \to CH_3OCH_3 + Y^- - \Delta H$

Y:	$OCOCF_3$	>	Br	>	Cl	>	OPh	>	F	>	OCH_3
$-\Delta H$(kJ/mol)	251		199		165		123		45		0
Rel. rate	58		54		34		12		1		<0.07

Part A of Table 53 indicates that the thermodynamic order is different from the kinetic order. Of special interest are the relative low reactivities of hydride, methylthiolate, and cyanide. It is hypothesized that it is the diffuseness of charge in these anions that causes them to be weak nucleophiles in the gas phase and good ones in solution (because of weaker solvation). Part B of Table 53 shows that with methoxide the order of kinetic reactivity is the same as the order of reaction enthalpies (exothermicities). Apparently the ease of breaking the CH_3-Y bond plays a dominant part in the rates of reaction. Only when exothermicities become very high, or rather similar, secondary factors become of importance. Thus the order of the three very fast leaving groups can change. Toward OH^- and F^- we have $CH_3Br > CH_3Cl > CH_3O_2CCF_3$ and $CH_3O_2CCF_3 > CH_3Cl > CH_3Br$, respectively. With F^-, CH_3S^-, and CN^-, whose reactions with methyl halides have rather similar exothermicities, symbiotic effects take over: the hard F^- reacts faster with CH_3Cl than with CH_3Br, whereas the reverse is true for the soft anions CH_3S^- and cyanide.

It is remarkable that relative nucleophilicities deviate from the corresponding Swain-Scott parameters n (Table 47).

The same is observed in dipolar-aprotic solvents (see above) and in the presence of *crown ethers* (see below) (Liotta et al. *Tetrahedron Lett.* 1975, 4205). Surprisingly, the order of relative nucleophilicities of negative ions in gas phase displacement reactions with *acid halides* parallels that observed in *protic* solvents.[236b]

This contrast with alkyl halides is ascribed to different interactions of the substrate with the solvated nucleophiles in the respective transition states in solution. The strength of attachment of the nucleophile to an alkyl halide is much smaller than the nucleophile-solvent bond strength. Desolvation therefore contributes a large part to the S_N2 activation energy. The transition state of carbonyl additions, however, must have much of the character of an alkoxide ion. Therefore it responds to changes of solvent in a manner similar to that of anionic nucleophiles and the influence of these changes on relative reactivity is small.

Because of their importance in synthesis, great interest has been generated recently in "naked" anions (e.g., $RCOO^{-237}$, F^{-238}, CH_3O^{-239}), easily obtainable in benzene or acetonitrile by strong binding of the countercation (especially K^+, see **201**) by so-called *crown ethers*[237a] (e.g., "18-crown-6" in **201**).

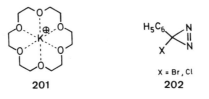

| **201** | **202** |

$X = Br, Cl$

These anions show extreme reactivity in S_N2 and E2 reactions; even substitutions in unactivated aryl halides were observed[239] (*229*):

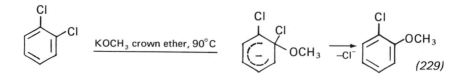

(*229*)

(This result accords strikingly with results of gas phase reactions of various nucleophiles (e.g., methoxide) with phenyl acetate.[236c] Nucleophilic aromatic substitution (e.g., formation of anisole, expulsion of acetate) is the only reaction observed. However, if only a *single* protic solvent molecule is attached to the nucleophile, the reaction pathway is drastically changed to the one expected from the enormous body of solution data: addition to the carbonyl group and expulsion of phenoxide. This change may occur because in the gas phase charge-localized transition states of carbonyl addition have higher energies than charge-dispersed transition states of nucleophilic aromatic substitution. A single protic solvent molecule attached to the nucleophile apparently is sufficient to disperse the charge on the carbonyl oxygen to such a degree that the normal behavior is regained. Methyl esters too, in the gas phase, show S_N2 at methyl instead of carbonyl attack[236d]).

α-Elimination from benzal halides leads in such systems to free phenylhalocarbenes with the same competition constants (with respect to cyclopropane formation with pairs of olefins) as those of species obtained from diazirines (*202*) by photolysis *in solution*.[240]

In the absence of crown ethers selectivities were different; that is, under these conditions the reactive species are carbenoids containing KHal or KO−*t*−Bu. In addition, the results with crown ether show that the reactive species in diazirine photolysis is not vibrationally excited diazirine or the isomeric diazo compound.

In recent years gas phase acidities have also been studied intensively.[236e] For carbon acids in nonaqueous polar solvents the order of relative acidities is largely the same as in the gas phase. However structural effects on pK are much larger in the latter, since these effects operate with full strength in the gas phase, whereas their anion-stabilizing task in solution is partly taken over by the solvent. Comparison of gas phase acidities with solution acidities impressively demonstrates the enormous effect of solvation of the anions of oxygen and nitrogen acids on the solution acidities of these acids: whereas in polar solvents the acidity of methanol is much higher than that of diphenylmethane, the reverse is true in the gas phase. In the gas phase the enthalpy of HO−H ionization is practically the same as that of $CH_2=CHCH_2-H$

ionization! Enthalpies of acid dissociation in the gas phase serve as starting point for the calculation of various other important thermochemical data (Scheme 46A). Gas phase heats of formation of anions can yield methyl cation affinities ΔH°_{MCA} or in the case of hydrogen-bearing anions (e.g., alkoxides, enolates, $Ph\bar{C}HCH_3$, HCO_2^-) hydride affinities ΔH°_{redn} of unsaturated compounds (e.g., ketones, α,β-unsaturated ketones, styrene, CO_2). Electron affinities of radicals $(EA(A\cdot))$ or bond dissociation energies $(D(A-H))$ (and from these heats of formation of radicals $\Delta H^\circ_f(A\cdot)$) can be calculated if the value of one of the two parameters is known from other sources.

$$AH \rightarrow A^- + H^+ - \Delta H^\circ_{acid}$$

Heats of formation of anions:
$$\Delta H^\circ_f(A^-) = \Delta H^\circ_{acid}(AH) - \Delta H^\circ_f(H^+) + \Delta H^\circ_f(AH)$$

Methyl cation affinities of anions (see Table 37a):
$$ACH_3 \rightarrow A^- + CH_3^+ - \Delta H^\circ_{MCA}(A^-)$$

$$\Delta H^\circ_{MCA}(A^-) = \Delta H^\circ_f(A^-) + \Delta H^\circ_f(CH_3^+) + \Delta H^\circ_f(ACH_3)$$

Hydride affinities of unsaturated compounds:
$$HXY^- \rightarrow H^- + XY - \Delta H^\circ_{redn}(XY)$$

$$\Delta H^\circ_{redn}(XY) = \Delta H^\circ_f(H^-) + \Delta H^\circ_f(XY) - \Delta H^\circ_f(HXY^-)$$

Electron affinities of radicals and bond dissociation energies
$$EA(A\cdot) = IP(H\cdot) - \Delta H^\circ_{acid}(AH) + D(AH)$$

Scheme 46A[236e] Gas phase thermodynamic data derivable from gas phase enthalpies of acid dissociation ΔH°_{acid} ($\Delta H^\circ_f(H^+)$ = 367.2 kcal/mol, $\Delta H^\circ_f(CH_3^+)$ = 262 kcal/mol, $\Delta H^\circ_f(H^-)$ = 33.2 kcal/mol, $IP(H\cdot)$ [=ionization potential] = 313.6 kcal/mol).

The same methodology has also been used to investigate gas phase equilibria involving carbocations,[82b] for example,

$$R'-X + R^+ \rightleftharpoons R'^+ + X-R$$

where X = Hal or H

From their heats of formation secondary carbocations were found to be about 67 kJ/mol less stable than tertiary carbocations. A similar value was found in superacids. Abnormal stability was found for the 2-norbornyl cation, being only about 25 kJ/mol above 2-methylnorbornyl.

Gas phase reactivity and stability data are particularly suited for comparison with results of quantum mechanical calculations, which can be done only for the gas phase.

3.2.4.2 Correlations

Solvent Activity Coefficients. A quantitative description of solvent effects (presented above in a qualitative way) is possible if one makes use of "solvent activity coefficients" $^0\gamma_S^L$. Their logarithms are proportional to free energy changes that occur when a substrate (S) is transferred from a reference solvent (0) to another solvent (L).* The methods used to determine these properties are described by Parker and Kolthoff[241] [e.g., by means of solubility products K^0 and K^L of a salt (AgY) in the two solvents; log K^0/K^L (AgY) = log $^0\gamma_{Ag^+}^L$ + log $^0\gamma_{Y^-}^L$]. Some values are given in Tables 54 and 55.

The increasingly negative values of log $^0\gamma^L$ of simple cations (e.g., H^+, Table 54; K^+, Tables 54 and 55) in the series methanol, water, dimethylformamide (DMF), dimethylsulfoxide (DMSO), hexamethylphosphoric triamide (HMPA) show the decrease in free energy of the positively charged species in this solvent series. Free energy minima for

TABLE 54 Logarithms of Solvent Activity Coefficients $^M\gamma_{ion}^L$ for Transfer of Ions from Methanol to Other Solvents[242]

Ion	log $^M\gamma_{ion}^W$	log $^M\gamma_{ion}^{DMF}$	log $^M\gamma_{ion}^{DMSO}$
$As(C_6H_5)_4^+$	4.1	-2.6	-2.4
Na^+	-1.5	-3.1	-3.9
K^+	-1.8	-3.5	-3.9
Rb^+	-1.8	-3.6	-3.7
Cs^+	-1.7	-3.3	-3.9
Ag^+	-1.3	-4.3	-7.2
Tl^+	-0.7	-2.7	-4.3
H^+	-1.9	-4.4	-5.2
Pic^-	0.8	(-0.7)	
ClO_4^-	-1.0		
Cl^-	-2.2	5.9	4.6
Br^-	-2.0	4.1	2.5
I^-	-1.2	2.1	0.4
SCN^-	-1.0	1.9	0.5
N_3^-	-1.9	4.1	2.6
NO_3^-	-2.2		
CH_3COO^-	-2.8	8.6	7.5
$C_6H_5COO^-$	-1.3	6.5	
$B(C_6H_5)_4^-$	4.1	-2.6	-2.4

M   = Methanol
W   = Water
Pic^- = Picrate

*$\Delta G = RT \ln {}^0\gamma_S^L$.

TABLE 55 Logarithms of Solvent Activity Coefficients $^{CH_3CN}\gamma^L_{ion}$ for Transfer of Ions from Acetonitrile to Other Solvents[243]

Ion	CH₃CN	TFE	H₃CNO₂	PC	Acetone	TMS	CH₃OH	C₂H₅OH	H₂O	HCONH₂	DMF	DMA	NMePy	DMSO	HMPT
Ag⁺	0.0	12.4	9.1	6.9	5.4	5.1	4.6	4.4	3.0	1.0	0.7	-0.4	-0.5	-2.6	-3.9
K⁺	0.0	4.6	2.0				0.2	1.0	-3.2	-1.8	-3.3	-4.2		-4.2	-4.7
(H₅C₆)₄As⁺	0.0		0.3	-0.5	-0.3	-0.1	0.1	0.8	6.5	0.2	-1.7	-1.2	-1.6	-0.1	-2.0
(H₅C₆)₄C	0.0		0.2	0.0	-0.4	-0.5	0.5	0.5		1.7	-0.7	-0.8	-1.2	-0.8	-0.9
(C₅H₆)₄B⁻	0.0	-1.0		-1.6	0.2	2.4	1.4 (-0.5)	2.5	5.7 (1.2)	1.8	-1.1	-0.9	-2.1	0.2	-0.1
Cl⁻	0.0	-8.4	-2.3	-0.1	2.6	0.1	-4.4	-3.5	-6.4	-4.7	0.9	1.8	1.5	0.0	3.0
Br⁻	0.0	-5.9	-0.4	0.5	2.5	0.4	-2.3	-1.5	-3.9	-2.7	1.4	2.0	1.5	0.2	3.4
I⁻	0.0	-3.5	-0.9	-0.5	2.1	-0.6	-0.5	0.2	-1.5	-0.9	0.9	0.9	1.0	-0.3	3.1
N₃⁻	0.0		-1.4	-0.3	2.4	-0.1	-3.0	-2.3	-4.3	-2.9	0.7	1.6	2.2	-0.4	3.4
SCN⁻	0.0		-1.3	-0.8			-0.7		-1.4	-1.3	0.8	0.9	1.4	-0.2	1.9
Pic⁻	0.0						-0.5	-1.6	1.8		-0.4				

TFE = F₃C—CH₂OH

PC = Propylencarbonate

TMS = Tetramethylenesulfone
DMF = Dimethylformamide
DMA = Dimethylacetamide

NMePy = N-Methylpyrrolidone
DMSO = Dimethylsulfoxide
HMPT = Hexamethylphosphoric triamide
Pic⁻ = Picrate

195

anions are observed in alcohols (trifluoroethanol), water, and forma-mide (Table 55).

Using solvent activity coefficients Müller[244] was able to describe the *solvent dependence* (log k_S/k_{CH_3CN}) of some S_N2 reactions *(230)* (k_S = rate constant in solvent S, k_{CH_3CN} = rate constant in acetonitrile).

$$X^\ominus \quad + \quad A-Y \quad \longrightarrow \quad X-A \quad + \quad Y^\ominus$$

$$\lg \frac{k_S}{k_{CH_3CN}} = a \lg \left[{}^{CH_3CN}\gamma^S_{X^\ominus} \right] + b \tag{230}$$

A–Y	$X^\ominus$	a	b
CH_3OTos	$Br^\ominus$	0,788	0,108
C_4H_9Br	$N_3^\ominus$	0,612	– 0,8
CH_3I	$Cl^\ominus$	0,813	0,33
CH_3I	$SCN^\ominus$	0,538	0,16

Similar, but less precise correlations of solvent dependence for other reactions involving nucleophilic attack have been given by Parker.[245]

Solvolysis of Halides, Tosylates, and Similar Esters When the solvent itself functions as the nucleophile *(231)*, the rate of solvolysis is expected to obey a four-parameter equation *(232)*.

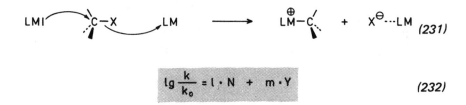

$$\lg \frac{k}{k_0} = l \cdot N + m \cdot Y \tag{232}$$

where N and Y, respectively represent (relative) nucleophilicities, and (relative) ionizing power (i.e., the possibility of accomplishing the heterolysis of C−X) of the two solvents; l and m are the corresponding susceptibility (reaction) constants of the substrate; k_0 is the rate constant in a solvent chosen as standard, k is the rate constant in sol-vent LM *(231)*.

One can distinguish two extreme situations.

1 In *tertiary substrates* backside attack by the nucleophile is sterically hindered (the extreme case is attack at a bridgehead ester). In these instances, termed by Winstein[246] "limiting," one may neglect the $l \cdot N$ term. The same is true for a series of solvents where only Y is

changing. For the resulting Winstein-Grunwald equation[247] [(*233*); see also Fig. 44] by definition

$$\log \frac{k}{k_0} = m \cdot Y \qquad\qquad (233)$$

where $m = 1$ for t-butyl chloride and $Y = 0$ for 80:20 (vol) ethanol-water; m values (susceptibility coefficients) for some representative tertiary, secondary, and primary substrates are given in Table 56. Y values, which increase with the water content of various solvent mixtures may be obtained from Fig. 43.

TABLE 56 *m* Values in Ethanol-Water Mixtures at 25°C[248]

Substrate	m
1—Adamantylbromide	1,20
$\langle\text{phenyl}\rangle-\overset{\text{H}}{\underset{\text{Cl}}{\text{C}}}-CH_3$	0,966
1—Bicyclo [2.2.2] octylbromide	1,03
$t-C_4H_9Cl$	1,00
$t-C_4H_9Br$	0,94
2—Adamantyltosylate	0,91
$H_2C=CH-\overset{\text{H}}{\underset{\text{Cl}}{\text{C}}}-CH_3$	0,89
$(H_5C_6)_2CHCl$	0,75
exo—2—Norbornyltosylate	0,75
$H_3C-O-\langle\text{phenyl}\rangle-\overset{\text{CH}_3}{\underset{\text{CH}_3}{\text{C}}}-CH_2-OTos$	0,42
$i-C_3H_7Br$	0,43
C_2H_5Br	0,34 (55°C)
CH_3Br	0,22

The complex nature of the empirical parameter Y representing the ionizing power of the solvent may be inferred from the dispersion of m values observed for numerous compounds. Data for 2-methyl-2-phenylpropylchloride (neophyl chloride) (Table 57) and the Y-dependence of solvolysis rates of 1-bromoadamantane and

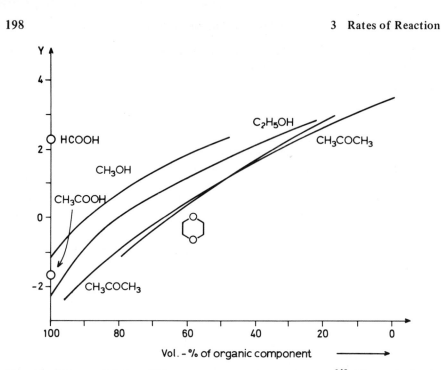

Fig. 43 Values of Y for different aqueous solvent mixtures.[249] [From A. Streit-wieser, Jr., *Solvolytic Displacement Reactions*, McGraw-Hill Book Co., New York, 1962, p. 45 (ref. 249).]

TABLE 57 Dispersion of *m* Values in Solvolysis of 2-Methyl-2-phenylpropyl Chloride (50°C)[250]

Solvent	m
Dioxane-water (20-70 vol-%)	0.961
Acetic acid-formic acid (0-100%)	0.837
Ethanol-water (20-80 vol-%)	0.833
Methanol-water (30-80 vol-%)	0.790
Water in acetic acid (0-16 m)	0.733

1-adamantyl tosylate (Fig. 44) show that to obtain rigorous corre-lations, one must use different *m* values for various solvent types.

The scattering observed for adamantane derivatives in trifluoro-ethanol, acetic acid (for 1-AdOTos) and phenol-benzene mixtures may be due to increased hydrogen bonding between these solvents and the leaving group.

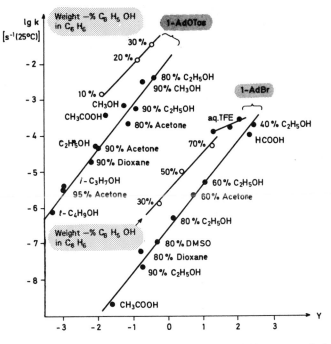

Fig. 44 Winstein-Grunwald correlation of the solvolysis rates of 1-adamantyl tosylate (1-AdOTos) and 1-adamantyl bromide (1-AdBr)[251] (aq. TFE = aqueous trifluoroethanol). [From K. Okamoto, K. Matsubara, and T. Kinoshita, *Bull. Chem. Soc. Jap.* **45**, 1191 (1972) (ref. 251).]

Additional factors that may have a different impact on R–X, R^+X^-, and/or transition states leading to these species in different solvent families are

(*a*) Covalent nucleophilic interactions (see below).

(*b*) Solvent polarizability.

(*c*) Differences in the chemistry of ion pairs (different degrees of internal return).

(*d*) Change in solvent structure.

The effect of the solvent on the *rate of addition of bromine to olefins* may also be described by (*233*) ($m_{1\text{-pentene}}$ = 1.16; m_{styrene} = 0.96; m_{stilbene} = 1.20).[251 a]

2 The lowest sensitivity toward ionizing power is found for *methyl bromide* and *methyl tosylate* ($m_{\text{CH}_3\text{OTos}}$ = 0.3). This indicates a second limiting situation: in these sterically unhindered substrates where ionization would lead to the least stabilized carbocations the $l \cdot N$ *term takes preponderance*. Consequently, for methyl tosylate

Schleyer defined $l = 1.$[252]. Using $m = 0.3$ and known Y values one can use (232) to determine N values for various solvents [(234) and Table 58] [because of the definition of Y, ethanol-water (80:20) is also the zero point k_0 of the N scale].

TABLE 58   Values for the Nucleophilicity Parameter N for Various Solvents, Calculated by Means of (234)

Solvent	N		N_{KL}[a]
Isopropanol	+0.09		
Ethanol	+0.09		+0.46
Methanol	+0.01		+0.58
Ethanol-Water (80:20, vol.)	0.00		0.00
Acetone-Water (56:44, weight)	−0.47		−0.44
Ethanol-Water (50:50, vol.)	−0.20		−0.42
Water	−0.26		−0.87
Dioxane-Water (50:50, weight)	−0.41		−0.41
Acetic acid	−2.05		−1.34
Formic acid	−2.05		−1.61
Trifluoroacetic acid	−4.74	97% Trifluoroethanol	−2.22
Fluorosulfonic acid	~−4		

[a]Subsequently derived by a procedure designed to minimize errors caused by the choice of the value of the mY term in (232).[252a]

$$N = \log \left(\frac{k}{k_0}\right)_{CH_3OTos} - 0.3Y \qquad (234)$$

So far very little is known about l values. Schleyer gave approximate values for a series of *tosylates*. He formulated the solvent dependence of solvolysis rates of tosylates as a *linear combination* of the solvent dependence of methyl tosylate solvolysis controlled by nucleophilicity and that of 2-adamantyl tosylate controlled exclusively by the ionizing power of the solvent (235a) and Table 59:

$$\log \frac{k}{k_0} = (1-Q) \log \left(\frac{k}{k_0}\right)_{CH_3OTos} + Q \log \left(\frac{k}{k_0}\right)_{2\text{-AdOTos}} \qquad (235a)$$

where k = solvolysis rate in a given solvent
$\quad k_0$ = solvolysis rate in ethanol-water (80:20, vol).
$\qquad 0 < Q < 1$

Definition: $Q = 0$ for methyl tosylate; $Q = 1$ for 2-adamantyl tosylate.

TABLE 59 Q Values (*235a*) and l Values (*235b*) for Some Primary and Secondary Alkyl Tosylates, Calculated from Relative Solvolysis Rates k/k_0 in a Given Solvent and in Ethanol-Water (80:20, vol)[a]

Tosylate	Q	l	l_{KL}	m_{KL}
Methyl	0.00^b	1	1.20	0.50
Ethyl	0.16	0.84	1.00	0.55
Benzyl	0.22	0.78	0.90	0.67
Isopropyl	0.56	0.44	0.50	0.55
Isobutyl	0.64	0.36		
Cyclopentyl	0.67	0.33	0.25	0.56
2-Pentyl	0.69	0.31		
3-Pentyl	0.69	0.31		
4-Heptyl	0.76	0.24		
Cyclohexyl	0.75	0.25	0.18	0.60
2-Adamantyl	1.00^b	0.00	0.00	0.81

[a] The l_{KL} and m_{KL} are for use in (*232*) with N_{KL} and Y.[252a]
[b] Defined.

With the approximation $\log(k/k_0)_{2\text{-AdOTos}} = Y$ (actually $\log(k/k_0)_{2\text{-AdOTos}} = 0.91\ Y$, see Table 56), substituting (*234*), in (*235a*), and comparing coefficients with (*232*), one obtains

$$l = (1 - Q) \qquad\qquad (235b)$$

$$m = (0.3 + 0.7Q) \qquad\qquad (235c)$$

The continuous change in Q and l (Table 59) may be interpreted as follows: only a single factor changes within the reaction series: the extent to which the departure of the leaving group is facilitated by simultaneous nucleophilic attack by solvent. It diminishes regularly within this series, and in 2-adamantyl tosylate nucleophilic solvent participation has completely vanished because of unfavorable steric interaction (*203*).

203

	Role of Nu^- in Transition State	Species Formed After Passage of Transition State	Terminology
Primary substrates	$Nu^{\delta-}\cdots CH_2R \cdots Y^{\delta-}$ Strong nucleophile attachment → strong coupling (synchronicity) of bond breaking and bond formation	$Nu-CH_2R + Y^-$ Substitution product[a]	S_N2
Secondary substrates (borderline cases)	$Nu^-\cdots\cdots CHR_2^{\delta+}\cdots\cdots Y^{\delta-}$ Nucleophile attachment less strong → bondbreaking more advanced, finished before full bond formation	$(1-\delta)^-$ $(1-\delta)^+$ $Nu\cdots\cdots CHR_2$ Y^- Nucleophilically solvated ion pair[b] (Doering-Zeiss intermediate)	"S_N2 (intermediate)"[252]
Limiting substrates	$Nu^- \quad CR_3^{\delta+}\cdots\cdots Y^{\delta-}$ No nucleophile attachment, bond breaking only	$Nu^- \quad CR_3^+ \quad Y^-$ Intimate ion pair	S_N1

[a] Complete bond between Nu and carbon → kinetic order of Nu^- in rate equation: unity.
[b] Partial bond between Nu and carbon → kinetic order of Nu^- in rate equation < 1, cf. M. L. Bird, E. D. Hughes, and C. K. Ingold, *J. Chem. Soc.* **1954**, 634, p. 640.

Scheme 46B   The continuous modes of achieving nucleophilic substitution.

This picture is that of a *continuum of solvolysis mechanisms* encompassing as one extreme, the S_N2 reaction (methyl tosylate), followed by a borderline region (see Section 5.2.6.3), which includes "normal" secondary substrates (i.e., substrates solvolyzing without neighboring group participation or resonance stabilization, p. 30). and as the other extreme, the S_N1 reaction (2-adamantyl tosylate and tertiary substrates), see Scheme 46B.

Extensive work by Schleyer and co-workers has shown that among normal secondary tosylates *only* those of the 2-adamantyl type (see also pp. 129, 131) undergo solvolysis with rate-determining formation of a carbocation, that is, uninfluenced by the nucleophile,[253] as do tertiary substrates of the "limiting" type. In solvolyses of other "normal" secondary substrates such cationic intermediates are excluded; this also excludes the hypothesis of concurrent S_N2 and S_N1 processes (*236*), proposed as an alternative to the continuum of solvolysis mechanisms.

$$R-X \quad \begin{cases} R^+ + X^- \xrightarrow{\text{LM}} R-\overset{+}{M}L \text{ (racemic)} \\[2mm] \xrightarrow[-X^-]{\text{LM}} R-\overset{+}{M}L \text{ (inverted)} \end{cases} \qquad (236)$$

3.3 FURTHER APPLICATIONS OF CORRELATION EQUATIONS

3.3.1 Analysis of Reaction Mechanisms

Correlation equations are used not only to calculate relative reaction rates for new members of known reaction series [containing other (known) substituents, or involving other (known) solvents or nucleophiles] but also to *analyze reaction mechanisms*.

When in the cases discussed above $k_{calc} \simeq k_{exp}$ this is a rather strong indication that the compound in question reacts according to the mechanism on which the correlation is based and that no new effects or mechanisms occur. On the other hand, significant discrepancies between calculated and experimental values of k strongly indicate the appearance of such effects or mechanisms. A few examples representing both situations are discussed below. Further examples showing how to use correlation equations in mechanistic studies [by determining and comparing susceptibility (reaction) constants and substituent (reagent, solvent) constants] are treated in Section 4.2.5.2.

3.3.1.1 Solvolysis of Pinacolyl Tosylate

A Taft analysis of the solvolysis rate of pinacolyl tosylate in trifluoroacetic acid showed good agreement between the calculated and experimental values (Fig. 45).[254]

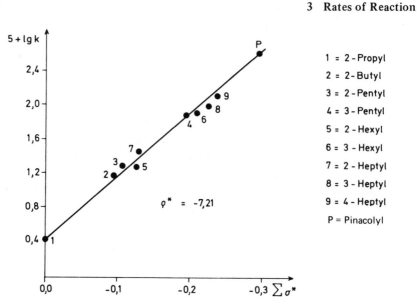

Fig. 45   Dependence of rate constants k $(25°C, s^{-1})$ for trifluoroacetolysis of secondary tosylates $R^1R^2CHOTos$ on $\Sigma\sigma^*$ of the two alkyl groups R^1 and R^2. [Values of σ^* from: R. W. Taft, Jr., in M. S. Newman (Ed.), *Steric Effects in Organic Chemistry*, John Wiley & Sons, New York, 1956, p. 591.] [From J. Slutsky, R. C. Bingham, P. von Ragué Schleyer, W. C. Dickason, and H. C. Brown, *J. Amer. Chem. Soc.* **96**, 1970 (1974) (ref. 254).]

Isopropyl derivatives were hypothesized to solvolyze relatively slowly because of "hidden return," $k_{-1} \gg k_2$ (*237a*). The much higher solvolysis rate of pinacolyl derivatives was explained by the prevention of this process by very fast methyl migration ($k_2 \gg k_{-1}$) (*237b*).[255]

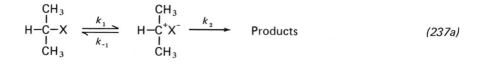

(237a)

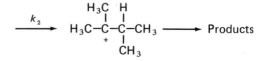

(237b)

However, the Taft analysis shows that the higher rate of pinacolyl derivatives is due exclusively to increased inductive-hyperconjugative stabilization of the transition state for solvolysis and that there are no mechanistic differences that can be attributed to "hidden return" or methyl migration in these two types of compounds.

3.3.1.2 Mechanism of the General Acid-Catalyzed Hydrolysis of Ethyl Orthocarbonate

The general acid-catalyzed hydrolysis of *ethyl orthocarbonate* could conceivably proceed via the A-S_E2 mechanism* or the A2 mechanism[†][256] (Scheme 47).

In the A2 mechanism addition of strong nucleophiles should lead to deviations from the Brönsted correlation. The magnitude of the expected deviation on addition of NaI at constant pH, in dimethylarsinate buffer $NaO-\overset{\overset{O}{\|}}{As}(CH_3)_2/HO-\overset{\overset{O}{\|}}{As}(CH_3)_2$ was evaluated by means of the Swain-Scott equation (222). k_{H_2O} and $k_{NaO-\overset{\overset{O}{\|}}{As}(CH_3)_2}$ were obtained from the buffer concentration dependence of the rate constant for hydrolysis (Fig. 46; see also p. 170).

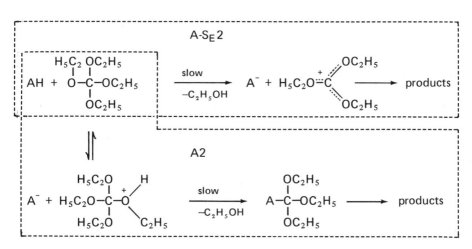

Scheme 47 Possible mechanisms for the rate-determining step in the general acid-catalyzed hydrolysis of ethyl orthocarbonate.

*Bimolecular electrophilic substitution (in this instance, of a OC_2H_5 group) by acid "A".

[†]Bimolecular nucleophilic substitution in the conjugate acid of the substrate.

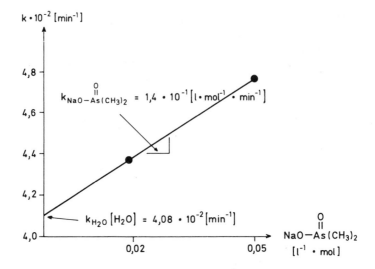

Fig. 46 Dependence of the hydrolysis rate (k, 30°C, pH 5.67) of ethyl orthocarbonate on the concentration of sodium dimethylarsinate (see also p. 000).

$$\lg \frac{k_{NaO-\overset{O}{\underset{\|}{As}}(CH_3)_2}}{k_{H_2O}} = \lg \frac{1,4 \cdot 10^{-1}}{4,08 \cdot 10^{-2}/55,5} = s \, n_{NaO-\overset{O}{\underset{\|}{As}}(CH_3)_2} \qquad (238)$$

Correlation of n values of anions of other oxygen acids with their pK_a's gave $n_{NaO-\overset{O}{\underset{\|}{As}}(CH_3)_2} = 3.4$ by way of the pK_a of dimethylarsinic acid.

This enabled the calculation of $s = 0.67$. Knowing n_{I^-} (5.04) it was possible to calculate k_{I^-}:

$$\lg \frac{k_{I^\ominus}}{k_{H_2O}} = \lg \frac{k_{I^\ominus}}{7,3 \cdot 10^{-4}} = 0,67 \cdot 5,04 \; ; \quad k_{I^\ominus} = 1,7 \; l \cdot mol^{-1} \cdot min^{-1} \qquad (239)$$

The contribution of the I^- catalyzed reaction to the total rate constant (in 0.03 m NaI) should be $k_{I^-}[NaI] = 0.051 \, min^{-1}$. However, the actual value of the total rate constant under these conditions was only $0.042 \, min^{-1}$ without showing any deviation from the Brönsted correlation. This result excludes the A2 mechanism.

3.3.1.3   Deviations from Correlation Equations

Some situations where deviations from correlation equations were observed are given in Table 60, together with the interpretations given to these deviations.

TABLE 60 Deviations from Correlations and Their Interpretation

Reaction	Correlation Equation	Nature of Deviation	Interpretation	Ref.
Acetolysis of	Foote-Schleyer	$\log k_{exp} - \log k_{calc} = 13$ $\log k_{exp} - \log k_{calc} = 4.4$	Formation of delocalized ions and	P. v. R. Schleyer, *J. Amer. Chem. Soc.* **86**, 1856 (1964)
Hydrolysis of aromatic radical anions	$\log k = -a \cdot \Delta E_\pi + b$	Nonlinear relation	In the transition state of the reaction there is resonance between the "no bond" structure (A) and the "charge transfer" structure (B). In addition to ΔE_π ionization potentials of the radical anions and changes in hybridization contribute to reactivity.	S. Hayano and M. Fujihira, *Bull. Chem. Soc. Jap.* **44**, 2046 (1971)
	Brønsted	For H_2N-NH_2 and $HONH_2$: $\log k_{exp} - \log k_{calc} = $ 2.4 and 4, respectively	These are examples of the *α-effect*. Nucleophiles with α-atoms bearing an unshared electron pair (e.g., also HOO^-, ClO^-) show increased reactivity toward electrophilic centers involved in π bonding[a]	T. C. Bruice, A. Donzel, R. W. Huffman, and A. R. Butler, *J. Amer. Soc.* **89**, 2106 (1967)

TABLE 60 (Continued)

Reaction	Correlation Equation	Nature of Deviation	Interpretation	Ref.
Isotope exchange with (cyclohexyl–NH$_2$ / cyclohexyl–NH–Li; resp. cyclohexyl–NH–Cs) D-cubane etc. (triptycene structure: j$_1$ j$_2$ j$_3$, T)	$\log k = a J^{13}\mathrm{CH} + b$	$\log k_{exp} - \log k_{calc} \approx 3$ (assuming that $\log k_{calc} \approx \log k_{cyclopropane}$, since $J^{13}\mathrm{CH}$-cyclopropane $\approx J^{13}\mathrm{CH}$-cubane) $\log k_{exp} - \log k_{calc} \approx 4.5$	In the highly strained cubane, anionization is accompanied by rehybridization: the anionic exocyclic orbital acquires more s-character, the skeletal C–C bonds acquire correspingly more p-character, resulting in a decrease in strain.[b] In addition to the (increased) s-character of the C$_i$–H bond the combined inductive effect of the benzene rings contributes to the kinetic acidity.[c]	T.-Y. Luh and L. M. Stock, *J. Amer. Chem. Soc.* **96**, 3712 (1974) A. Streitwieser, Jr., and G. R. Ziegler, *J. Amer. Chem. Soc.* **91**, 5081 (1969)

[a] See also: R. F. Hudson, in G. Klopman (Ed.), *Chemical Reactivity and Reaction Paths*, John Wiley and Sons, New York, 1974, p. 212; also: J. F. Liebman and R. M. Pollack, *J. Org. Chem.* **38**, 3444 (1973). Strong positive deviations for α-effect nucleophiles are shown in Fig. 40.

[b] Consequently this behavior contradicts what was said on page 139.

[c] Use of the correlation equation reflecting the inductive effect: $\log k/k_0 = 0.624 \; \Sigma \; 1/r_{ij} - 1.609$ (see p. 125, k_0 = rate of isotope exchange of benzene) gives a reaction rate corresponding to the excess value (4.5) of the J^{13}CH correlation. Accordingly, the kinetic acidity of triptycene ($\log k_{exp} \approx 7.3$) is determined to the extent of about $\frac{1}{3}$ ($\log k_{calc} \approx 2.8$) by the s-character of the C–H bond and about $\frac{2}{3}$ ($\log k_{exp} - \log k_{calc} \approx 4.5$) by the inductive effect.

3.3.1.4 Solvolysis with Formation of Phenonium Ions

The quantitative analysis of deviations from the Hammett correlation observed in the acetolysis of *threo*-3-aryl-2-butyl brosylates* and similar β-arylethyl derivatives is one of the corner stones of modern theories of solvolysis of secondary substrates (see also p. 203). The observations are shown in the plot in Fig. 47.[257]

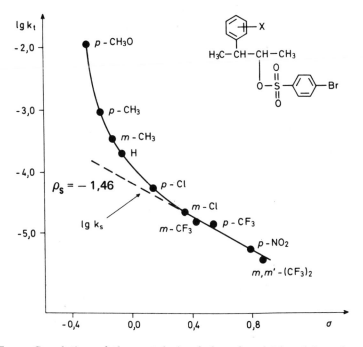

Fig. 47 $\rho\sigma$ Correlation of the acetolysis of *threo*-3-aryl-2-butyl brosylates: $k_t =$ titrimetric rate constant, $k_s =$ rate constants for the solvent-assisted portion of the reaction obtained by extrapolation of values for X = m,m'-(CF$_3$)$_2$, p-NO$_2$, p-CF$_3$, etc. [From H. C. Brown, C. J. Kim, C. J. Lancelot, and P. von Ragué Schleyer, *J. Amer. Chem. Soc.* **92**, 5244 (1970) (ref. 257).]

Titrimetric rate constants k_t for 3,5-(CF$_3$)$_2$, p-NO$_2$, p-CF$_3$, and m-CF$_3$ follow a $\sigma\rho$ correlation attributed to solvent-assisted solvolysis (k_s). Deviations starting with p-Cl are attributed to a new, competing mechanism, namely solvolysis via *phenonium ions* (k_Δ). Only fraction F ($F < 1$) of the phenonium ions reacts further to give *threo* products, whereas fraction 1-F reforms the brosylate ("internal return") (Scheme 48).

*Brosylate = 4-bromobenzene sulfonate.

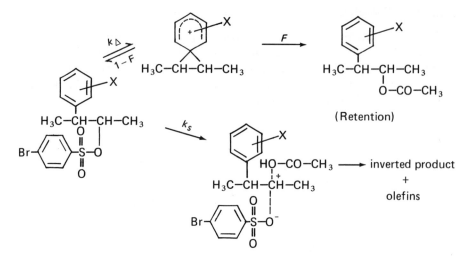

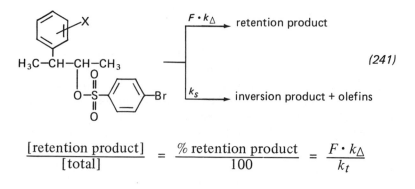

Scheme 48 Acetolysis of *threo*-3-aryl-2-butyl brosylate. Competition between formation of phenonium ions (k_Δ) and the solvent-assisted process (k_s).

The total rate constants is

$$k_t = k_s + F \cdot k_\Delta \qquad (240)$$

If one assumes that the $\sigma\rho$ correlation for substrates that solvolyze without aryl participation also represents the contribution of k_s to the solvolysis rate of substrates reacting via phenonium ions (dashed line in Fig. 47) one can compute $F \cdot k_\Delta$ as the difference $k_t - k_s$. The ratio $F \cdot k_\Delta/k_t$ obtained in this way is in good agreement with the value obtained from the percentage of products with retained configuration. (Inverted acetate and all other products were ascribed to the intermediate of the k_s pathway) [(*241*), Table 61].

$$\frac{[\text{retention product}]}{[\text{total}]} = \frac{\% \text{ retention product}}{100} = \frac{F \cdot k_\Delta}{k_t}$$

TABLE 61   Analysis of the Acetolysis Rate of *threo*-3-Aryl-2-Butyl Brosylates
(for symbols see text; %Ret. = $100 \cdot Fk_\Delta/k_t$)

X	$10^5 \times k_t$	$10^5 \times k_s$	$10^5 \times Fk_\Delta$	k_t/k_s	Retention Product (%)	
					Calculated	Experimental
p-CH$_3$O	1060	14.9	1045	71	99	100
p-CH$_3$	81.4	10.7	70.7	7.6	87	88
m-CH$_3$	28.2	7.66	20.5	3.7	73	68
H	18.0	6.08	11.9	3.0	66	59
p-Cl	4.53	2.85	1.68	1.6	37	39

Data derived from the initial stage of the microscopic transformation (ionization, k_t, k_s → column 6, Table 61) and data for the final
phase (product formation from intermediates → column 7, Table 61)
give the same result. This can mean only that from beginning to end
the transformation proceeds via two reaction paths (mechanisms)
independent of each other. Crossover between the two paths, which
would invalidate the quantitative correlation between rate data and
product data, does not take place. However, such crossover would be
expected if the intermediates in the k_s path were "normal" carbocations (204) [(*242*)].

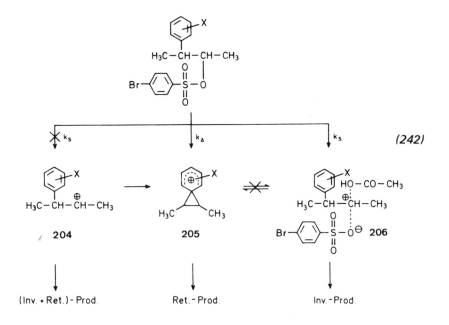

Consequently, the conclusion was drawn[258] that **204** is not formed and that the k_S path yields species **206** in which the participating solvent is so strongly attached to the backside that the transition **206** → **205** does not occur within the lifetime of **206**. In "normal" secondary substrates without steric hindrance by the phenyl group, solvent participation must be even stronger. In β-aryl substrates the contribution of the k_S path increases when the ability of X to provide electrons decreases (column 5, Table 61).

A similar dissection of the reaction into a k_Δ path and competing *bromonium ion* formation was performed for the addition of bromine to [259] It could be shown that on addition of bro-

mine to stilbenes, processes involving carbocations (p. 16) and those involving bromonium ions take place simultaneously.[259a]

3.3.1.5 Competing Mechanisms of Charge Delocalization

The $\rho\sigma^+$ plot for the hydrolysis of *syn*-7-aryl-*anti*-7-norbornenyl-4-nitrobenzoate[260] (Fig. 48) shows an inflection at X = OCH$_3$. This means that for this electron-donating substituent, interaction **207** with very little positive charge on C-7 ($\rho = -2.30$) [as it exists for X = 3,5-(CF$_3$)$_2$, 4-CF$_3$, and H] is replaced *abruptly* by interaction **208**.

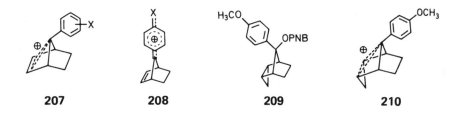

| **207** | **208** | **209** | **210** |

The new ρ (-5.27) now shows high positive charge at C-7, corresponding to that in the 7-norbornyl derivative. Rates of norbornyl and norbornenyl derivatives are also equal in this region. The stereochemistry of product formation supports this interpretation: when X belongs to the first group of substituents the only product is the *anti* alcohol (p. 31); for X = OCH$_3$ and N(CH$_3$)$_2$ the product contains 8% *syn* alcohol.[260a]

(The cyclopropane derivative **209** shows interaction **210** even for X = OCH$_3$[261]: **209** reacts 10^3 times faster than 7-(4-methoxyphenyl)-7-norbornyl-4-nitrobenzoate. Delocalization of the positive charge into the cyclopropane ring in the orientation shown is stronger than

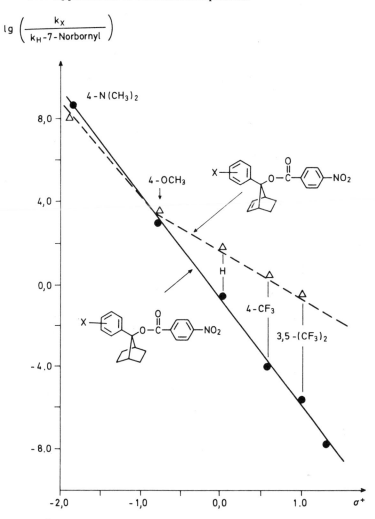

Fig. 48 $\rho\sigma^+$- Correlation of the hydrolysis of 7-aryl-7-norbornyl and *syn*-7-aryl-*anti*-7-norbornenyl-4-nitrobenzoates (70% dioxane, 25°C); k_x is the rate constant of the corresponding norbornane or norbornene derivative. The reference compound in both cases is 7-phenyl-7-norbornyl-4-nitrobenzoate (k_H-7-norbornyl). [From P. G. Gassman and A. F. Fentiman, Jr., *J. Amer. Chem. Soc.* **92**, 2549 (1970) (ref. 260).]

delocalization into 4-methoxyphenyl or the double bond of the norbornene derivative; see also Section 5.2.3.1).

3.3.1.6 ρ-Values in 1,3-Dipolar Additions[262]; Concave Hammett Plots

Reactions with 4-substituted phenyl azides show positive ρ values for cyclopentene (+0.9), norbornene (+0.88), and 1-pyrrolidinocyclo-

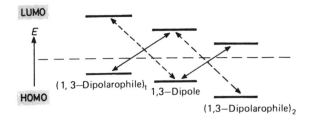

Fig. 49 Reactivity determining HOMO-LUMO interactions in 1,3-dipolar additions. Because of smaller energy differences, HOMO-LUMO interactions indicated by solid arrows dominate over those represented by broken arrows. Depending on the relative position of the orbitals of the 1,3-dipole and the 1,3-dipolarophile, both the combinations $HOMO_{1,3-dipolarophile}$-$LUMO_{1,3-dipole}$ and $HOMO_{1,3-dipole}$-$LUMO_{1,3-dipolarophile}$ may prevail.

hexene (+2.54) and negative values for maleic anhydride (−1.1) and N-phenylmaleinimide (−0.8). These reactions are *orbital controlled* and the combination with the lowest HOMO-LUMO separations determines reactivity (see Fig. 49, also pp. 147-150 and Section 5.2.4.3.).

Substituent effects may be explained by an increase or decrease in the decisive separation between highest occupied molecular orbital (HOMO) and lowest unoccupied molecular orbital (LUMO). An electron donor (e.g., *p*-OCH$_3$) *raises* orbital levels, an electron acceptor (e.g., *p*-NO$_2$) *lowers* them. In reactions controlled by $HOMO_{1,3-dipole}$-$LUMO_{1,3-dipolarophile}$ (maleic anhydride and others) substitution of the dipole by *p*-OCH$_3$ increases the rate with respect to a constant dipolarophile, whereas *p*-NO$_2$ lowers it; ρ is negative. The situation is reversed (ρ = +) for $LUMO_{1,3-dipole}$-$HOMO_{1,3-dipolarophile}$ interactions, for example, with cyclopentene and other olefins.

If dipole and dipolarophile have approximately equal HOMO-LUMO separations, both electron donors and electron acceptors enhance the reaction rate, resulting in a curved Hammett plot (Fig. 50).

It is to be expected that these effects will be relatively small, since the intensification of one HOMO-LUMO interaction (solid arrows in Fig. 50) is counterbalanced by the weakening of the other (dashed arrows in Fig. 50).

Absolute values for ρ depend on the HOMO-LUMO separation. If it is very large substituent effects are *nil*. Concave Hammett plots like the one in Fig. 50 generally are observed in *dual mechanism* situations where one type of interaction in the transition state (e.g., $HOMO_{dipole}$-$LUMO_{dipolarophile}$) prevails in one (contiguous) part of the reaction series and another one (e.g., $HOMO_{dipolarophile}$-$LUMO_{dipole}$) takes over in the following part. Another example is provided by the formation of macrocyclic lactones through cyclization of ω-hydroxy-2-

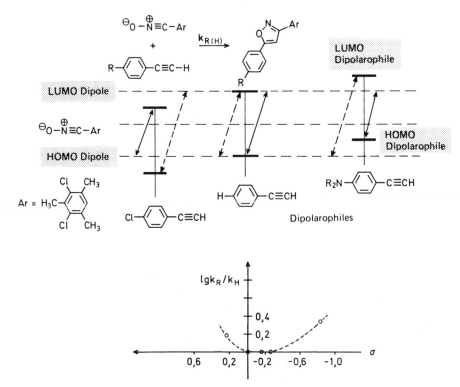

Fig. 50   Variation of the reaction rate for a constant 1,3-dipole (3,5-dichloro-2,4,6-trimethylbenzonitrile oxide) and a variable 1,3-dipolarophile where HOMO-LUMO spacings of 1,3-dipole and 1,3-dipolarophile are supposed to be approximately equal; k_R is the rate constant of the para-substituted phenylacetylene, k_H is the rate constant of the unsubstituted phenylacetylene taken as reference.[262 a]

pyridyl thiolesters (Fig. 50a).[262 b] For $\sigma_m < 0.12$, N–H bond formation dominates in the transition state ($\rho = -4.34$), whereas for $\sigma_m > 0.12$ C–O bond formation is more advanced, leading to a negatively charged side chain of the pyridine ring ($\rho = 1.77$). In the rate minimum at $\sigma_m = 0.12$ both interactions are of same strength, and they counterbalance each other.

3.3.1.7 Multistep Reactions, Convex Hammett Plots

Convex Hammett plots result when a reaction mechanism can consist of an equilibrium followed by an irreversible step and when within the reaction series there is a change of the rate-determining step. An example is the formation of a Schiff base with substituted benzaldehydes (*243*)[263] (Fig. 51).

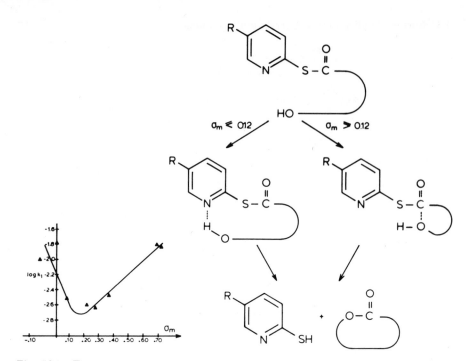

Fig. 50A Formation of macrocyclic lactones through cyclization of ω-hydroxy-2-pyridyl thiolesters.

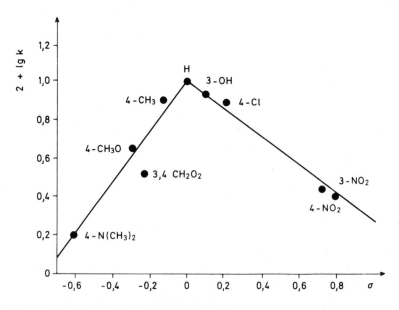

Fig. 51 ρσ Plot for reaction (243). [From J. O. Schreck, J. Chem. Educ. 48, 103 (1971) (ref. 263).]

The maximum at X = H is considered to be the point at which the rate-determining step is changing. When X is an electron donor the rate-determining step is addition. Decrease in electron-donating capacity of X enhances the reaction rate. For electron acceptors the addition is fast and the elimination is rate determining. Its rate is reduced when the electron-accepting character of X increases.[264]

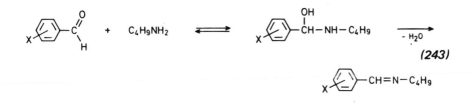

$$(243)$$

3.3.2 Characterization of Substituents

Properties of new substituents may be determined by involving them in known reaction series. For instance, the inductively electron-withdrawing character of the cyclopropane ring became apparent from its positive σ^* value. This value was obtained by placing the pK of cyclopropyl acetic acid on the pK-σ^* plot for other acetic acids. On the other hand placement of the solvolysis rate of cyclopropane-substituted 2-phenylethyl chlorides on the appropriate log k-σ^+ plot showed the pronounced tendency of cyclopropane rings toward mesomeric electron donation[265]:

$$\sigma_p^+ (\Delta) = -0.45$$

Placing cyclic tertiary substrates 211 on the known $\rho^* \Sigma \sigma^*$ plot for the solvolysis of acyclic tertiary substrates Harris and McManus[266] were able to obtain σ^* values for structural units 212; these so-called σ_t^* values (Table 62) are supposed to represent the *sum of steric, hyperconjugative, and inductive effects* of 212.

Using ρ^* values for acetolysis and alcoholysis of *acyclic* secondary substrates, which take place with nucleophilic solvent participation, it was possible to calculate for the first time rates of acetolysis and alcoholysis of "normal" *cyclic* secondary substrates 213. For com-

TABLE 62 Substituent Constants
(σ_I^*) for the Structural Unit 212 in
Alicyclic Systems

212 in	σ_t^*
Cyclobutyl	0,35
Cyclopentyl	−0,48
Cyclohexyl	0,10
Cycloheptyl	−0,47
Cyclotridecyl	0,00
Cyclopentadecyl	0,06
7-Norbornyl	1,31
7-Norbornadienyl	−0,12
anti-7-Norbornenyl	−0,45
exo-2-Norbornyl	−0,74
endo-2-Norbornyl	0,20
2-Adamantyl	−0,21

pounds solvolyzing *without* solvent participation (2-adamantyl tosylate, 7-norbornyl tosylate) the calculated rates were *too high*; for substrates ionizing with *neighboring group participation* (e.g., *anti*-7-norbornenyl tosylate, cyclobutyl tosylate) the values obtained were *too low*. Both *exo*- and *endo*-2-norbornyl tosylates gave correct values. This means that *endo*-2-norbornyl tosylate solvolyzes with solvent participation and that in *exo*-2-norbornyl tosylate the participation of the C_1-C_6 bond contributes the same amount to the reaction rate as does solvent in the *endo* isomer.

3.3.3 Estimation of Relative Reactivities of New Compounds; Values of Susceptibility Factors

To estimate effects of substituent, solvent, or reagent changes on new reactions one may use, as a first approximation, some general rules with respect to values of susceptibility factors.

3.3.3.1 *Effect of the Nature of the Reaction*

If in going from ground state to transition state in a given reaction, effects of substituents, solvents, or reagents (quantitatively expressed by σ, pK, n, Y, N_+, etc.) on the reacting system are large, numerical values of the corresponding susceptibility factors (reaction constants) are high.

Examples

Hypothesis with Respect to the Transition State	Expectation for the Susceptibility Factor
Little change in charge	Low ρ values

Proton transfer insignificant   Low values for α or β
Predominant soft-soft interactions   Low values for β
Reduced $S_N 1$ character in solvolysis   Low values for m

3.3.3.2 Effect of Reactivity

As a very approximate rule of thumb one may state that *highly reactive systems* are often *less selective* and when amenable to correlations such systems show *lower* values for susceptibility factors than similar less reactive systems. This is known as the *"reactivity-selectivity relationship"* or the *"stability-selectivity relationship."* It is mentioned here only briefly. A more detailed treatment is given in Section 5.2.3.3.

Values of ρ for electrophilic aromatic substitutions decrease when the electrophilicity ("reactivity") of the reagent increases[61] (Table 63; see also p. 111); the reagent becomes less capable of discriminating between two differently substituted benzene derivatives.

TABLE 63 ρ Values for Some Electrophilic Aromatic Substitutions

Electrophile: Br_2	Cl_2	CH_3CO^+	Br^+	"$C_2H_5^+$"
ρ -12.1	-10	-9.1	-6.2	-2.4

Similarly, ρ_π values in the Hammett correlation (*244*) of polarographic half-wave potentials ($E_{1/2}$) in reductions of aromatic compounds decrease with increasing reactivity (Fig. 52).[267]

$$\left(E_{\frac{1}{2}}\right)_X - \left(E_{\frac{1}{2}}\right)_H \sim \ln\frac{(k_f^\circ)_X}{(k_f^\circ)_H} = \rho_\pi \sigma_x \qquad (244)$$

where k_f° = rate constant of the irreversible electrode reaction
            $(E_{1/2})_X$ = half-wave potential of a substituted compound
            $(E_{1/2})_H$ = half-wave potential of the parent compound (X = H)

Methyl perchlorate reacts with Br^- about 10^3 times faster than methyl iodide and its s value is lower[268]:

$$CH_3ClO_4 : s = 0,92 \ (H_2O, \ 0°C)$$
$$CH_3I : \quad s = 1,27 \ (H_2O, \ 0°C)$$

Reactivity-(stability-)selectivity relationships are also observed in situations that are not amenable to quantitative correlations. Known

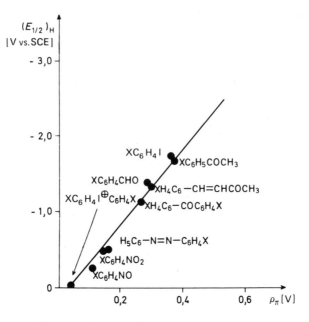

Fig. 52 Polarographic reduction of various types of aromatic compounds at pH 5-8. The plot shows the relationship between the half-wave potential $(E_{\frac{1}{2}})_H$ (in volts with respect to the standard calomel electrode) of the parent compound (X = H) as measure of the ease of reduction ("reactivity") and the susceptibility factor ρ_π for the individual reaction series. [From P. Zuman, *Collect. Czech. Chem. Commun.* 25, 3225 (1960) (ref. 267).]

examples of decreasing selectivity with decreasing stability are found in the chemistry of short-lived intermediates: carbocations, carbenes, and radicals (Fig. 53 and Tables 64, 64a and 65). The selectivities $\log(k_i/k_{isobutene})_{CXY}$ of a series of carbenes CXY toward a pair of olefins ("i" and isobutene) were found[270a] to correlate with the calculated

TABLE 64 Relative Abilities of Phenylcarbene ($R = C_6H_5$) and Halocarbenes ($R = Cl, Br$) to Insert into Primary and Secondary C—H Bonds in Pentane[270]: Halocarbenes Are Considered to Be More Stable than Phenylcarbene

$$R-\overset{..}{\underset{|}{C}}-H \ + \ H-\overset{|}{\underset{|}{C}}- \ \xrightarrow{k_{prim} + k_{sec}} \ H-\overset{\overset{R}{|}}{\underset{\underset{H}{|}}{C}}-\overset{|}{\underset{|}{C}}-$$

Mode of Generation of the Carbene	Carbene	k_{sec}/k_{prim}
$C_6H_5CHN_2, h\nu$	C_6H_5CH	7.3
$ClCHN_2, h\nu, -50°C$	$ClCH$	20
$BrCHN_2, h\nu, -50°C$	$BrCH$	25

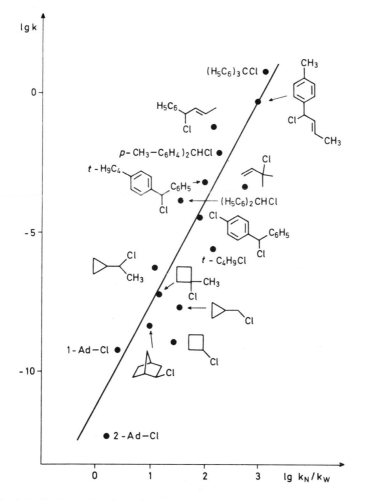

Fig. 53 Selectivities of carbocations toward azide ion and water measured as the ratio of alkyl azide (RN_3) and alcohol (ROH). In general the slower solvolyzing systems form the higher energy (therefore less selective) carbocations (k = hydrolysis rate constant in acetone-water, 80:20, 25°C, s^{-1}: $k_N/k_W = (\% \; RN_3[H_2O])/(\% \; ROH[NaN_3])$.[269] [From D. J. Raber, J. M. Harris, R. E. Hall, and P. von Ragué Schleyer, *J. Amer. Chem. Soc.* **93**, 4821 (1971) (ref. 269).]

energies of stabilization of CXY by X and Y. Carbenes more stabilized than CCl_2 (arbitrarily taken as standard) show relative selectivities (m, Table 64A) greater than 1; less stabilized carbenes have m less than 1.

The reasons for reactivity-selectivity relationships are unclear. An interesting possibility is that a differential decrease in selectivity is a consequence of a differential increase in reactivity (see Section 5.2.2.2). However, other factors and points of view also must be taken into consideration (see Section 5.2.2.3).

TABLE 64A   Carbene Selectivity Indices m^{270b}

Carbene:	CF_2	CFCl	CCl_2	MeSCCl	PhCCl	PhCBr	CBr_2	MeCCl	$BrCCO_2Et$
m^a	1.48	1.28	1.00	0.91	0.89	0.70	0.65	0.50	0.29

$^a m = \log(k_i/k_{\text{isobutene}})_{CXY} \cdot \log(k_i/k_{\text{isobutene}})_{CCl_2}$.

TABLE 65   Relative Reactivities of Cl·, C_6H_5·, and Br· with Respect to Abstraction of Primary, Secondary, and Tertiary Hydrogen Atoms in Alkanes[271]: Stability Is Considered to Increase in the Order Cl, C_6H_5, Br

	k_{CH_3-H}	$k_{CH_3CH_2-H}$	$k_{(HC_3)_2CH-H}$	$k_{(HC_3)_3C-H}$
Cl·, 40°C	~0,004	1 (Definition)	4,3	6,0
C_6H_5·, 60°C		1 (Definition)	7,8	37
Br·, 77°C	~0,0007	1 (Definition)	220	19400

3.3.3.3   Effects of Structure, Solvent, and Temperature

In conformity with its role as a *measure for changes in charge* "felt" by a substituent in the transition state of a reaction, ρ is strongly dependent on all factors affecting charge density. Examples of the weakening effect of charge delocalization on ρ are provided by ρ values for solvolysis rates of compounds **214-216** (Table 66).[272]

Similar effects are found in anionic systems $(245)^{272a}$ and $(246)^{272b}$:

$$(p\text{-}R\text{--}C_6H_4)_2CH\text{--}CCl_3 \xrightarrow[\text{slow}]{OR^-/HOR} (p\text{-}R\text{--}C_6H_4)_2\bar{C}\text{--}CCl_3 \xrightarrow[-Cl^-]{\text{fast}} \qquad (245)$$

$$\rho\,(30°C) = 2.5\text{-}3.0 \qquad\qquad (p\text{-}R\text{--}C_6H_4)_2C\text{=}CCl_2$$

$$p\text{-}R\text{--}C_6H_4\text{--}CH_2\text{--}CH_2\text{--}\overset{+}{N}(CH_3)_3Br^- \xrightarrow[\substack{-N(CH_3)_3 \\ -Br^-}]{OR^-/HOR} p\text{-}R\text{--}C_6H_4\text{--}CH\text{=}CH_2 \qquad (246)$$

$$\rho\,(30°C) = 3.8$$

The low value of ρ in system (245) (in spite of the postulated formation of a *full* negative charge) must be ascribed to increased delocalization (part of the negative charge may also be taken over by the chlorine atoms). However, since several factors change at the same time, such conclusions must be made with caution (the same is true for cationic systems in Table 66).

TABLE 66   Decrease in Absolute Value of ρ with Increasing Possibilities of Positive Charge Delocalization

Compound	Solvent	Temperature (°C)	ρ
214	80% Acetone	25	-2.71
215	50% Acetone	125	-1.9
216	80% Acetone	25	-0.9

It is also obvious that for equal charges at the reaction center substituents further removed from it will have less influence than immediate neighbors. The lowering of ρ by insulating groups is described by the *transmission coefficient* ϵ (Table 67).

Nucleophilic solvent participation in solvolysis also results in decreased numerical values for ρ (Table 68).

Similar charge dispersion effects show up in m, *which is a measure of changes in charge "felt" by solvent at the C—X group.* Low values for m are found in substrates belonging to the limiting class (see pp. 196 and 201) whenever these compounds ionize with *delocalization* of the positive charge [(247)-(249)] .

TABLE 67   Transmission Coefficients $\epsilon (= \rho_{RXZ} : \rho_{RZ})$ for Some Insulating Groups X

X	ϵ^{273}
$-CH_2-$	0,4
$-CH_2-CH_2-$	0,2 (= ca. $0,4^2$)
$-CH=CH-$	0,4
	0,3

TABLE 68   Decrease in Absolute Value of ρ in Solvolysis with Increasing Solvent Nucleophilicity

System[274]	ρ	System[275]	ρ^*
$C_6H_5CH_2Cl$, HCOOH, 0,38 M H_2O	$-7,4$	$R_1R_2CHOTos$, CF_3COOH	$-7,1$
$C_6H_5CH_2Cl$, 80% C_2H_5OH–H_2O	$-1,7$	$R_1R_2CHOTos$, HCOOH	$-3,5$
		$R_1R_2CHOTos$, CH_3COOH	$-2,6$
		$R_1R_2CHOTos$, C_2H_5OH	ca. 0

In addition, the ability of the *leaving group* to delocalize *negative* charge leads to a lowering of *m* in the order $Cl > Br > OTos$.

For the same substrate *m* values in solvent systems of relatively high nucleophilicity are lower than in systems of low nucleophilicity.

Numerical values of all susceptibility factors decrease with increasing temperature.

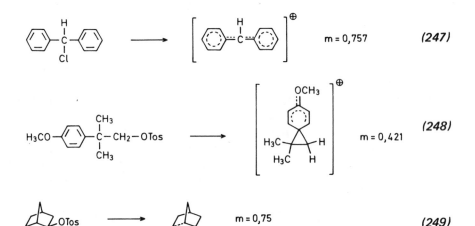

$$m = 0,757 \qquad (247)$$

$$m = 0,421 \qquad (248)$$

$$m = 0,75 \qquad (249)$$

TRANSITION STATES

CHAPTER 4

Reactivity in Terms
of Transition
State Theory

4.1 ESSENTIALS OF THE TRANSITION STATE THEORY

The *transition state theory*, developed by Eyring and Evans and
Polanyi, plays a major role in the interpretation of chemical reactivity.[276] The theory postulates that in a one-step reaction the combination of reaction partners assumes a continuous series of interatomic
configurations between the states of "independent reaction partners"
and "products that have just become independent of each other." At
some point in the intermediate region the system achieves the *critical
configuration* of the "activated complex" $M^{\ddagger}$. This is the highest point
on the most favorable reaction path of the energy hypersurface relating
the potential energy of the interacting system and the positions of all
the atoms (see Section 5.1.1). In a schematic representation of the
situation (Fig. 54) $M^{\ddagger}$ corresponds to the top of the *activation barrier*
that must be overcome by the interacting system. One speaks about
reactants as being in the ground state and activated complexes as being
in the *transition state*.

The *reaction rate* of elementary reactions (one-step transformations
without intermediates; see Fig. 54) is the product of the concentration
of $M^{\ddagger}$ and the rate constant $k^{\ddagger}$ of its collapse to products:

Macroscopic

$$A + B \xrightarrow{k_{exp}} products$$

$$\text{Reaction rate} = k_{exp} [A][B] \qquad (250a)$$

Microscopic

$$A + B \underset{}{\overset{K^{\ddagger}}{\rightleftharpoons}} M^{\ddagger} \xrightarrow{k^{\ddagger}} \text{products}$$

$$\text{Reaction rate} = k^{\ddagger} \, [M^{\ddagger}] \tag{250b}$$

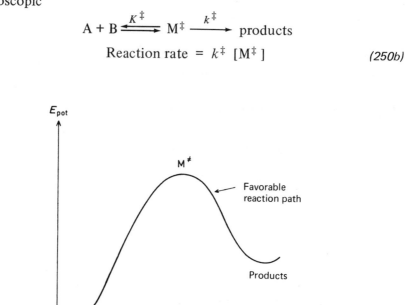

Fig. 54 Schematic representation of the continuous series of atomic configurations traversed in a one-step transformation $A + B \rightarrow$ products. Abscissa: all $3N - 6$ geometric coordinates are encompassed in the coordinate "structural changes" (reaction coordinate); some or all of these geometric coordinates may change. Ordinate: corresponding changes in potential energy.

As shown in (250b) $M^{\ddagger}$ must be in thermal equilibrium with the reactants throughout the course of the reaction [(251), (252)]:

$$[M^{\ddagger}] = K^{\ddagger} \, [A] \, [B] \tag{251}$$

$$k_{\exp} = k^{\ddagger} \, K^{\ddagger} \tag{252}$$

$K^{\ddagger}$ may be calculated, like any ordinary equilibrium constant, from the partition functions Q and the difference in potential energy, ΔE_0, between $M^{\ddagger}$ and $A + B$ (253)

$$K^{\ddagger} = \frac{Q_{M^{\ddagger}}}{Q_A \, Q_B} \, e^{-\Delta E_0 / RT} \tag{253}$$

The only difference between the activated complex and a normal molecule* is that excitation of one of the vibrational modes of the former must lead to collapse to products. $k^{\ddagger}$ is taken to be equal to the frequency $\nu^{\ddagger}$ of this particular vibration (which constitutes the reaction coordinate). From (252) and (253):

$$k_{exp} = \nu^{\ddagger} \; \frac{Q_{M^{\ddagger}}}{Q_A \, Q_B} \; e^{-\Delta E_0/RT} \qquad\qquad (254)$$

If $kT/h\nu^{\ddagger}$ is the partition function of the vibrational mode leading to collapse (for $\nu^{\ddagger} \ll kT/h$; k = Boltzmann's constant; h = Planck's constant) $Q_{M^{\ddagger}}$ can be written as the product of the partition functions for all degrees of freedom:

$$Q_{M^{\ddagger}} = \frac{kT}{h\nu^{\ddagger}} \cdot Q'_{M^{\ddagger}} \qquad\qquad (255)$$

from which:

$$k_{exp} = \frac{kT}{h} \cdot \frac{Q'_{M^{\ddagger}}}{Q_A Q_B} \; e^{-\Delta E_0/RT} \qquad\qquad (256)$$

where $Q'_{M^{\ddagger}}$ is the partition function of all the other "normal" degrees of freedom of $M^{\ddagger}$ and one can define an equilibrium constant $K'^{\ddagger}$ that, based on known thermodynamic relationships, may be derived from the differences in free energy, enthalpy, and entropy of the partners A, B, and $M^{\ddagger}$ in equilibrium (257):

$$\frac{Q'_{M^{\ddagger}}}{Q_A Q_B} \; e^{-\frac{\Delta E_0}{RT}} = K'^{\ddagger} = e^{-\frac{\Delta G^{\ddagger}}{RT}} = e^{\frac{\Delta S^{\ddagger}}{R}} \cdot e^{-\frac{\Delta H^{\ddagger}}{RT}} \qquad (257)$$

Substituting (257) into (256) yields the Eyring equation (258):

Eyring equation [§]

$$k_{exp} = \frac{kT}{h} \; e^{-\Delta G^{\ddagger}/RT} = \frac{kT}{h} \, e^{\Delta S^{\ddagger}/R} \cdot e^{-\Delta H^{\ddagger}/RT} \qquad (258)$$

---

*The similarity of activated complexes and molecules implicit in this formulation is treated in Section 5.2.1.

[§] $k/h = 2.0836 \times 10^{10} s^{-1} \cdot K^{-1}$; $R = 8.3143$ J mol$^{-1} \cdot$K^{-1}

The *activation parameters* $\Delta S^{\ddagger}$ (*entropy of activation*) and $\Delta H^{\ddagger}$ (*enthalpy of activation*) can be calculated from the temperature dependence of the rate constant, for example, by rewriting (258) as

$$\log \frac{k_{exp}}{T} = 10{,}319 + \frac{\Delta S^{\ddagger}}{19{,}173} + \frac{\Delta H^{\ddagger}}{19{,}173 T} \qquad (259)$$

($\Delta H^{\ddagger}$ and $\Delta S^{\ddagger}$ in J/mol and J/mol·K; k_{exp} in $(mol/1)^{-n}$ s^{-1} when the order of the reaction is $n + 1$) and plotting

$$\log \frac{k_{exp}}{T} \text{ versus } \frac{1}{T}$$

Alternatively, one can plot $\log k_{exp}$ versus $1/T$ and obtain the *Arrhenius parameters* A (frequency factor, preexponential factor) and E_a (the Arrhenius activation energy) (260-263):

Arrhenius equation:
$$k_{exp} = A \, e^{-E_a/RT} \qquad (260)$$

$$\log k_{exp} = \log A - \frac{E_a}{2.3RT} \qquad (261)$$

$$E_A = \Delta H^{\ddagger} + nRT \qquad (262)$$

(For reactions in the gas phase n is equal to the molecularity of the reaction; for reactions in solution $n = 1$.)

$$A = \frac{kT}{h} \cdot e^{\frac{\Delta S^{\ddagger}}{R}} \cdot e^{-(\Delta n - 1)} \qquad (263)$$

(For reactions in the gas phase: unimolecular, $\Delta n = 0$; bimolecular, $\Delta n = -1$; termolecular, $\Delta n = -2$. For reactions in solution, approximately: $\Delta n = 1$.)

Rigorous expressions for k_{exp} contain in (258) the factor κ—the *transmission coefficient*. This factor takes into account the possibility that for particular shapes of the energy hypersurface part of the systems that have already crossed the barrier may return to reactants and not lead to products ($\kappa < 1$). Since values for κ are insufficiently known, the value 1 is generally used. The possibility that systems with lower energy than that corresponding to the activation barrier may cross it ("tunneling effect", see p. 261) may manifest itself in values of $\kappa > 1$.

The assumption of equilibrium partitioning of the energy over all degrees of freedom in the system A, B, $M^{\ddagger}$ lies at the basis of the transition state theory. It must be valid also for interactions of the activated complex with the medium (solvent effects).*

In gas phase reactions proceeding at low pressure, or in photochemical reactions, there is the possibility that primary products formed with excess energy may react further, prior to the equilibration of the energy. [In the case of dimethylcyclopropane the lifetime of thermally activated molecules with an average energy of ≈ 331 kJ/mol is $\approx 10^{-6}$ s; the lifetime of molecules "chemically activated" by the exothermicity of addition of CH_2 to 2-butene (average energy ≈ 461 kJ/mol) is, on the other hand, only 10^{-8} s).[278] The product ratio of further (intramolecular) reactions of such species (labeled) differs from that obtained in the corresponding thermally equilibrated system. An example is the photochemical and thermal decomposition of 3-ethyl-3-methylaziridine (264).[279] In the photochemical reaction in the *gas phase* further reaction of the intermediate methylethylcarbene takes place before $K_A^{\ddagger}$ and $K_B^{\ddagger}$, which determine the ratio of products A and B, achieve values corresponding to complete energy equilibration (see however, p. 192).

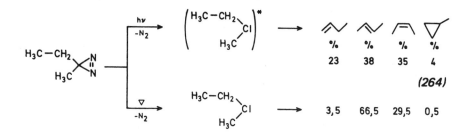

(264)

4.2 RELATIVE REACTIVITIES

4.2.1 General Aspects

The relative reactivity of two systems A and B may be expressed in terms of transition state theory as follows:

$$\lg \frac{k_A}{k_B} = \frac{\Delta G_B^{\ddagger} - \Delta G_A^{\ddagger}}{2,3 \, RT} = \frac{-\Delta\Delta G^{\ddagger}}{2,3 \, RT} \qquad (265)$$

The various possibilities are shown in fig. 55.

*Avery's results[277] show that energy flow in molecules is very fast, which allows us to assume equilibrium partitioning over the $3N-6$ possible degrees of freedom with their multitude of excited levels.

Situation I $\Delta\Delta G^{\ddagger}$ of two systems (e.g., with identical or isomeric reactive partial structures*) are due to differences both in *ground state* and *transition state*.

The Foote-Schleyer equation (Section 3.1.2.2) is a good illustration of the general situation. The second term takes into account the free energy difference, due to torsional effects, between cyclohexyl tosylate and the tosylate to be compared to it. The other terms take into account mainly the differences in the transition states leading to the carbocations produced in acetolysis. (In the formulation of the Foote-Schleyer equation, differences in properties of transition states were replaced by differences in properties of the carbocations themselves. For the approximate description of transition states by means of model systems, see Section 5.2).

4.2.2 Intramolecular Competition; The Curtin-Hammett Principle; The Winstein-Holness Equation

Situation II, that is, differences in reactivity as a consequence of identical ground state free energies and different transition state free energies, corresponds to intramolecular competition (Section 2.2.1), for instance, in conformationally rigid systems such as in aromatic substitution or in cycloadditions (*18, 19*). In such systems ratios of reaction rates depend solely on free energy differences between the corresponding transition states.

Systems of type 1, where A and B are in equilibrium and this equilibrium is established *faster* than A and B undergo further reaction behave *as if situation II were prevailing* (Scheme 49).

This *Curtin-Hammett* principle,[280] relevant only for kinetically controlled reactions and mainly for the analysis of product ratios, can be easily derived [(*266-268*) and Fig. 56].

Equation (*268*) shows that the ratio of product derived from A to product derived from B depends exclusively on the properties of the corresponding transition states $A^{\ddagger}$ and $B^{\ddagger}$.

---

*Aside from special steric, polar, electronic, or association effects (which then become actually responsible for $\Delta\Delta G^{\ddagger}$) the differences in the residual structures should not contribute to $\Delta\Delta G^{\ddagger}$; that is, substitution of H in A by CH_3 ($\rightarrow$ B) should contribute to the difference in ground states, $G_B - G_A$, by the same amount as to the difference in transition states, $G_B^* - G_A^*$. In diagrams, this allows a displacement (by this amount) on the G scale making comparisons much easier.

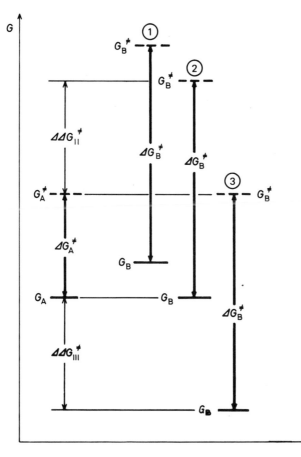

Fig. 55 Possible relative free energies *at constant relative reactivity* ($\Delta\Delta G^{\ddagger}$) of ground (solid lines) and transition states (dashed lines) of two systems A (constant in figure) and B (variable in figure). In addition to various possible forms of situation (I) (ground *and* transition states differ in their free energies) two special cases are possible: (II) ($\Delta\Delta G^{\ddagger}$ is determined only by differences in transition states) and (III) [$\Delta\Delta G^{\ddagger}$ is determined only by differences in ground states (transition states are identical or similar)].

The Curtin-Hammett principle applies primarily to relatively slow reactions of conformationally mobile substrates (e.g., Schemes 50-53).

The exclusive formation of the *trans* olefin **218** in the Claisen-Schmidt reaction (Scheme 50) is due to the sterically unfavorable phenyl-phenyl interaction in the transition state leading to the *cis* iso-mer **217**. Such an interaction results in the distortion of the π-electron system and loss of coplanarity.

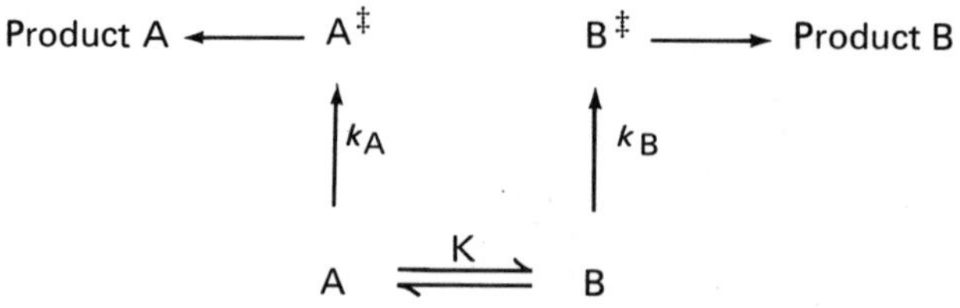

Product A ←――― A‡ B‡ ――――→ Product B

$$\Big\uparrow k_A \qquad\qquad \Big\uparrow k_B$$

$$A \underset{\longleftarrow}{\overset{K}{\rightrightarrows}} B$$

$$\frac{[\text{Product A}]}{[\text{Product B}]} = \frac{k_A[A]}{k_B[B]} = \frac{k_A}{k_B}\cdot\frac{1}{K} \tag{266}$$

$$\frac{[\text{Product A}]}{[\text{Product B}]} = \exp\frac{\Delta G_B^{\ddagger} - \Delta G_A^{\ddagger} + \Delta G}{RT} \tag{267}$$

$$\frac{[\text{Product A}]}{[\text{Product B}]} = \exp\frac{G_B^{\ddagger} - G_B - G_A^{\ddagger} + G_A + G_B - G_A}{RT} = \exp\frac{-(G_A^{\ddagger} - G_B^{\ddagger})}{RT} \tag{268}$$

Scheme 49 The Curtin-Hammett principle. Product formation from equilibrating reactants.

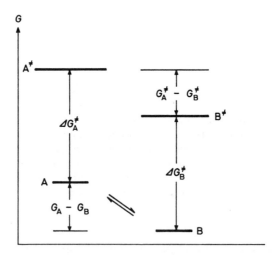

Fig. 56 The Curtin-Hammett principle. Ground and transition states in the system A ⇌ B and the differences in free energy of the two transition states controlling the *product ratio*.

234

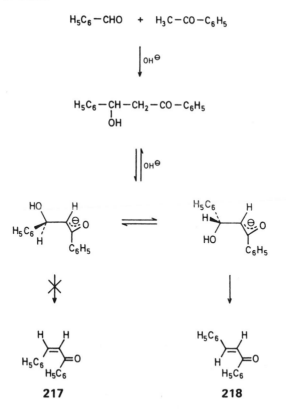

Scheme 50 Formation of *cis-trans* isomeric olefins on elimination of OH⁻ from equilibrating rotamers of hydroxy enolates in the Claisen-Schmidt reaction.[281]

The same situation pertains in E2 reactions where *cis-trans* isomeric olefins may be formed. The determination of relative migratory aptitudes of various groups by intramolecular competition experiments (p. 55) must be performed under the conditions of the Curtin-Hammett principle.

The inversion of the nitrogen atom in 1-benzyl-4-phenyl-piperidine proceeds much faster than quaternization with methyl iodide. The transition state for quaternization with the benzyl group equatorial and the unshared pair on nitrogen axial (→**219**) is more favorable than the one in which the benzyl group is axial and the unshared pair equatorial (→**220**). (By contrast, the sulfur analogues, thianes, are alkylated, preferably through equatorial approach of the alkylating agent.[282a])

Important applications of the Curtin-Hammett principle can be found with acyclic substrates reacting via *cyclohexane-like* transition states (Schemes 52 and 53 and Fig. 57).

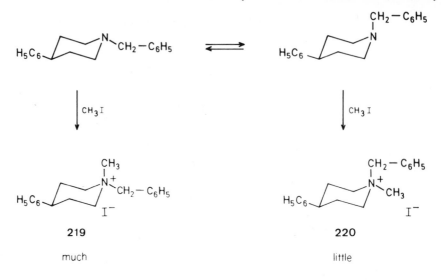

Scheme 51 Methylation of 1-benzyl-4-phenylpiperidine.[282]

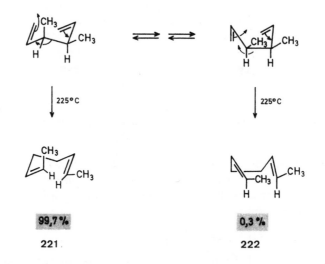

Scheme 52 Cope rearrangement of *meso*-3,4-diemthylhexadiene-1,5.[283] Favoring of the quasi-chair transition state.

The ratio of *cis-trans-* (221) and *trans-trans*-octa-2,6-diene (222) formed in the *Cope rearrangement* (p. 432) of *meso*-3,4-dimethylhexadiene-1,5 (Scheme 52) shows that the quasi-chair transition state (223) leading to 221 is preferred over the quasi-boat transition state (224) leading to 222 by 24 kJ/mol.

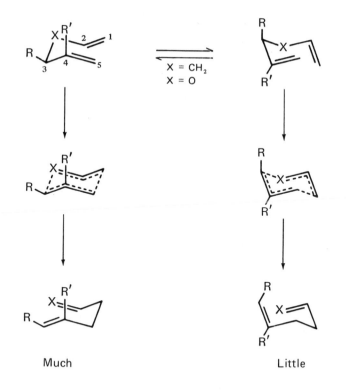

Scheme 53 Cope and Claisen rearrangements. Transition states with "equatorial" R are favored.

In spite of the great similarity with the free energy difference between chair and twist-boat forms of cyclohexane, it is by no means clear that the same or similar steric factors are responsible for the free energy difference between **223** and **224**. Orbital interactions may play a role as well (see p. 407).

The chair model for the transition state of the Cope rearrangement and the very similar nonaromatic *Claisen* rearrangement (Scheme 53, X = O) proved to be very fruitful for the prediction of the stereochemical outcome of these reactions.

The *E/Z* ratio[16d] of the olefins formed corresponds in many instances to the cyclohexane conformation energy (Table 10) of the R group[284] (Scheme 53).

The Claisen rearrangement, a method for carbon-carbon bond formation, therefore became an important tool in synthetic chemistry because of its predictable stereochemistry [e.g., in syntheses of trialkyl olefins[285] via sequence (269) or in generation of chiral centers when substituents at C-5 are different[286] (Scheme 53)].

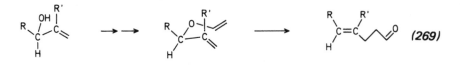

(269)

In a similar vein the high preference for *trans*-alkene formation observed in the thermal allyl sulfenate-allyl sulfoxide interconversion (Scheme 53A)[287] can be ascribed to the lower free energy of the five-membered ring transition state containing R in a pseudoequatorial position.

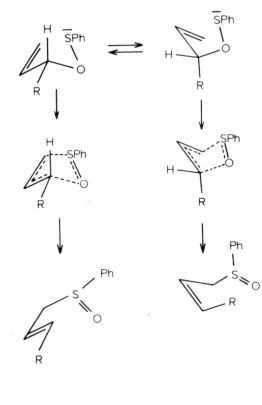

major   minor

Scheme 53A The allyl sulfenate-allyl sulfoxide interconversion.

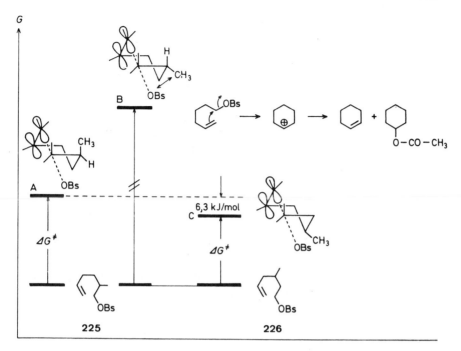

Fig. 57 Transition states in the acetolysis of isomeric methyl-5-hexenyl brosylates.[288] Simultaneous operation of steric effects of several substituents in the cyclohexane-like transition states. (For clarity **B** was derived from the epimer of the reactant molecule leading to **A**).

$$Bs \ = \ O_2S\text{—}\langle\ \rangle\text{—}Br$$

Cationic cyclizations of 5-hexenyl derivatives to cyclohexyl cations (Scheme 18 and Fig. 57) proceed via the more favorable chair transition state.[288]. The 6.3 kJ/mol difference in free activation energy of acetolysis of 2-methyl-5-hexenyl brosylate (**225**) and its 3-methyl isomer (**226**) may be ascribed to the cyclohexane conformation energy of the methyl group: in the transition state **C** for solvolysis of **226** this group can occupy the equatorial position. On the other hand, in the case of **225** equatorial orientation would lead to an eclipsed interaction with the leaving group (**B**), so that the orientation **A** with the methyl group axial is preferred.

Examples of systems undergoing fast *structural* isomerizations ($\Delta G^{\ddagger}$ for interconversion of isomers $\approx$ 42 kJ/mol) are the *valence isomerization* equilibria between cycloheptatrienes and bicyclo[4.1.0]hepta-2,4-dienes ("norcaradienes") (*270*).[289]

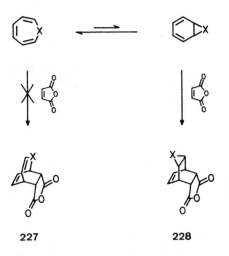

$$X = CR_2 , NR , O \qquad\qquad (270)$$

In general the monocyclic isomers are highly favored*, but reaction products are often derived from the bicyclic form. The best known example is the exclusive formation of **228** in the Diels-Alder reactions shown. Favoring of the corresponding transition state over that leading to **227** agrees with behavior observed in simple cyclic dienes: Diels-Alder reactivity of cyclic dienes (starting with cyclopentadiene) decreases with increasing ring size and vanishes completely in cycloocta-1,3-diene.[290] (The addition of the dienophile to the 1,3-cyclohexadiene system complies with the "*endo*" rule (Sections 1.3.1.5 and 5.2.4.6) and takes place from the sterically more accessible side, opposite to the three-membered ring.)

Eight-membered ring polyenes show a behavior similar to that of the seven-membered ring polyenes (*271*).[291]

The operation of the Curtin-Hammett principle in a prototropic equilibrium is shown in Scheme 55.[292]

227   **228**

Scheme 54   Valence isomerization and Diels-Alder reaction of cycloheptatriene derivatives with maleic anhydride.

$$(271)$$

*It is useful to point out that even for $\Delta G = 21$ kJ/mol the equilibrium is practically completely "fixed" on one side (see Table 7). However, for $\Delta G^{\ddagger} = 42$ kJ/mol there still exists a very fast transformation of the equilibrium partners into each other.

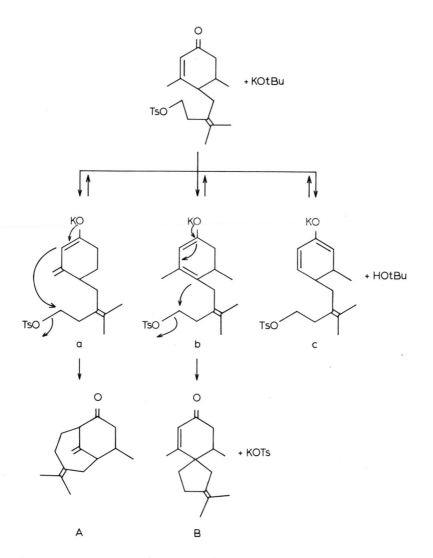

Scheme 55 Operation of the Curtin-Hammett principle in a prototropic equilibrium system of tosyloxydienolates.

DMSO (ml)	H$_2$O (ml)	% A	% B
10.0	0.05	100	0
8.8	1.2	49.7	50.3
7.5	2.5	6.5	93.5

In an attempted synthesis of the mixture of diastereomers B (one of which is the sesquiterpene ketone β-vetivone) A was obtained exclusively, when pure DMSO was used as solvent. Under these conditions dienolate a is formed fastest and does not equilibrate with its isomers b and c. Equilibration between a, b, and c is induced by the addition of water. Ultimately B is formed predominantly, because from the rapidly interconverting mixture of dienolates a-c the transition state of five-membered ring formation is the one that is reached most easily.

Under the conditions of the Curtin-Hammett principle mole fractions γ of A and B remain constant throughout the duration of the reaction and the overall rate can be expressed by:

$$\frac{-d([A] + [B])}{dt} = (k_A \cdot [A] + k_B \cdot [B]) \cdot [\text{reaction partner}] \qquad (272)$$

$$= (k_A \cdot \gamma_A + k_B \cdot \gamma_B) \cdot ([A] + [B]) \cdot [\text{reaction partner}]$$

$$(273)$$

and correspondingly:

Winstein-Holness equation[293]

$$k_{\text{total}} = k_A \cdot \gamma_A + k_B \cdot (1 - \gamma_A) \qquad (274)$$

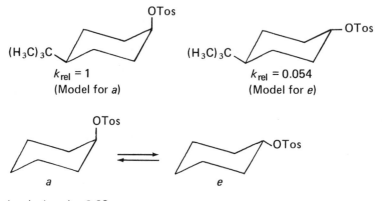

$k_{\text{rel}} = 1$ $k_{\text{rel}} = 0.054$
(Model for a)   (Model for e)

a   e

$k_{\text{rel}}\ (= k_{\text{total}}) = 0.28 \qquad\qquad\qquad\qquad\qquad\qquad\qquad (275)$

$k_{\text{total}} = 0.28 = 1 \cdot \gamma_a + 0.054\ (1 - \gamma_a) \qquad\qquad\qquad \gamma_a = 0.24$

Scheme 56 Relative rate constants of S_N2 reactions of cyclohexyl tosylates and $NaSC_6H_5$ in 87% ethanol at 25°C.[294] Analysis by means of the Winstein-Holness equation (274). Because of the high conformation free energy of the t-butyl group (23.5 kJ/mol) the equilibrium mixtures of conformers of cis-trans isomeric 4-t-butylcyclohexyl tosylates contain exclusively ($\gamma \approx 1$) the conformer in which the t-butyl group is equatorial. The reactivity of the tosyloxy group is considered *not* to be affected by the presence of the t-butyl group.

If k_A and k_B are known (or if approximate values can be derived from appropriate models for A and B) the determination of k_{total} makes it possible to calculate the composition of a fast equilibrating mixture A $\rightleftarrows$ B.

Example

Conformer equilibrium in cyclohexyl tosylate[294] (Scheme 56). The equilibrium constant $a \rightleftarrows e$ turns out to be 3.16 and the free conformation energy of the tosyloxy group, 2.9 kJ/mol. "Conformation energies" of other groups were derived in a similar manner (Table 10).

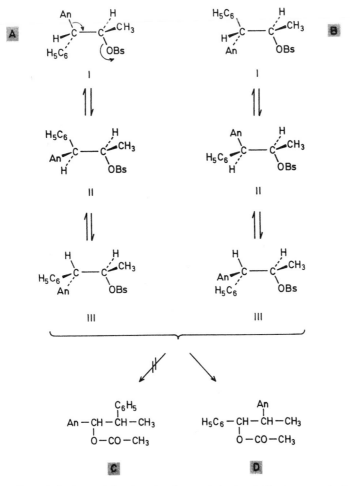

Scheme 57 *p*-Anisyl participation in the acetolysis of diastereomeric 1-phenyl-1-(*p*-anisyl)-2-propylbromobenzene sulfonates.

The Winstein-Holness equation must be used when relative reactivities of conformationally mobile systems are to be compared:

Comparison between p-anisyl and phenyl participation.[295] Acetolysis (25°C) of the diastereomeric 1-phenyl-1-(*p*-anisyl)-2-propylbromobenzene sulfonates A and B yields exclusively product D resulting from *p*-anisyl migration (Scheme 57).

Assuming that only *trans*-coplanar arrangements are involved:

$$k_{total}^{A} = k_I^A \gamma_I^A \ (= 157.6 \times 10^{-7} \ s^{-1}) \gg k_{II}^A \gamma_{II}^A + k_{III}^A \gamma_{III}^A \qquad (276)$$

$$k_{total}^{B} = k_{II}^B \gamma_{II}^B \ (= \ 86.7 \times 10^{-7} \ s^{-1}) \gg k_I^B \gamma_I^B + k_{III}^B \gamma_{III}^B \qquad (277)$$

Because it was impossible to determine the aptitude to participate of the *p*-anisyl group versus that of the phenyl group as the quotient [D] : [C] one had to measure the acetolysis rate of a model, 1,1-diphenyl-2-propylbromobenzene sulfonate E (Scheme 58).

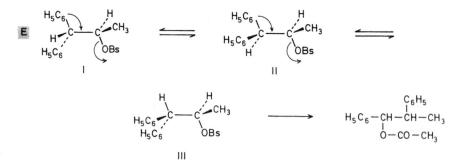

Scheme 58 Phenyl participation in the acetolysis of 1,1-diphenyl-2-propylbenzene sulfonate (E).

$$k_{total}^{E} = k_I^E \gamma_I^E + k_{II}^E \gamma_{II}^E \ (= 3,94 \times 10^{-7} \ s^{-1} \ [25°C]) \qquad (278)$$

To deal with the complicating factor that this compound has two rotamers (I and II), both able to participate in the reaction, the following simplifying assumptions were made: substitution of phenyl by *p*-anisyl does not affect the rotamer population (i.e., γ_I, γ_{II}, and γ_{III} have the same values in A, B, and E); the effect of interchanging positions of (nonparticipating) H and C_6H_5 on the relative reactivity of the rotamers is the same in both the *p*-anisyl and the phenyl system (i.e., $k_I^E : k_{II}^E = k_I^A : k_{II}^B$). One then can write

$$\frac{k_I^E \gamma_I^E}{k_{II}^E \gamma_{II}^E} = \frac{k_I^A \gamma_I^A}{k_{II}^B \gamma_{II}^B} = \frac{157,6}{86,7} \qquad (279)$$

[see (276) and (277)] which means that rotamer I of A and rotamer II of B are model systems for rotamers I and II of E. However, in contrast to the *t*-butylcyclohexyl system, in A, B, and E there is no clear favoring of a single equilibrium partner (for which one can set $\gamma \approx 1$) and one can only compare the products $k \cdot \gamma$.

Combination of (*279*) and (*278*) gives the values of the two terms in (*278*). They are given along with data from (*276*) and (*277*) in Table 69. It can be seen that ionization of rotamer I of A and rotamer II of B involving *p*-anisyl participation proceeds 62 times faster than ionization of the corresponding rotamers I and II of E involving phenyl participation.

If equilibration is slower than reaction, in the extreme case even the slowest reacting component of the system will be consumed before rearrangement to the faster reacting form can occur. In this instance the ratio of reaction products will reflect the ratio of individual components in the reacting system. The very fast hydroboration of thujopsene[296] (complete after a few seconds at 0°C) is thought to represent such a "non-Curtin-Hammett" situation (Scheme 59).

TABLE 69 Comparison of *p*-Anisyl Participation and Phenyl Participation [see Schemes 57 and 58, (*276*)-(*278*)]

System	$\gamma_I k_I$ ($\times 10^{-7}$ s^{-1}	Relative Value)	$\gamma_{II} k_{II}$ ($\times 10^{-7}$ s^{-1}	Relative Value)
A	157.6	62		
B			86.7	62
E	2.54	1	1.40	1

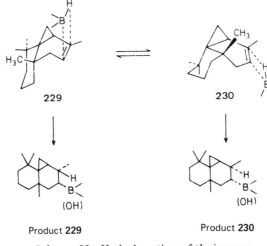

Scheme 59 Hydroboration of thujopsene.

Of the possible interconvertible conformations **229** and **230**, **229** should be attacked from the β-side, while **230** should be attacked from the α-side[16d], because in this conformer the β-side of the double bond is shielded by the angular methyl group. Since only the product derived from **229** is formed, it was concluded that **229** is the only conformation present.

The rotamer distribution in the ground state of cyclopropylmethyl ketones (Scheme 4), known from nmr and electron diffraction data, is relatively closely matched by the ratio of enol acetates obtained on reduction of the ketones with Li/NH$_3$, followed by treatment of the enolates with acetic anhydride.[297] Electron transfer and irreversible ring opening must proceed faster than equilibration of the rotamers of the ketone or of the ketyl formed from it (Scheme 60; see also Scheme 4).

Understandably, similar situations are encountered in reactions involving typical reactive intermediates (cf. the reactions of the 9-decalyl cation, scheme 18 and ref. 288). Extrusion of nitrogen (and ring expansion) in diazonium betaines formed on addition of diazoalkanes to

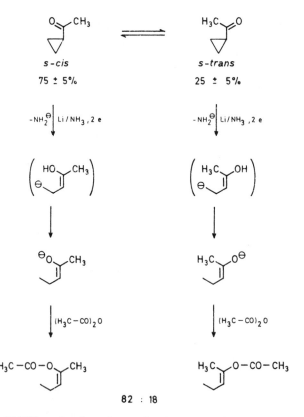

Scheme 60   Li/NH$_3$ reduction of acetylcyclopropane [From W. G. Dauben and R. E. Wolf, *J. Org. Chem.* **35**, 2361 (1970) (ref. 297).]

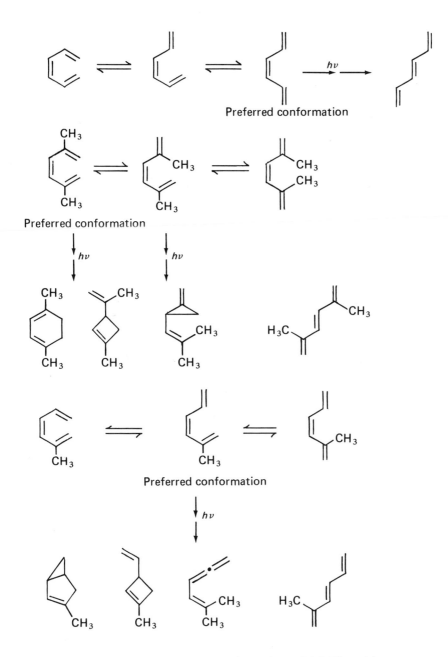

Scheme 61 Photochemical transformations of 1,3,5-hexatrienes.

247

cyclopropanones (Scheme 14) is probably faster than equilibration of the betaine rotamers. Typical products formed on photolysis of various 1,3,5-hexatrienes are best derived from preferred ground state conformations. Because of this it is assumed that both the excitation ("Franck-Condon principle"*) and further reaction of the excited species represent a "non-Curtin-Hammett" situation[298] (Scheme 61).

The change from a "non-Curtin-Hammett" situation to the corresponding Curtin-Hammett situation is nicely illustrated by the Diels-Alder reactions of the equilibrium mixture (established by sigmatropic [1,5]-trimethylsilyl migration) of the three isomeric trimethylsilylcyclopentadienes (Scheme 61A).[298a]

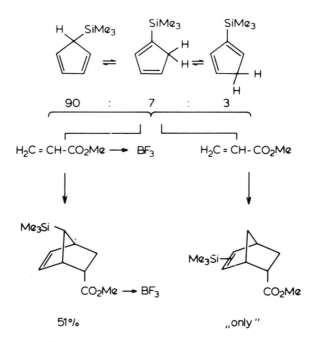

Scheme 61A   Diels-Alder reactions of the trimethylsilylcyclopentadienes.[298a]

The extremely fast reaction of the boron trifluoride complex of methyl acrylate yields 51% of the adduct derived from the major component. Methyl acrylate alone, however, is not a reactive enough dienophile to surpass in rate the 1,5-trimethylsilyl migration: in this

*The Franck-Condon principle states that transitions between various electronic states occur so fast that nuclei of the corresponding atoms do not change their positions. Immediately after a transition, the atoms involved still occupy positions corresponding to the equilibrium structure (energy minimum) of the initial state. Only later does reorganization to the characteristic equilibrium structure of the new state occur (with emission of energy).

case the only product is the one derived from the least abundant isomer.

4.2.3   Ground State Effects

Situations of type III (Fig. 55), that is, differences in reactivity as a consequence of same transition states and different free energies of ground states, are more difficult to denote than examples of situation II. Assuming that the transition between product and reactant states takes place within a continuum of structures and is accomplished by the mutation of a specific vibration of the activated complex into a translation, *each* structurally discrete *reactant-product pair* (even if differing only in one partner) has its *own* transition state with *its characteristic geometry, product-forming vibration* (reaction coordinate), and *free energy*. Consequently, reactions of isomers leading to the same product can at best be expected to proceed via similar transition states and only approximately mimic situation III.

Assuming that the free energy of transition states leading to unstable intermediates is very similar to that of the intermediates themselves (see p. 328) it can be argued that the high sensitivity of 3-methylenecyclohexadiene-1,4 toward acid[299] is due mainly to lack of *aromatic stabilization* in its ground state, compared to its isomer toluene, where such stabilization is present (Fig. 58).

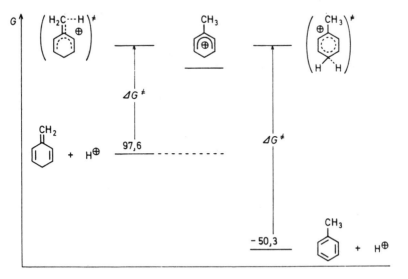

Fig. 58 Free energies (kJ/mol) of ground states of toluene and 3-methylenecyclohexadiene-1,4 and the free energies of activation $\Delta G^{\ddagger}$ for protonation of these substrates.

Differences in free energy of isomeric transition states and variations in these differences on changing the leaving group can be observed on studying the cationic reactivity in the *anti*-7-norbornenyl system (*280*).[300]

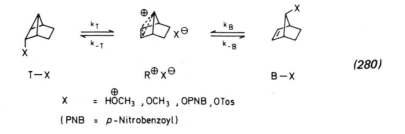

$$(280)$$

$$X \quad = \quad H\overset{\oplus}{O}CH_3 \ , OCH_3 \ , OPNB \ , OTos$$

$$(PNB \ = \ p\text{-Nitrobenzoyl})$$

The "reactivity ratio" $R = k_T/k_B$ gives the difference in activation barriers of the tricyclic structure T—X and the bicyclic isomer B—X, the "partition factor" $P = k_{-B}/k_{-T}$ shows the difference in activation barriers surrounding the intermediate ion pair $R^+ X^-$ and leading to products B—X and T—X. The product

$$P \cdot R \ = \ \left(\frac{k_{-B}}{k_{-T}} \right) \left(\frac{k_T}{k_B} \right) \ = \ K \qquad (281)$$

is the equilibrium constant [BX]/[TX].

The rate constants for hydrolysis of T—OCH$_3$ and B—OCH$_3$ via R^+X^- (probably) involving specific acid catalysis ($X^- = HOCH_3$) are given by (*285*):

$$H^\oplus \ + \ T\text{—}OCH_3 \ \underset{}{\overset{K_H}{\rightleftharpoons}} \ T\text{—}\overset{\oplus}{O}HCH_3 \ \overset{k_T}{\longrightarrow} \ R^\oplus HOCH_3 \ \longrightarrow \ \longrightarrow$$

$$B\text{—}OH \ + \ T\text{—}OH \qquad (282)$$

$$- \frac{d\,[T\text{—}OCH_3]}{dt} \ = \ k_T[T\text{—}\overset{\oplus}{O}HCH_3] \ = \ k_T K_H [H^\oplus] [T\text{—}OCH_3] \qquad (283)$$

$$k_{exp} \ = \ k_T K_H [H^\oplus] \qquad (284)$$

$$\left(\frac{k_{exp}}{[H^\oplus]} \right)_{T\text{—}OCH_3} \ : \ \left(\frac{k_{exp}}{[H^\oplus]} \right)_{B\text{—}OCH_3} \ = \ \frac{k_T}{k_B} \ = \ \frac{9,7}{1,4 \cdot 10^{-6}} \qquad (285)$$

(It was assumed that the equilibrium constants K_H for protonation of B—OCH$_3$ and T—OCH$_3$ are equal.)

If the partition factor P is chosen in such a way as to correspond to the ratio $[B-OCH_3]:[T-OCH_3] = 300$, as was found in the methanolysis of B–OTos, $K = 2 \times 10^9$.

With respect to *ground states* this means that the free energy of $T-\overset{+}{O}HCH_3$ (and $T-OCH_3$) is 53 kJ/mol higher than that of $B-\overset{+}{O}HCH_3$ (and $B-OCH_3$), probably because of ring strain in the bicyclo[2.1.0]-pentane moiety present in the tricyclic substrate. However, according to the value of P, the tricyclic transition state is also *higher* (by 14 kJ/mol) than the bicyclic one; (Fig. 59a).

In the presence of sodium methoxide methanolysis of B–OTos gives increasing amounts of $T-OCH_3$ (see p. 160); consequently, it must be assumed that for the system $R^+OTos^- + {}^-OCH_3$ $P \leqslant 1$. If the reaction of $R^+OCH_3^-$ happened to yield a similar value, then the tricyclic transition state in this system (Fig. 59b) should be roughly at the *same height* as that for the bicyclic one.

Finally, if one assumes that the ground states of *p*-nitrobenzoates T-OPNB and B-OPNB differ by the same amount as those of (protonated) methyl ethers, it follows from the ratio of their hydrolysis rate constants

$$R = \frac{k_T}{k_B} = 8 \times 10^{11} \qquad (286)$$

that $P = 2.5 \times 10^{-3}$; that is, the tricyclic transition state must now be 14.7 kJ/mol *lower* than the bicyclic one (Fig. 59c).

According to Rüchardt,[98, 301a] the *increased tendency toward formation of radicals* observed when the radicalic carbon atom becomes more highly alkylated (Table 70), as well as the corresponding decrease

TABLE 70 Relative Half-Lives of Radical Formation from Peroxy Esters (a) and azoalkanes (b)

(a)	$R_1R_2R_3C-COO-OC(CH_3)_3$	$\xrightarrow{60°C}$	$R_1R_2R_3C\cdot + CO_2 + \cdot OC(CH_3)_3$
(b)	$R_1R_2R_3C-N=N-CR_1R_2R_3$	$\xrightarrow{300°C}$	$R_1R_2R_3C\cdot + N_2 + \cdot CR_1R_2R_3$

$R_1R_2R_3$	$t_{1/2}$ (rel.)	
	(a)	(b)
H,H,H,	(1670)	13000
H,H,CH$_3$	(700)	1170
H,CH$_3$,CH$_3$	333	117
CH$_3$,CH$_3$,CH$_3$	1	1 (Definition)

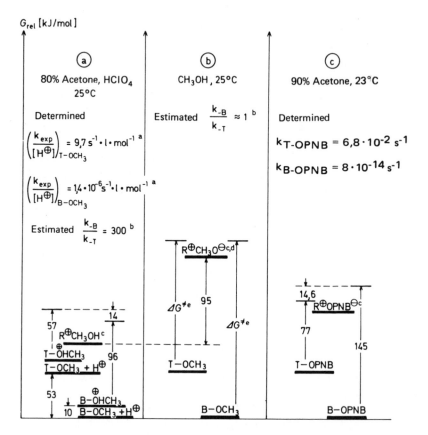

Fig. 59 Free energies of activation and relative free energies of ground states (thick lines) and transition states (thin lines) in solvolysis of bicyclic and tricyclic compounds (*280*). Rate constants were transformed into $\Delta G^{\ddagger}$ by means of (*258*). The following notes correspond to the superscript indices in the figure:

(*a*) From these values one can calculate the rate constant for ionization of the protonated substrate via (*284*). For K_H one used the value 0.014, derived from pK of protonated ether (p$K_{(C_2H_5)_2\overset{+}{O}H}$ = −3.59)[301] and pK of H_3O^+ (−1.74).

(*b*) See text.

(*c*) The free energy of these states is unknown.

(*d*) G_{H^+} = 0 was chosen (ground levels of *a* and *b* are equal). Considering pK_{CH_3OH} = 16.35 this state turns out to be 95 kJ/mol higher than the state R^+CH_3OH.

(*e*) The height of this barrier is unknown.

TABLE 71   CH Dissociation Energies (kJ/mol)[98]

CH_4 $\longrightarrow$	$CH_3\cdot + H\cdot$	436
H_3C-CH_3 $\longrightarrow$	$H_3C-CH_2\cdot + H\cdot$	412
$H_3C-CH_2-CH_3$ $\longrightarrow$	$H_3C-\overset{\cdot}{C}H-CH_3 + H\cdot$	395
$H_3C-\underset{\underset{CH_3}{\mid}}{CH}-CH_3$ $\longrightarrow$	$H_3C-\underset{\underset{CH_3}{\mid}}{C}\cdot-CH_3 + H\cdot$	382

in CH dissociation energies (Table 71) are not due to the stabilizing effect of alkyl groups in radical **232** and in the transition state leading to it, as is generally assumed, but rather to relief of the steric destabilization present in the ground state **231** (*287*). These conclusions are based on the following arguments:

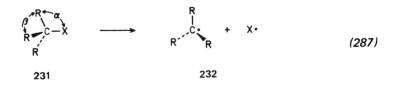

$$(287)$$

231   **232**

1   Electron-spin resonance (esr) studies of alkyl radicals have shown that spin density at the central carbon atom is only slightly decreased by methyl groups. Since this effect must be even weaker in the transition state, it cannot lead to appreciable stabilization of the transition state and one must look for other reasons to account for the higher rate of formation of alkyl radicals when the degree of substitution increases.

2   Substitution of H in **231** (R = H) by alkyl groups results in an increase in β ($\beta_{propane}$ = 111.5°) and a corresponding decrease in α. The resulting unfavorable steric interactions between the three ligands and X are relieved on dissociation to **232** (relief of back strain). The effect is largest for X = H; this C—X bond is extremely short and repulsion between its electrons and those of bonds to ligands is at its maximum (see also p. 143).

The free energy of activation for dissociation of many hydrocarbons into a pair of radicals (*287a*) is linearly dependent on the ground state strain (front strain) of these hydrocarbons[301a].

TABLE 72 Relative Hydrolysis Rates of t-Alkyl p-Nitrobenzoates[302]

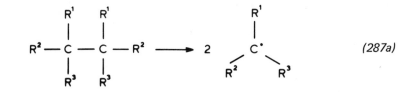

R_1	R_2	R_3	k_{rel}
CH_3	CH_3	CH_3	1.0 (Definition)
t-C_4H_9	t-C_4H_9	t-C_4H_9	13,500
$(H_3C)_3C-CH_2$	t-C_4H_9	t-C_4H_9	19,400
$(H_3C)_3C-CH_2$	$(H_3C)_3C-CH_2$	t-C_4H_9	68,000

$$R^2 - \underset{\underset{R^3}{|}}{\overset{\overset{R^1}{|}}{C}} - \underset{\underset{R^3}{|}}{\overset{\overset{R^1}{|}}{C}} - R^2 \longrightarrow 2 \quad \underset{R^2 \qquad R^3}{\overset{R^1}{\overset{|}{C^\cdot}}} \qquad (287a)$$

Similar situations—increase in reactivity with increase in volume of subtituents—are found in heterolyses of t-alkyl esters[302] (Table 72).

However, the situation here is less clear, since in addition to steric effects in the ground state there may be differences in solvation of transition states leading to charged species. This manifests itself in the acid-catalyzed dehydration of the corresponding alcohols.[303] Systems most reactive in solvolysis react the most slowly, probably because of hampered solvation (Table 73).

In solvolysis reactions of p-nitrobenzoates relief of steric interactions due to the bulkier leaving group seems to be stronger than loss of solvation.

TABLE 73 Rate Constants (s^{-1}) of Sulfuric Acid Catalyzed Dehydration of Tertiary Alcohols in Acetic Acid at $25°C$[303]

$$R_1R_2R_3C-OH \xrightarrow{H^+} R_1R_2R_3C-\overset{+}{O}H_2 \longrightarrow R_1R_2R_3C^+ + H_2O$$

R_1	R_2	R_3	$10^5 \cdot k$
CH_3	t-C_4H_9	t-C_4H_9	39.5
t-C_4H_9	t-C_4H_9	t-C_4H_9	3.38
$(H_3C)_3 C-CH_2$	t-C_4H_9	t-C_4H_9	1.47

When one of the R groups is phenyl, increased steric effects result in lower rate constants for the ionization of the *p*-nitrobenzoates; this is ascribed to inhibition of resonance in the transition state and in the carbocation,[304] since large R groups can no longer be coplanar with the phenyl ring.

The higher *oxidation rates* of axial cyclohexanols with chromic acid correspond to differences in steric energies of axial and equatorial epimers[305] (Fig. 60).

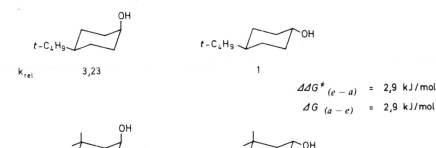

k_{rel} 3,23   1

$$\Delta\Delta G^{\ddagger}_{(e-a)} = 2,9 \text{ kJ/mol}$$
$$\Delta G_{(a-e)} = 2,9 \text{ kJ/mol}$$

k_{rel} 33,5   1

$$\Delta\Delta G^{\ddagger}_{(e-a)} = 9,0 \text{ kJ/mol}$$
$$\Delta G_{(a-e)} = 8,2 \text{ kJ/mol}$$

Fig. 60 Relative rates and $\Delta\Delta G^{\ddagger}$ in oxidations of axial and equatorial cyclohexanols by chromic acid and (steric) *ground state differences* ΔG of the corresponding epimeric pairs.

The relationship (*287b*) has been shown to yield the relative rates of oxidation by chromic acid of a large number of epimeric pairs of alcohols.[305a] This implies that all oxidations studied have very similar transition states.

$$\Delta G_{a \rightleftarrows e} = 0.8 RT \ln \left(\frac{k_a}{k_e}\right) \qquad (287b)$$

As shown in Section 3.2.4.1 activation of anionic nucleophiles by dipolar aprotic solvents is largely due to destabilization of ground states. (If the change of the rate constant of an $S_N 2$ reaction $Y^- + RX \rightarrow YR + X^-$ on going from reference solvent 0 to solvent L is written as

$$\log k^L - \log k^0 = \log {}^0\gamma_Y^L + \log {}^0\gamma_{RX}^L - \log {}^0\gamma_{YRX}^L{}^{\ddagger}$$

the difference $\log {}^0\gamma_{RX}^L - \log {}^0\gamma_{YRX}^L{}^{\ddagger}$ is fairly constant).[245]

4.2.4   Kinetic Isotope Effects

Relative reactivities of substrates differing only in isotopic substitution correspond most closely to situation III (Fig. 55): energy hypersurfaces, reflecting the dependence of the potential energy of reacting systems on the positions of all atoms involved and thus the geometry and potential energy of the transition state are *not* affected by isotopic substitution. (According to the *Born-Oppenheimer approximation*, interatomic forces depend only on attractions and repulsions of electronic and nuclear charges, not on nuclear mass). However, through its influence on the vibrational energy of a reacting X—H bond substitution of H by D does have an effect on the reaction rate, k_H/k_D (*288a, 288b*) known as the *kinetic isotope effect* (in this instance, the deuterium isotope effect).

$$X-H \ + \ Y \ \xrightarrow{\ k_H\ } \ X \ + \ H-Y \qquad (288a)$$

$$X-D \ + \ Y \ \xrightarrow{\ k_D\ } \ X \ + \ D-Y \qquad (288b)$$

One speaks of a *primary isotope effect* when isotopic substitution involves an atom attached to a bond formed or broken during the reaction, as in (*288a*) and (*288b*). The generally weaker effects of isotopic substitutions at other positions are known as *secondary isotope effects*. Effects of isotopic substitution in solvents used in reactions are known as *solvent isotope effects*.[306] There are also *equilibrium isotope effects*. Only primary H/D isotope effects, clearly identifiable when they are larger than the largest secondary isotope effects ($k_H/k_D \approx 2.5$) are discussed in some detail.

Primary kinetic isotope effects are mainly due to differences in zero-point vibrational energies of the isotopically substituted substrates.

According to transition state theory (for identical transmission coefficients for H and D species):

$$\frac{k_H}{k_D} = \frac{Q'^{\neq,H} \cdot Q_{XD}}{Q'^{\neq,D} \cdot Q_{XH}} \cdot e^{\frac{-\Delta E_0}{RT}} \qquad (289)$$

$$\Delta E_0 = \Delta E_{0,H} - \Delta E_{0,D} \qquad (290)$$

$\Delta E_{0,H}$ ($\Delta E_{0,D}$) is the energy difference between the transition state belonging to $X-H$ ($X-D$) and the ground state $X-H+Y$ ($X-D+Y$). It consists of the differences in potential energies and zero-point energies [(*291*) and Fig. 61]:

$$\Delta E_{0,H} = \Delta E_{pot} + E_{0,vibr.}^{\neq,H} - E_{0,vibr.}^{X-H} \qquad (291)$$

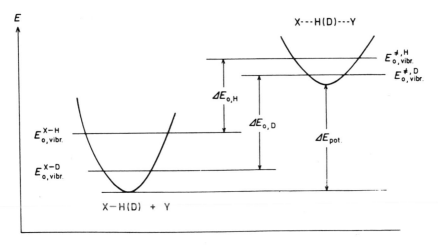

Fig. 61 Primary kinetic H/D isotope effect. Schematic representation of potential curves for the ground state species $X-H(D)$ and the transition state $X \cdots H(D) \cdots Y$ and of zero-point energies of H and D species.

Since ΔE_{pot} is the same in both cases:

$$\Delta E_o = E_{o,vibr.}^{\neq,H} - E_{o,vibr.}^{X-H} + E_{o,vibr.}^{X-D} - E_{o,vibr.}^{\neq,D} \qquad (292)$$

Assuming[307] $Q'^{\neq,H} = Q'^{\neq,D}$ and $Q_{X-H} = Q_{X-D}$ one can write (293):

$$\frac{k_H}{k_D} \approx e^{-\frac{(E_{o,vibr.}^{\neq,H} - E_{o,vibr.}^{\neq,D})}{RT}} \cdot e^{\frac{(E_{o,vibr.}^{X-H} - E_{o,vibr.}^{X-D})}{RT}} \qquad (293)$$

Zero-point energies are sums of the contributions $\frac{1}{2}h\nu_i$ of the $3N-6$ vibrational modes with frequencies ν_i of normal molecules ($3N - 5$ for linear molecules) and of the $3N^{\neq} - 7$ vibrational modes of activated complexes ($3N^{\neq} - 6$ for linear activated complexes), for example, (294):

$$E_{o,vibr.}^{X-H} = 1/2\, h \sum_{i}^{3n-6} \nu_i^{X-H} \qquad (294)$$

In the extreme case, *when* **one** *of the vibrations of the ground state disappears* **completely** *when the transition state is reached, whereas the* **sum** *of the* **other** *frequencies* **remains the same**, one obtains an isotope effect that is determined exclusively by the different contribution of the zero-point energy of the disappearing vibration to the ground states of the H and D substrates (situation III, Fig. 55).

This situation arises when the forces binding H(D) to X and Y, respectively, in the symmetric vibration $\overleftrightarrow{X}\cdots H(D)\ldots\overleftrightarrow{Y}$ of the activated complex are balanced, so that H(D) is no longer involved in a stretching vibration as in the ground state and when at the same time the sum of the other frequencies on reaching the activated complex remains the same (the second, asymmetric vibration $\overleftrightarrow{X}\cdots H(D)\cdots\overleftrightarrow{Y}$ having degenerated into the product-forming translation). [For vibrations of parts of the molecule further removed from the reaction center this is understandable, but there is experimental evidence that deformation frequencies $X\diagup^H_{\downarrow}$ (in the ground state) and $\overset{\uparrow}{X}\cdots\overset{\downarrow}{\underset{\downarrow}{H}}\cdots\overset{\uparrow}{Y}$ (in the transition state) are quite similar].[308]

The ratio k_H/k_D is obtained through (295):

$$-\Delta E_o \;=\; \frac{N_L}{2}\,h\,(\nu^{X-H} - \nu^{X-D}) \;=\; \frac{N_L}{2}\,h\nu^{X-H}\left(1 - \frac{\nu^{X-D}}{\nu^{X-H}}\right) \qquad (295)$$

where N_L = Avogadro's number

Since the force constants of the vibration involved in X–D and X–H are the same, the quotient $\nu^{X-D}:\nu^{X-H}$ is equal to $(\mu_{X-H}/\mu_{X-D})^{1/2}$ (μ = reduced mass). For large masses of X the value obtained should be $\sqrt{0.5} = 0.707$. An average value of 0.741 is generally found (296). The effects of lower masses of X and slight anharmonicity of the vibrations involved apparently contribute to this value.

$$-\Delta E_o \;=\; \frac{N_L}{2}\,h\nu^{X-H}\,(1-0{,}74) \;=\; 0{,}13\,N_L \cdot h \cdot \nu^{X-H} \qquad (296)$$

$$\frac{k_H}{k_D} \;=\; e^{\dfrac{0{,}13\,N_L \cdot h \cdot \nu^{X-H}}{RT}} \;=\; e^{\,0{,}628 \cdot 10^{-3} \cdot \bar{\nu}} \qquad (297)$$

Numerical values for k_H/k_D (at 298 K) for the limiting case are given for various X–H vibrations in Table 74.

Effects of isotopic substitution of carbon, oxygen, and sulfur are given below. Because vibrational frequencies of most bonds are lower and because the ratios of reduced masses for these heavy nuclei are close to unity, these isotope effects are only slightly larger than 1 (Table 75).

Kwart et al.[309] found that H/D effects in reactions (298) and (299) can be ascribed exclusively to differences in zero-point energies of the C–H and C–D bonds and accordingly assumed balance of force constants in a "symmetric" transition state A with a *linear* arrangement of the migrating hydrogen atom and atoms C-1 and C-5.

TABLE 74 Primary H/D Isotope Effects of Some X—H Bonds Calculated According to (297) for the Limiting Case, Which Assumes That Only the Considered Vibration Is Involved and That It Completely Vanishes in the Transition State

Bond	$\bar{\nu}\,(cm^{-1})$	k_H/k_D(298 K)
C—H	2900	6.2
N—H	3100	7.0
O—H	3300	7.9
S—H	2500	4.8

TABLE 75 Calculated Primary Kinetic Isotope Effects of Heavier Elements

Bond $A-{}^1B/A-{}^2B$	$\bar{\nu}$ (cm^{-1})	k_1/k_2(298 K)
$H-{}^{12}C/H-{}^{14}C$	2900	1.041
$H-{}^{12}C/H-{}^{13}C$	2900	1.022
${}^{12}C-{}^{12}C/{}^{12}C-{}^{14}C$	1000	1.092
$C-{}^{16}O/C-{}^{18}O$	1100	1.063
$C-{}^{32}S/C-{}^{34}S$	700	1.015

$$\xrightarrow{\;400\;-\;500\,^\circ C\;}$$

(298)

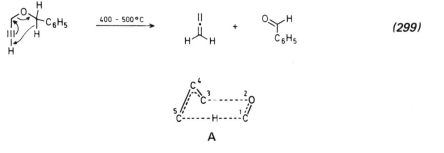

$$\xrightarrow{\;400\;-\;500\,^\circ C\;}$$

(299)

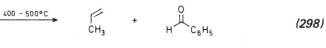

A

In acid-base reactions such transition states (manifested by maximum values of k_H/k_D) should occur when pK values of reaction partners are equal (see p. 333 and p. 371).[310]

This simplest model for the primary kinetic isotope effect is widely used in mechanistic analyses of hydrogen transfer reactions, but only rarely for transfer of other atoms (or groups of atoms). In this context it should be mentioned that in calculations[309a] maximum values for the $^{12}C/^{14}C$ isotope effect of transfer of carbon CH_2R in the transition state of an S_N2 reaction were obtained when it was assumed that the partial bonds to the nucleophile and the leaving group are almost equal in strength (bond order). This criterion was used to distinguish between the possibilities of classical S_N2 reactions and S_N2 reactions proceeding via reversibly formed contact ion pairs (see p. 450).

Values of H/D effects given in Table 74 become *lower* when one has to take into account *zero-point energy differences* in the *activated complex*—that is, when as a consequence of stronger bonding to one of the partners, X or Y, the vibration of H(D) in $X \cdots H(D) \cdots Y$ no longer vanishes completely ("unsymmetric transition states").

Increased values of k_H/k_D (compared to those in Table 74) result when in addition to the complete disappearance of the valence vibration in the transition state, one also has a reduction of the sum of isotope-dependent deformation frequencies. A reduction by a total of 750 cm^{-1}, considered to be extreme, results in "semiclassical" maximum H/D effects[308] (Table 76) (see Section 4.2.5).

TABLE 76   Maximum "Semiclassical" H/D Effects[308]

Bond	$\dfrac{k_H}{k_D}$ (298° K)
C–H	10
N–H	11
O–H	13
S–H	8

Higher H/D effects are also known,[310] for example, (*300*):

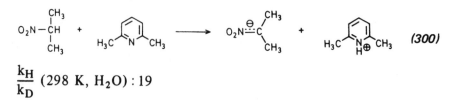

$\dfrac{k_H}{k_D}$ (298 K, H_2O) : 19

They are attributed to the *tunneling effect*.[311] Heisenberg's uncertainty principle allows the delocalization of particles via the wave-particle dichotomy over a region corresponding to their de Broglie wavelength λ_x:

$$\lambda_x = \frac{h}{m v_x}$$

where h = Planck's constant
m = mass
v_x = velocity in the x direction.

If λ_x is equal to the distance that must be travelled, for instance, by a hydrogen atom, on decomposition of the activated complex this travel can occur even if the energy of the system is *below* that of the transition state. In this instance one observes reaction rate constants higher than those predicted by the uncorrected transition state theory. Since the lighter isotope always has the longer wavelength, it shows larger deviations from classical behavior, and the ratio of transmission coefficients κ_H/κ_D reflecting the tunneling effect is always larger than unity. Its specific value depends on the form of the energy hypersurface in the region of the activated complex (i.e., on the width and form of the activation barrier).

Lower k_H/k_D values (compared to those in Table 76) can also result when in the activated complex X, H (D), and Y are no longer *linear*, but *bent* (233).[312] In these cases bending vibrations (whose frequencies are lower than those of stretching vibrations) change on going to the transition state.

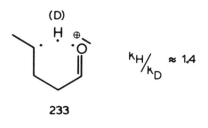

233

Assumedly bent transition state of hydrogen migration in a radical cation[312a]

It has been pointed out[312b] that k_H/k_D may be large in quite asymmetric transition states if motion of X and Y (Fig. 61) dominates in the symmetric stretching vibration X $\cdots$ H(D) $\cdots$ Y of the transition state. The mass of the central hydrogen or deuterium then has relatively little effect and the zero-point energies of this mode are the same for both isotopes, irrespective of the relative degree of bonding of H (D) to X and Y, respectively.

An important external factor affecting the values of primary isotope effects may be "internal return," for example, in base-catalyzed tritium exchange:

$$R-T+B^- \underset{k_{-1}}{\overset{k_1}{\rightleftharpoons}} R^- + T-B \xrightarrow{k_2} R^- + H-B \rightarrow product \qquad (300a)$$

When anionization is accompanied only by small changes in structure (e.g., in the formation of phenyl anions) $k_{-1} \gg k_2$ and $k_{exp} = K k_2$. Since primary isotope effects of k_1 and k_{-1} are roughly equal, the equilibrium isotope effect of K is about 1. Since k_2, the rate constant for diffusion of T−B away from R⁻, is not dependent on an isotope effect, the isotope effect controlling k_{exp} is small. A similar situation pertains when a proton has to be removed from a strong intramolecular hydrogen bond.[312c] [Internal return is insignificant (k_{-1} is small) when formation of the anion results in drastic changes in structure, e.g., in fluorene. See also: A. Streitwieser et al., *J. Amer. Chem. Soc.* **95**, 4254 (1973).]

In summary, the large number of factors affecting individually or in combination the ratio k_H/k_D of primary (and secondary) H/D effects does not in general, allow one to draw definitive conclusions with respect to the structure of the activated complex ("symmetric," "unsymmetric," "productlike," see p. 327) from a single k_H/k_D value. However, some possibilities appear when one studies the *variation* of the kinetic isotope effect within a reaction series containing very similar members (see also p. 371).

The existence of even a small primary kinetic isotope effect (as long as it can be readily distinguished from a secondary isotope effect or from generally small equilibrium isotope effects, e.g., $k_H/k_D > 2.5$), or its absence, is a sure indication that the corresponding bond is either formed or broken in the rate-determining step, or not at all affected. In all fields of mechanistic studies in organic chemistry extensive use is made of this technique. A few illustrations follow.

The absence of an H/D effect in some electrophilic aromatic substitutions (e.g., nitration, bromination, alkylation)[313] is an indication that the rate-determining step is the first one in which the isotopically substituted bond is not broken (*301*):

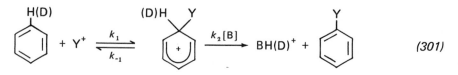

where [B] = [base (solvent)]

Assuming stationary states (Section 2.3.3):

$$k = \frac{k_1 k_2 [B]}{k_{-1} + k_2 [B]} \qquad\qquad (302)$$

the condition $k_1 \ll k_2[B] \gg k_{-1}$ (first step slow and irreversible) gives $k = k_1$.

Deviation from the condition above, for example, $k_2[B] \ll k_{-1}$ (first step reversible) or $k_1 \gg k_2[B] \gg k_{-1}$ (consecutive reaction, Section 2.3.2) leads to expressions for k containing k_2. In such cases a primary isotope effect is to be expected.

A value of $k_H/k_D \approx 7$ is found, for example, in sulfonation of the 9-position of anthracene. Neighboring hydrogen atoms at C-1 and C-8 make deprotonation strongly sterically hindered and thus rate determining.[314]

Base-catalyzed halogenations of CH acidic compounds, where the rate-determining step is the *formation* of the enolate[315] ($k_2[BH] \ll k_3[E^+] \gg k_1[CH][B]$)

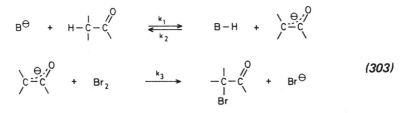

$$(303)$$

show primary kinetic H/D effects, as do E2 reactions and E1cB reactions of the irreversible type.

A review of observed and possible isotope effects in elimination was compiled by Fry (Table 77).[316]

Table 77, shows that a combination of isotope effects observed on isotopic substitution of *several* positions of the same system allows substantially sharper mechanistic interpretations than a single isotope effect.

Primary $^{12}C/^{14}C$ kinetic isotope effects (k/k^*) measured by Benjamin and Collins[317] show that in the 1,3-dipolar addition of N,α-diphenyl nitrone to styrene *both* carbon atoms of styrene participate in the rate-determining step (Scheme 62; see also Table 75).

Even the smallest of the three isotope effects is larger than any known secondary $^{12}C/^{14}C$ isotope effect. This result strongly supports the concertedness of this and similar cycloadditions. The stereospecificity of many such reactions (Section 1.3.1.4) is not an unequivocal criterion for concertedness, since it cannot rule out a two-step mechanism in which the second-step, ring closure, is faster than internal rotation.

TABLE 77 Mechanisms for Elimination and Possible Isotope Effects

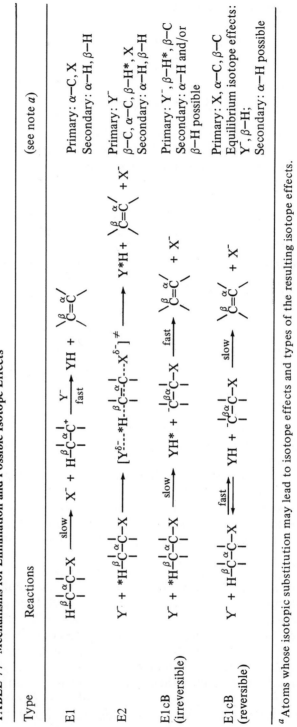

Type	Reactions	(see note a)
E1	$H\overset{\beta}{-}C\overset{\alpha}{-}C-X \xrightarrow{\text{slow}} X^- + H\overset{\beta}{-}C\overset{\alpha}{-}C^+ \xrightarrow[\text{fast}]{Y^-} YH + \ \overset{\beta}{\diagdown}C=C\overset{\alpha}{\diagup}$	Primary: α–C, X Secondary: α–H, β–H
E2	$Y^- + {}^*H\overset{\beta}{-}C\overset{\alpha}{-}C-X \longrightarrow [Y^{\delta-}\cdots{}^*H\cdots\overset{\beta}{C}\overset{\alpha}{=}C\cdots X^{\delta-}]^{\neq} \longrightarrow Y^*H + \ \overset{\beta}{\diagdown}C=C\overset{\alpha}{\diagup} + X^-$	Primary: Y^- β–C, α–C, β–H*, X Secondary: α–H, β–H
E1cB (irreversible)	$Y^- + {}^*H\overset{\beta}{-}C\overset{\alpha}{-}C-X \xrightarrow{\text{slow}} YH^* + {}^-\overset{\beta}{C}\overset{\alpha}{-}C-X \xrightarrow{\text{fast}} \ \overset{\beta}{\diagdown}C=C\overset{\alpha}{\diagup} + X^-$	Primary: Y^-, β–H*, β–C Secondary: α–H and/or β–H possible
E1cB (reversible)	$Y^- + H\overset{\beta}{-}C\overset{\alpha}{-}C-X \underset{\text{fast}}{\rightleftarrows} YH + {}^-\overset{\beta}{C}\overset{\alpha}{-}C-X \xrightarrow{\text{slow}} \ \overset{\beta}{\diagdown}C=C\overset{\alpha}{\diagup} + X^-$	Primary: X, α–C, β–C Equilibrium isotope effects: Y^-, β–H; Secondary: α–H possible

^{a} Atoms whose isotopic substitution may lead to isotope effects and types of the resulting isotope effects.

264

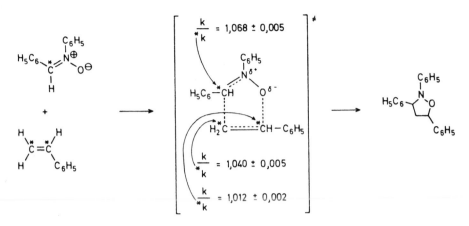

Scheme 62 Primary kinetic $^{12}C/^{14}C$ isotope effects in the 1,3-dipolar addition of N,α-diphenylnitrone to styrene.[317] Isotopic substitution of each of the three carbon atoms participating in the formation of the two new bonds results in what is considered to be a substantial isotope effect; this is interpreted as evidence for *concerted* formation of the C–C and C–O bonds.

4.2.5 Linear Free Energy Relationships

4.2.5.1 General Aspects

According to transition state theory, log-log correlations (e.g., the Hammett equation) are consequences of a linear relationship between the effects of substitution of H by X [in other instances: change of nucleophile, Brönsted acid (or base), solvent] on the free energy of activation of the reaction under consideration on one hand, and the free energy of reaction (or free activation energy) of the model reaction (here: dissociation of benzoic acid):

$$\log \frac{k_X}{k_H} = \rho \cdot \sigma = \rho \log \left(\frac{K_X}{K_H} \right)_{\text{diss. benzoic acids, } H_2O, 25°C} \qquad (304)$$

$$\triangle\triangle G^{\ddagger} = \rho \cdot \triangle\triangle G \qquad (305)$$

(This example was chosen because we are primarily interested in reaction rates. The $\sigma\rho$ correlations of equilibrium constants[114] show that the same is true for equilibria.) In general, *linear free energy relationships* can be formulated as:

$$\delta_x \varDelta G_1 = a \cdot \delta_x \varDelta G_2 \qquad (306)$$

The operator δ_x expresses the following:

The influence of changing the variable X on ΔG (or $\Delta G^{\ddagger}$) in reaction series 1 is proportional to its influence in reaction series 2.

4.2.5.2 Analysis of Reaction Mechanisms by Means of Linear Free Energy Relationships

The fact that logarithms of rate constants [or $\delta_X \Delta G^{\ddagger}$, see (306)] within-in a reaction series can be correlated with those of another series shows that the same factors are essential for relative reactivities in both series. However, the existence of correlations of logarithms of dissociation constants of substituted benzoic acids and logarithms of rate constants of reactions mechanistically belonging to completely different types clearly illustrates that such correlations *per se* do not allow conclusions with respect to the reaction mechanism; they only indicate that reactivity is affected by polar effects of the type observed in *m*- and *p*-substituted benzene derivatives. But if the main characteristics of the mechanism of a given reaction are known from independent sources, the knowledge of the nature of the substituent constant leading to the best linear correlation may allow us to draw conclusions with respect to special properties, for example, of the intermediate involved. The same is true for values of the susceptibility factor (reaction constant), which reflect in quantitative terms the contribution of a given factor represented by the respective substituent constant to the reaction under consideration. For instance, experiment shows that the rate of bromine addition to styrenes obeys a σ^+ correlation ($\rho = -4.5$). This suggests that the intermediates are resonance-stabilized ions, not bromonium ions (p. 16).

On the other hand, solvomercuration of styrenes is better described by simple σ constants[318] ($\rho = -2.25$), suggesting the occurrence of mercurinium ions (307):

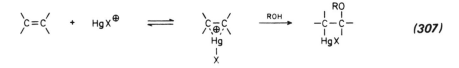

$$ (307) $$

The extremely negative value of ρ [in the Yukawa-Tsuno correlation (133)] for the cycloaddition of tetracyanoethylene to *p*-substituted styrenes, as well as its solvent dependence show that this cycloaddition occurs stepwise via intermediate dipolar ions [(308) and Fig. 62].[319]

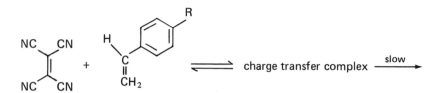

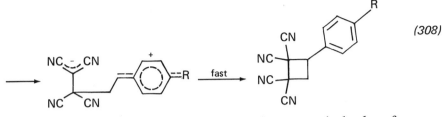

(308)

Examples of less polar reactions with low numerical values for ρ are ester pyrolyses and radical reactions:

(309)

$$\lg\left(\frac{k_R}{k_H}\right)_{600\,°C} = -0,66 \cdot \sigma^+ \quad \text{Lit. 320}$$

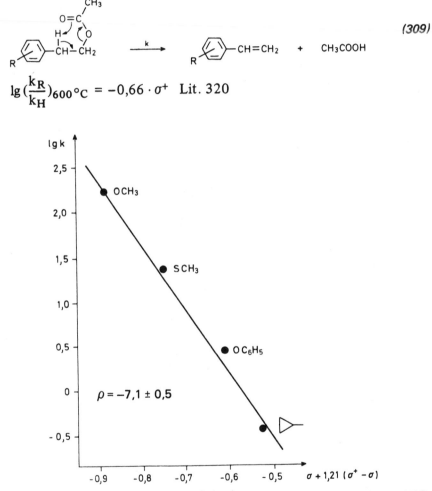

Fig. 62 Reaction rate constants (1 mol^{-1} s^{-1}) for the reaction of $(NC)_2C=C(CN)_2$ and R—⟨benzene⟩—CH=CH$_2$ (R = CH$_3$O, CH$_3$S, C$_6$H$_5$O, cyclopropyl). [From P. D. Bartlett, *Quart. Rev.* **24**, 473 (1970) (ref. 319).]

$$C_{11}H_{23}\cdot \quad + \quad \text{(aryl)}-CH_3 \quad \xrightarrow{k} \quad C_{11}H_{24} \quad + \quad \text{(aryl)}-CH_2\cdot \qquad (310)$$

$$lg\left(\frac{k_R}{k_H}\right) = 0{,}50 \cdot \sigma \quad \text{Lit. 321}$$

Resonance hybrids of the transition state for benzylic hydrogen abstraction by radicals $R\cdot$:

$$R\cdot \;\overset{..}{H}\cdot CH_2-C_6H_5 \quad\longleftrightarrow\quad R\overset{\ominus}{:}\overset{..}{H}\;\overset{\oplus}{C}H_2-C_6H_5 \quad\longleftrightarrow\quad \overset{\oplus}{R}\;\overset{..}{H}\overset{\ominus}{:}CH_2-C_6H_5$$

$$\mathbf{234} \qquad\qquad\qquad \mathbf{235} \qquad\qquad\qquad \mathbf{236}$$

Corresponding to the definition of σ, a positive (negative) ρ shows accumulation of negative (positive) charge in the transition state. Consequently, alkyl radicals may be classified as nucleophilic (contribution of structure **236** is greater than that of **235**). Halogen atoms, amino-, acyloxy-, and trichloromethyl radicals are electrophilic.[322]

Addition of cycloheptatrienylidene to styrenes has a positive ρ. This means that in contrast to the generally electrophilic character of carbenes and carbenoids we are dealing here with a *nucleophilic* carbene: contribution of structure **237** to the transition state is considerable. The reason for the particular behavior of this carbene may be the incorporation of the empty $2p$ orbital of C-7 into the stable tropylium system **238**.[323]

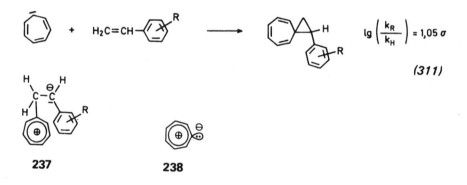

$$lg\left(\frac{k_R}{k_H}\right) = 1{,}05\,\sigma$$

$$(311)$$

237   **238**

An interesting example showing how aromatic substituent constants can be used to interpret reactivities of nonaromatic compounds is provided by the analysis of the effect of aluminum chloride on the Diels-Alder reaction of substituted butadienes with methyl acrylate (*312*).[324] This effect manifests itself in a substantial increase in rate (by a factor of 10^5) and a change in the ratio of "*meta*" (**239**) to "*para*" product (**240**):

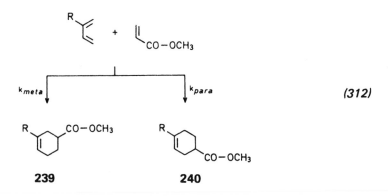

239   **240**

Whereas the ratio **239**:**240** cannot be described by means of σ^+ constants either for the uncatalyzed, nor for the catalyzed reaction, the ratio of partial rate factors (with respect to R = H) for the catalyzed and uncatalyzed reaction obeys (*313*):

$$\log \left(\frac{k_{\text{R-meta or para}}}{k^H/2} \right)_{\text{catalyzed}} - \log \left(\frac{k_{\text{R-meta or para}}}{k^H/2} \right)_{\text{uncatalyzed}} \qquad (313)$$

$$= \rho \cdot \sigma^+_{\text{meta or para}} (R) \qquad (\rho = -3.07)$$

that is, the *modification* of substituent effects by transforming the dienophile $H_2C=CH-CO-OCH_3$ into dienophile $H_2C=CH-CO-OCH_3 \rightarrow AlCl_3$ corresponds to substituent effects observed in the S_N1 solvolysis of aryldimethylcarbinyl (*cumyl*) chlorides.

This suggests that in both the solvolysis reaction and the transition state of the Diels-Alder reaction the substituent effect is transmitted by a *benzene-like* conjugated system (see also Section 5.2.4.4). Equation (*313*) can be rewritten so that the roles of the free energies of ground states (G°) and transition states ($G^\ddagger$) become apparent (*314b*):

$$\log (k_{p\text{-R}})_{\text{cat}} - \log (k_{p\text{-R}})_{\text{uncat}} = -3.07\, \sigma^+_{p\text{-R}} + C_1 \qquad (314a)$$

$$\frac{-(G^\ddagger_{p\text{-R}})_{\text{cat}} + (G^\circ_{p\text{-R}})_{\text{cat}} + (G^\ddagger_{p\text{-R}})_{\text{uncat}} - (G^\circ_{p\text{-R}})_{\text{uncat}}}{2.3\, RT} =$$
$$-3.07\, \sigma^+_{p\text{-R}} + C_1 \qquad (314b)$$

For all possible R's $(G^\circ)_{\text{cat}}$ differ from $(G^\circ)_{\text{uncat}}$ by the constant contribution of the free energy of coordination of aluminum chloride and methyl acrylate so that both these terms may be incorporated into the constant:

$$\frac{(G_{p\text{-R}}^{\ddagger})\text{uncat} - (G_{p\text{-R}}^{\ddagger})\text{cat}}{2.3\,RT} = -3.07 \cdot \sigma_{p\text{-R}}^{+} + C_2 \qquad (315)$$

Since

$$\sigma_{p\text{-R}}^{+} = -\frac{1}{4.54}\,\log\left(\frac{k_{p\text{-R}}}{k_\text{H}}\right)_{\text{hydrolysis of cumyl chlorides}} \qquad (316a)$$

and

$$\sigma_{p\text{-R}}^{+} = -\frac{1}{4.54} \cdot -\left(\frac{G_{p\text{-R}}^{\ddagger} - G_{p\text{-R}}^{\circ}}{2.3\,RT}\right)_{\text{hydrolysis of cumyl chlorides}} + C_3 \qquad (316b)$$

(315) and $(316b)$ yield (317):

$$(G_{p\text{-R}}^{\ddagger})\text{uncat} - (G_{p\text{-R}}^{\ddagger})\text{cat} = \frac{3.07}{4.54}\,(G_{p\text{-R}}^{\circ} - G_{p\text{-R}}^{\ddagger})_{\text{hydrolysis}} + C_4 \qquad (317)$$

Equation (317) shows that the effect of R on the free energy change of the activated complex accompanying the transition from the uncatalyzed to the catalyzed Diels-Alder reaction is proportional to its effect on the activation process of cumyl chloride hydrolysis. In a structural sense the latter corresponds to the transition **241** → **242**. On the basis of these similar substituent effects structures **243** and **244**, the analogues of **241** and **242**, implying *synchronous* formation of the two new C—C bonds become plausible for the transition states of the uncatalyzed and catalyzed Diels-Alder reactions. It also becomes clear that *no σ^{+} correlations* corresponding to (313) are to be expected when one considers separately the properties of the catalyzed and the uncatalyzed reaction, since in such systems one deals with relationships between ground states *without* electron delocalization (**245**) and transition states *with* delocalization (**243**).

241   **242**

Activation process in cumyl chloride ionization

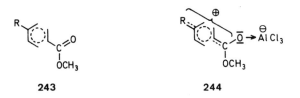

243   **244**

Transition states for the uncatalyzed and catalyzed Diels-Alder reaction

245   **243**

Activation process for the uncatalyzed Diels-Alder reaction

An example of an analysis of a reaction mechanism based not on known substituent parameters (which in this instance are unavailable) but on a known model reaction closest to the reaction under investigation is the isomerization of olefins catalyzed by $KOtC_4H_9$ in DMSO (*318*), (*319*).[325]

(318)

(319)

The nonlinearity of the correlation of the effect of R on the rate of reaction and the Taft parameters (Fig. 63) shows that the reaction is sensitive not only to polar but also to steric effects.

An adequate model reaction turned out to be the base-catalyzed bromination of ketones [(*320*) and Fig. 64], where the rate determining step is the formation of the enolate.

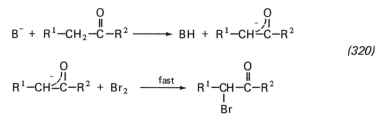

(320)

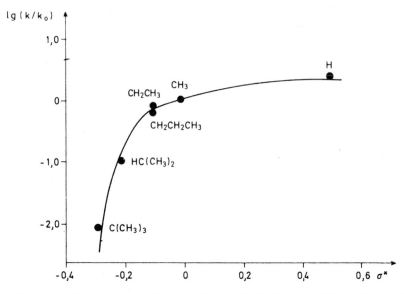

Fig. 63 Correlation of log (k/k_0) of *(319)* and σ^* (R = H, CH₃, C₂H₅, C₃H₇, i-C₃H₇, t-C₄H₉). [From A. Schriesheim, C. A. Rowe, Jr., and L. Naslund, *J. Amer. Chem. Soc.* **85**, 2111 (1963) (ref. 325).]

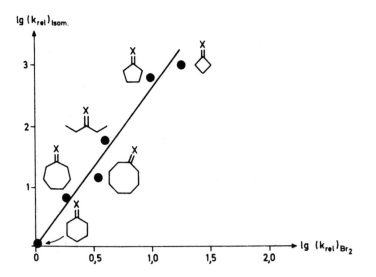

Fig. 64 Correlation of logarithms of relative rate constants for base-catalyzed olefin isomerization [X = CH₂, *(318)* and *(319)*] and base-catalyzed bromination of ketones [X = O, *(320)*]. Reference systems (k_{rel} = 1) are the cyclohexane derivatives. [From A. Schriesheim, C. A. Rowe, Jr., and L. Naslund, *J. Amer. Chem. Soc.* **85**, 2111 (1963) (ref. 325).]

In each $\overset{CH_2}{\bigwedge}$ / $\overset{O}{\bigwedge}$ pair structure-dependent effects (polar, stereo-electronic, steric—e.g., inhibition of attack by base) affect the reactivity of both systems to the same extent. Thus, it is plausible to assume that the rate-determining step in olefin isomerization is the formation of the allyl anions corresponding to the enolate ions.

The high rates of cyclobutane derivatives are probably stereoelectronic in origin: the protons to be abstracted are almost parallel to the neighboring $2p$ orbitals:

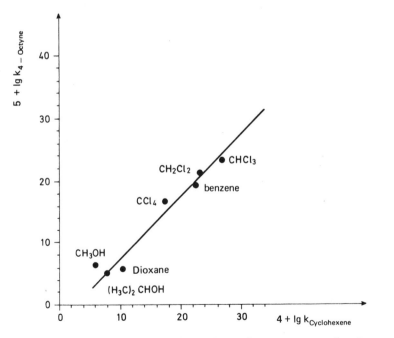

(321)

The effects of a series of very different solvents on the rate of reaction of m-chloroperbenzoic acid with cyclohexene and 4-octyne are identical [the slope of the correlation line is 1, (Fig. 65)].[326] This shows that the transition states are very similar and suggests the formation of the highly reactive *oxirene* as the nonisolable primary product in alkyne epoxidation (322).

Fig. 65 Linear relationship between logarithms of rate constants for the epoxidation of 4-octyne and cyclohexene in different solvents.[326] [From K. M. Ibne-Rasa et al., *J. Amer. Chem. Soc.* **95**, 7894 (1973) (ref. 326).]

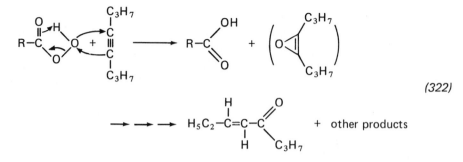

$$(322)$$

4.2.5.3 Origins of Linear Free Energy Relationships

Linear free energy relationships are *empirical* relationships between thermodynamic quantities; they are not derived from the laws of thermodynamics and are known as *extrathermodynamic* relationships. Their origins are not yet completely understood. There is no unequivocal answer to the question, "Why do substituent constants reflect *changes in free energy?*" It is commonly assumed, however, that they represent inductive and mesomeric effects that influence the *potential energies* of molecules (and activated complexes). Consequently, σ constants should provide an undistorted picture of these effects only *in vacuo*, at 0 K and without taking into account zero-point energies. In reality, this picture should be altered from *case* to *case* by contributions of all degrees of freedom (dependent on structure and solvent) as expressed by partition functions (Q). This situation is represented in (*323*) and (*324*) for dissociation equilibria (25°C, H_2O, $\rho = 1$) of benzoic acids (unsubstituted, HB; substituted, HBX):

$$RT \ln \left(\frac{K_{HBX}}{K_{HB}} \right) = -\Delta\Delta G = 2{,}3\, RT\, \sigma_x \tag{323}$$

$$RT \ln \left(\frac{K_{HBX}}{K_{HB}} \right) = RT \ln \frac{Q_{BX}\ominus Q_{HB}}{Q_B\ominus Q_{HBX}} - \left(E_{pot}^{BX\ominus} - E_{pot}^{HBX} - E_{pot}^{B\ominus} + E_{pot}^{HB} \right) \tag{324}$$

The best known situation where σ_x reflects $\Delta\Delta E_{pot}$ obtains when $\delta_x\, \Delta G \sim \delta_x\, \Delta H$, and when in addition $\delta_x\, \Delta H \sim \delta_x\, \Delta E_{pot}$. The latter is an unproved possibility.[327] $\delta_x\, \Delta G \sim \delta_x\, \Delta H$ results when the following conditions are obeyed [(*325*)-(*332*)]:

1 *Only a single interaction mechanism* may operate between substitutent and reaction center; that is, a single substituent constant must suffice to describe the reaction (*325*):

$$\delta_x\, \Delta H - T\delta_x\, \Delta S = -2.3RT \cdot \sigma \cdot \rho \tag{325}$$

2   ΔH, ΔS, and σ must be *temperature independent*.

3   The corresponding Hammett correlations must be valid at *any temperature*.

According to *(325)* ρ must be an inverse function of temperature:

$$\rho = \frac{A}{T} + B \tag{326}$$

where A and B are constants.

Combining *(325)* and *(326)* gives:

$$\delta_x \Delta H - T \delta_x \Delta S = -2.3 R \cdot \sigma \cdot A - 2.3 RT \cdot \sigma \cdot B \tag{327}$$

The validity of *(327)* at any temperature allows one to compare coefficients:

$$\delta_x \Delta H = -2.3 R \cdot \sigma \cdot A \tag{328}$$

$$\delta_x \Delta S = 2.3 R \cdot \sigma \cdot B \tag{329}$$

From *(328)* and *(329)*:

$$\frac{\delta_x \Delta H}{\delta_x \Delta S} = -\frac{A}{B} = \beta \tag{330}$$

This gives the proportionality between $\delta_x \Delta G$ and $\delta_x \Delta H$, respectively, $\delta_x \Delta S$:

$$\delta_x \Delta G = \delta_x \Delta H \left(1 - \frac{T}{\beta} \right) \tag{331}$$

$$\delta_x \Delta G = \delta_x \Delta S \left(\beta - T \right) \tag{332}$$

Equation *(330)* stating the proportionality of $\delta_x\Delta G$ and $\delta_x\Delta H$ (and possibly $\delta_x\Delta E_{pot}$) is often referred to as *isokinetic* (or isoequilibrium) relationship; β is the *isokinetic* (or isoequilibrium) temperature: substitution of $T = \beta$ and of *(330)* in *(325)* shows that at this temperature (which often has no real significance at all, e.g., $\beta = -1900$ K),[328] variation of substituents has *no* influence on ΔG. All compounds in the reaction series considered react equally fast at this temperature (all equilibrium constants of a reaction series have the same value at this temperature). Transgressing this temperature results in the reversal of the order of reactivities (Fig. 66).

Equation *(330)* is also known as the *compensation law*. It shows

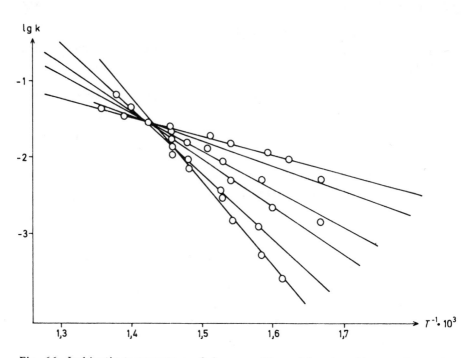

Fig. 66 Isokinetic temperature of decomposition of formic acid on various catalysts [From O. Exner, *Prog. Phys. Org. Chem.* **10**, 411 (1973) (ref. 328).]

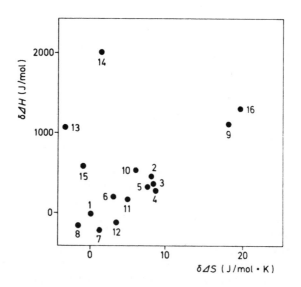

Fig. 67 Correlation of $\delta\Delta H$ and $\delta\Delta S$ of proton exchange between benzoic acids. [From T. Matsui, H. C. Ko, and L. G. Hepler, *Can J. Chem.* **52**, 2906 (1974) (ref. 330).]

that for positive β variations of ΔH and $T\Delta S$ partially compensate each other in their effect on ΔG.

If a family of reactions forming a reaction series belongs to the same isokinetic bundle of straight lines (i.e., has the same isokinetic temperature β; see Fig. 66), this is considered to be the strongest indication that all these reactions proceed by an identical mechanism.[328]

Only a vanishingly small percentage of the many reaction series obeying under certain conditions the Hammett equation or other log-log correlations obey the isokinetic (or isoequilibrium) relationship [(*330*) or (*331*) resp. (*332*), or one of the limiting cases of (*331*), resp. (*332*), $\beta = \infty$ (isoentropic series, only ΔH determines reactivity); $\beta = 0$ (isoenthalpic series, only ΔS determines reactivity).[329] Often plotting ΔH versus ΔS gives diagrams in which individual points are scattered without any regularity whatsoever. This is true even for reactions used for defining substituent constants or solvent parameters. The most important reaction, the dissociation of benzoic acids, obeys (*330*) only approximately, even if one rejects the anomalous points 13, 14, and 15 for p-hydroxy, p-methoxy, and p-methylbenzoic acid (Fig. 67).[330]

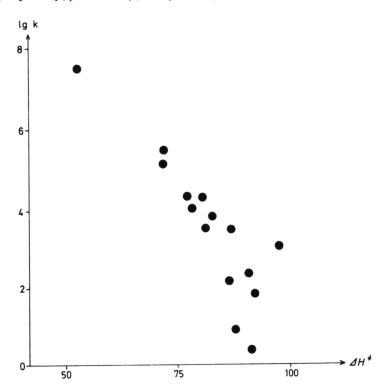

Fig. 68 log k (s^{-1}) versus $\Delta H^{\ddagger}$ (kJ/mol) for the hydrolysis of phenyldimethylcarbinyl chlorides. [From C. D. Johnson, *The Hammett Equation*, Cambridge University Press, Cambridge, 1973, p. 139 (ref. 114a).]

The reaction used to define σ^+ constants also shows only a very rough correlation between $\Delta G^{\ddagger}$ (log k) and $\Delta H^{\ddagger}$ (Fig. 68). Neither does the reaction used for defining Y obey the isokinetic relationships (Fig. 69).[331] However, the diagram nicely shows the compensating behavior of $\Delta H^{\ddagger}$ and $T\Delta S^{\ddagger}$.

Based on the examples above one may draw the conclusion that many reactions obeying (under certain circumstances) the Hammett equation or similar relationships do not adequately fulfill the conditions required to derive the isokinetic relationship. For instance, in the case of a four-parameter correlation (when two interaction mechanisms σ_1 and σ_2 are operative) and

$$\log \frac{k_X}{k_H} = \rho_1 \sigma_{1,X} + \rho_2 \sigma_{2,X} \qquad (333)$$

the quotient $\delta_x \Delta H : \delta_x \Delta S$ has a different value for each substituent because the ratio $\sigma_{1,x} : \sigma_{2,x}$ varies irregularly within a reaction series. The third assumption—the validity of the Hammett correlation at

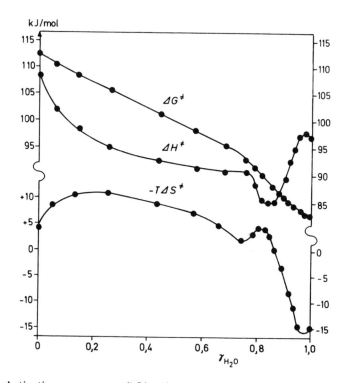

Fig. 69 Activation parameters (kJ/mol) for the solvolysis of t-butyl chloride in ethanol-water mixtures at 25°C [From S. Winstein and A. H. Fainberg, *J. Amer. Chem. Soc.* **79**, 5937 (1957) (ref. 331).]

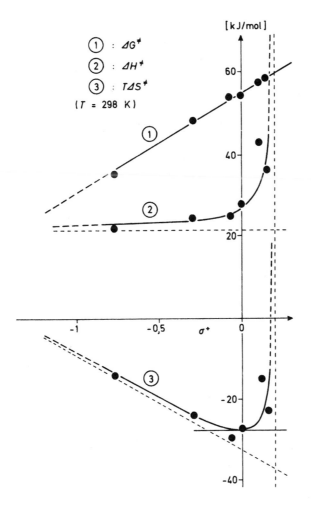

Fig. 70 Relationships between activation parameters for the addition of bromine to p-substituted styrenes and σ^+. Combination of (325) and (330) predicts linearity of (2) and (3). In reality

$$\Delta H^{\ddagger} = 5.29\ \frac{(\sigma^+ - 0.243)}{(\sigma^+ - 0.197)}$$

$$T\Delta S^{\ddagger} = -5.707\left[\frac{-0.234 + 1.184\ \sigma^+ + (\sigma^+)^2}{\sigma^+ - 0.197}\right]$$

Combination of these two expressions gives $\Delta G^{\ddagger} = 5.707\sigma^+ + 13.12$, while experiment gives $\Delta G^{\ddagger}_{exp} = 5.992\sigma^+ + 13.29$. [From J. E. Dubois and M. Marie de Ficquelmont-Loizos, *Tetrahedron Lett.* 635, **1976**, (ref. 332).]

any temperature—has been experimentally substantiated in only a few cases.

It has been shown[332] that the Hammett equation for the addition of bromine to styrenes (with σ^+) is valid for *several* temperatures and that the temperature dependence of ρ obeys (326). However $\Delta H^{\ddagger}$ and $T\Delta S^{\ddagger}$ turned out to be nonlinear with σ^+ (Fig. 70). A scheme depicting how enthalpy-entropy compensation leads to potential-energy-based conclusions has been outlined by Hepler.[333] Its basis is the partitioning of ΔH and ΔS into an *internal* term (dependent on electronic structure) and an *external* term (dependent on solvation) (334):

$$\delta_x \Delta G = \delta_x \Delta H_{int} + \delta_x \Delta H_{ext} - T\delta_x \Delta S_{int} - T\delta_x \Delta S_{ext} \qquad (334)$$

If one assumes that:

1 $\delta_x \Delta S_{int}$ can be neglected,
2 $\delta_x \Delta H_{ext} = \beta_{ext} \cdot \delta_x \Delta S_{ext}$ (i.e., the rough correlation of $\delta_x \Delta H$ and $\delta_x \Delta S$ is based on a much more precise correlation of $\delta_x \Delta H_{ext}$ and $\delta_x \Delta S_{ext}$),
3 β_{ext} is equal to the reaction temperature,

these assumptions lead to:

$$\delta_x \Delta G = \delta_x \Delta H_{int} \qquad (335)$$

In this connection it is also of interest to consider the following study.

Using the thermodynamic data for ionization of aqueous solutions of benzoic acids at 25°C,[330] it was possible to partition the "normal" substituent effect $\log(K_X/K_H) = \rho\sigma$ into an enthalpic contribution $\rho_H \sigma_H$, and an entropic contribution, $\rho_S \sigma_S$ [(336)-(339), and Table 78].[334]

$$\log\left(\frac{K_X}{K_H}\right)_{benzoic\ acids} = \rho_H \sigma_{H,X} + \rho_S \sigma_{S,X} \qquad (336)$$

$$-\frac{\Delta\Delta G}{2.3\ RT} = -\frac{\Delta\Delta H}{2.3\ RT} + \frac{\Delta\Delta S}{2.3\ R} \qquad (337)$$

Defining $\rho_H = \rho_S = 1$ in water at 25°C it follows from (336) and (337):

$$\sigma_H = -\frac{\delta_x(\Delta H)}{2.3\ RT} \qquad (338a)$$

TABLE 78 Thermodynamic Parameters for the Ionization of Aqueous Solutions of Benzoic Acids at $25°C$[330] and Enthalpic (σ_H) and Entropic (σ_S) Substituent Constants Derived from these Data[334]

Substituent	ΔG (kJ/mol)	ΔH (kJ/mol)	ΔS (J/mol K)	σ_H	σ_S	$f_H{}^a$	$f_S{}^b$
H	24.029	0.461	−79.06	0.000	0.000		
m-F	22.089	0.922	−70.97	−0.081	0.422	0.16	0.84
m-Cl	21.913	0.838	−70.68	−0.066	0.437	0.23	0.77
m-Br	21.779	0.754	−70.52	−0.051	0.446	0.10	0.90
m-I	22.039	0.796	−71.23	−0.059	0.409	0.12	0.88
m-OH	23.300	0.670	−75.92	−0.037	0.164	0.18	0.82
m-OCH$_3$	23.393	0.205	−77.76	0.045	0.068	0.39	0.61
m-CH$_3$	24.302	0.293	−80.53	0.029	−0.077	0.27	0.73
m-NO$_2$	19.777	1.592	−61.00	−0.198	0.942	0.18	0.82
p-Cl	22.781	1.005?	−73.03?	−0.095	0.315	0.23	0.77
p-Br	22.752	0.628	−74.20	−0.029	0.253	0.10	0.90
p-I	22.835	0.327	−75.50	0.023	0.186	0.11	0.89
p-OH	26.187	1.550	−82.62	−0.191	−0.186	0.51	0.49
p-OCH$_3$	25.643	2.514?	−77.55?	−0.359	0.079	0.81	0.19
p-CH$_3$	24.930	1.047?	−80.11?	−0.103	−0.055	0.65	0.35
p-NO$_2$	19.571	1.810	−59.58	−0.236	1.016	0.18	0.82

$$^a\quad f_H = \frac{|\sigma_H|}{|\sigma_H| + |\sigma_S|}$$

$$^b\quad f_S = \frac{|\sigma_S|}{|\sigma_H| + |\sigma_S|}$$

$$\sigma_S = \frac{\delta_x(\Delta S)}{2,3\,R} \qquad (338b)$$

In addition:

$$\sigma = \sigma_H + \sigma_S \qquad (339)$$

The data in Table 78 show that with the exception of "anomalous" substituents p-OCH$_3$, p-OH, p-CH$_3$, and m-OCH$_3$ the values of polar σ constants can be traced (to the extent of 70-90%) to *changes in interaction of the equilibrium partners (primarily the anions) with the medium* (σ_S) caused by the changes in substituents.

The predominance of the entropic effects manifests itself in the correlation equations:

$$\sigma_S = -3.6\sigma_H + 0.15 \qquad (340)$$

and

$$\sigma = 0.91\sigma_S - 0.07 \qquad (341)$$

In addition, there are relationships between σ_S and σ_I or σ_R [(342) and (343)]:

$$\sigma_{S(para)} = (1.2 \pm 0.2)\sigma_I + (0.8 \pm 0.2)\sigma_R + (0 \pm 0.1) \qquad (342)$$

$$\sigma_{S(meta)} = (1.2 \pm 0.1)\sigma_I + (0.4 \pm 0.1)\sigma_R + (0 \pm 0.1) \qquad (343)$$
$$(\sigma_I, \sigma_R : \text{see pp. 128 and 107})$$

They show that in this system changes in interaction between solvent and solute are ultimately caused by changes in the polar effect of the solute (contained in σ_I and σ_R). Consistent with this is the observation that even the dissociation constants K of benzoic acids in gas phase[335,236e] ($\delta_X \Delta S \approx 0$) can essentially be formulated via σ_S:

$$\log K = 8.2\sigma_S + 2.8\sigma_H \qquad (344)$$

In many acid-base equilibria ρ_S has almost the same value as the normal ρ! In addition, the existence of linear correlations between ρ_S and ρ_I calculated for the same reaction series [(134)] shows that the inductive effect contributes considerably to entropy effects in the ionization of aqueous benzoic acids [see also (342) and (343)] so that, in turn, σ_S and σ may be considered to be representative for the former.

A general analysis of the validity of the Hammett equation including systems where the proportionality between ΔH and ΔS is not observed can be found in the literature.[335a]

4.2.6 Polysubstitution; Cumulative Effects and Additivity

Reactivities of polysubstituted aromatic systems can often be computed, at least approximately, by multiplication of partial rate factors or by summation of σ constants for individual substituents (e.g., pp. 000 and 000). The reason for such so-called cumulative or additive effects is that ideally within a reaction series the contributions of individual substituents to the free energy of activation are *independent* of each other:

$$2.3\,RT\,\lg\frac{k_{X,Y}}{k_H} = 2.3\,RT \cdot \rho\,(\sigma_X + \sigma_Y) = \qquad (345)$$

$$- [(\Delta G_X^{\ddagger} - \Delta G_H^{\ddagger}) + (\Delta G_Y^{\ddagger} - \Delta G_H^{\ddagger})]$$

The *additivity principle* seems more plausible when polysubstitution occurs in different rings, rather than in the same ring, for example, 246-248:

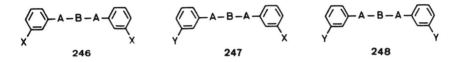

     246   247   248

Assuming the validity of the additivity principle in such systems, Simamura et al.[336] tried to determine whether certain reactions with peracids, where symmetric reaction centers yield unsymmetric oxidation products, involve in the rate-determining step symmetric intermediates (path a) or proceed directly (path b) (Scheme 63).

If the additivity principle is valid, formation of a symmetric intermediate (path a) requires that the effect of substituting both X by Y (246 → 248) be twice the effect of the simple substitution 246 → 247 (δ).

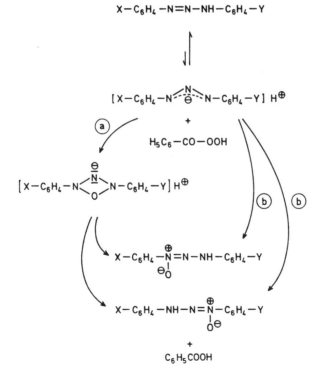

Scheme 63 Oxidation of diazoaminobenzenes.[336a]

$$247 : \quad \lg k_{XY,exp} = \delta + \lg k_{XX,exp} \qquad (346)$$

$$248 : \quad \lg k_{YY,exp} = 2\delta + \lg k_{XX,exp} \qquad (347)$$

Elimination of δ gives:

$$\frac{k_{XY,exp}}{(k_{XX,exp} \cdot k_{YY,exp})^{\frac{1}{2}}} = 1 \qquad (348)$$

In case (b) two competing processes take place at **247**:

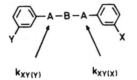

Attack at atom A closest to X is given by:

$$247 \qquad\qquad \log k_{XY(X)} = \delta_X + \log\frac{k_{XX,exp}}{2} \qquad (349)$$

where $k_{XX,exp}/2$ is the rate constant for the reaction at a *single* A in
246; δ_X reflects the effect of changing $(X \rightarrow Y)$, the more remote
substituent.

Attack at atom A closest to Y is given by:

$$247 \qquad\qquad \log k_{XY(Y)} = \delta_Y + \log\frac{k_{XX,exp}}{2} \qquad (350)$$

where δ_Y gives the effect of changing $(X \rightarrow Y)$, the closer substituent.

According to the additivity principle, attack at *one* A in **248** is given
by:

$$248 \qquad\qquad \log k_{YY} = \delta_X + \delta_Y + \log\frac{k_{XX,exp}}{2} \qquad (351)$$

The experimental rate constants of **247** and **248** are then given by:

$$k_{XY,exp} = k_{XY(X)} + k_{XY(Y)} = \frac{k_{XX,exp}}{2} \cdot (10^{\delta_X} + 10^{\delta_Y}) \qquad (352)$$

$$k_{YY,exp} = k_{XX,exp} \cdot 10^{(\delta_X + \delta_Y)} \qquad (353)$$

From *(352)* and *(353)* one forms again the quotient $k_{XY,exp}$:
$(k_{XX,exp} \cdot k_{YY,exp})^{1/2}$ contained in *(348)*:

$$\frac{k_{XY,exp}}{(k_{XX,exp} \cdot k_{YY,exp})^{1/2}} = \frac{1}{2}\left(10^{1/2(\delta_Y - \delta_X)} + 10^{-1/2(\delta_Y - \delta_X)}\right)$$

(354)

and since

$$\nu = \frac{k_{XY(Y)}}{k_{XY(X)}} = 10^{(\delta_Y - \delta_X)} = \frac{[\text{product} - XY(Y)]}{[\text{product} - XY(X)]}$$

(355)

it follows

$$\frac{k_{XY,exp}}{(k_{XX,exp} \cdot k_{YY,exp})^{1/2}} = \frac{\nu^{1/2} + \nu^{-1/2}}{2}$$

(356)

The determination of the rate constants showed that the oxidation
of diazoaminobenzenes proceeds via the rate-determining formation of
a symmetrical intermediate (path a) according to *(348)*.

On the other hand, rate constants for oxidation of azobenzenes to
azoxybenzenes[336b] *(357)* and for the Baeyer-Villiger rearrangement[336c]
[*(358)* and *(359)*, see also p. 57] are correlated with product ratios
via *(356)*; this eliminates in the former case the rate-determining formation of the symmetric oxadiaziridine **249**; in the latter case it shows
that the rate-determining step is the *rearrangement* (k_2, unsymmetrical
transition state) of the adduct **250**, not its *formation* (k_1, reaction of a
symmetrical species):

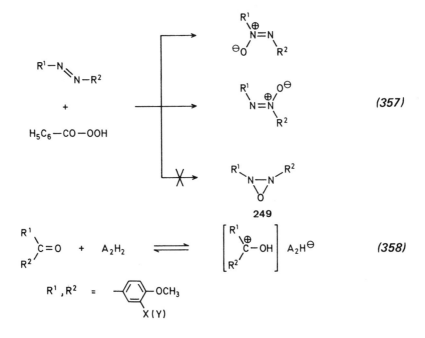

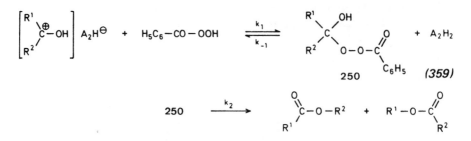

$$(359)$$

$$250 \xrightarrow{k_2}$$

A_2H_2 represents the dimer of dichloroacetic acid, used as medium; all ketones studied were almost completely protonated in this medium.

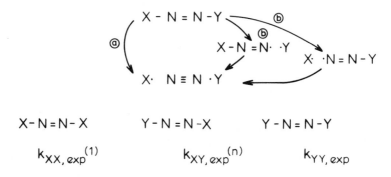

Scheme 63A Possible modes of decomposition of azo compounds.

The question of whether the thermal decomposition of azoalkanes occurs in a one-step process (path a, Scheme 63A) or via diazenyl radicals (path b, Scheme 63A) can be approached in a very similar manner. Using three differently substituted azoalkanes, additivity of the effects of substituent change leads to (348) in the case of path a. As before, in case b equations (349)-(353) apply. Simplifying by setting $k_{XX,exp} = 1$, $k_{XY,exp}/k_{XX,exp} = n$, and $\delta_X = 0$ (in case b, only a substituent change close to the breaking bond is assumed to affect the rate constant) transforms (348) into (360):

$$k_{YY,exp} = n^2 \quad \text{(path a)} \qquad (360)$$

whereas from (352) and (353) one obtains (361) for path b processes*:

$$k_{YY,exp} = 2n - 1 \quad \text{(path b)} \qquad (361)$$

*Application of (360) and (361) in the diagnosis of the one-step or two-step nature of processes like the azoalkane decomposition is called the Ramsperger criterion [H. C. Ramsperger, *J. Amer. Chem. Soc.* **51**, 2134 (1929).]

Experiments by Rüchardt[336d] for X = 1-norbornyl, Y = t-butyl gave values of $k_{YY,exp}$ much more in accord with (360) than with (361), thereby establishing path a for the decomposition of tertiary azoalkanes. In the case of highly unsymmetrical azo compounds (e.g., X = alkyl, Y = allyl), path b seems to be followed.

Additivity is observed in many other systems. Introduction of a 1-methyl group increases the free energy of activation of *thermolysis of cyclobutenes to butadienes* (362) by an average of 4.90 kJ/mol (Table 79: 252 and 251, 253 and 254, 255 and 256).

TABLE 79   Free Energy of Activation in
the Thermolysis of Cyclobutenes and
Methylcyclobutenes[337] (kJ/mol)

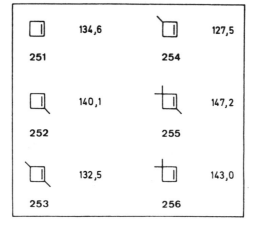

251	134,6	254	127,5
252	140,1	255	147,2
253	132,5	256	143,0

Introduction of a 3-methyl group decreases the free energy of activation by an average of 7.4 kJ/mol (Table 79: 251-254, 252-253).[337] On the other hand, introduction of two 3-methyl groups results in a relatively large increase in $\Delta G^{\ddagger}$ (Table 79: 255 and 252, 256 and 251), presumably because of unfavorable steric effects associated with ring opening (363):

$$H_3C-\!\!\!\square\!\!\!-H \quad\quad \longrightarrow \quad\quad H_3C-\!\!\!\langle\!\!\!\rangle\!\!\!-H \qquad\qquad (363)$$

The first two effects of a methyl group may be due, on one hand, to the stabilization of the ground state by alkyl groups (p. 44) and on the other to the stabilization of the transition state (where the 3-position adopts double bond character).

These increments for methyl groups give a calculated free energy of activation for the thermolysis of 1,2,3-trimethylcyclobutene of 134.6 + 2 · 4.90 − 7.4 = 137.0 kJ/mol. The experimental value is 138.7 kJ/mol.[338]

In certain *solvolyses* assumed to proceed with delocalization of the positive charge, introduction of methyl groups at different sites affects the rate constants by about the same factors. This experimental fact is considered to be evidence for a symmetric distribution of charge over

TABLE 80   Methyl Groups as Probes for Positive Charge

Substrate	Intermediate (Type)	Substituent	k_{rel}	Ref.
ONs R¹ R² (structure)	(structure) δ^+ δ^+ (see p. 000)	$R^1 = R^2 = H$ $R^1 = CH_3$ $R^2 = H$ $R^1 = R^2 = CH_3$	1 7 38.5	339

Substrate	Intermediate (Type)	R¹	R²	R³	R⁴	k_{rel}	Ref.
DNBO R³ R² R¹ R⁴ (structure)	(structure) δ^+ δ^+ δ^+ (see also p. 000)	H CH₃ H CH₃ CH₃ H CH₃ CH₃	H H CH₃ CH₃ H CH₃ H CH₃	H H H H H CH₃ CH₃ CH₃	H H H H CH₃ H CH₃ CH₃	1 11 8.2 80 124 82 290 1570	340

Substrate	Intermediate (Type)	Substituent	k_{rel}	Ref.
PNBO R¹ R² (structure)	(structure) δ^+ δ^+ δ^+ (see p. 000)	$R^1 = R^2 = H$ $R^1 = CH_3, R^2 = H$ $R^1 = R^2 = CH_3$	1 13.3 148	341

$$\left(N_S = O_2N - \!\!\! \bigcirc \!\!\! - SO_2-; \quad DNB = \begin{array}{c} O_2N \\ \bigcirc \\ O_2N \end{array} \!\!\! - C \!\!\! \begin{array}{c} O \\ \diagup \\ \diagdown \end{array} ; PNB = O_2N - \!\!\! \bigcirc \!\!\! - C \!\!\! \begin{array}{c} O \\ \diagup \\ \diagdown \end{array} \right)$$

these sites in the respective intermediates and the transition states leading to them (Table 80).

In view of the difficulty of examining the structures of short-lived ions, and the greater difficulty of examining transition states, the effect of substitution at different sites may be the best available criterion of the nature of the transition state in an ionization.[339]

Similar results in the *protolysis of cyclopropanes* show that the rate-determining step is the formation of a transition state leading to the delocalized "corner-protonated" cyclopropane intermediate **257** (Table 81). In contrast to the examples above, reactions that are supposed to proceed via *classical* carbocations with localized charge do not show cumulative methyl effects.

TABLE 81. Relative Reactivities in Reaction[342]

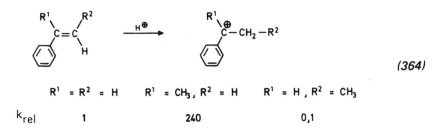

| Substituents | Mode of breaking | | k_{rel} |
	(%) a	(%) b	for breaking *a*
$R^1 = R^2 = H$			1^a
$R^1 = CH_3, R^2 = H$			9.2^a
$R^1 = H, R^2 = CH_3$	65	35	4.8
$R^1 = R^2 = CH_3$	60	40	37

[a]Corrected for two equivalent sites of cleavage

In such instances methyl groups are effective only at the atom where the positive charge is concentrated: *hydration of α-methylstyrene* proceeds 240 times faster than that of styrene,[343] whereas hydration of *trans-β-methylstyrene* is 10 times slower[344] (*364*):

(364)

	$R^1 = R^2 = H$	$R^1 = CH_3, R^2 = H$	$R^1 = H, R^2 = CH_3$
k_{rel}	1	240	0,1

In many instances one observes deviations from additivity, (toward lower reaction rates) when the number of substituents increases. A possible cause may be the *increase in steric effects* (e.g., Tables 80 and 81) resulting in weaker substituent interactions. Steric requirements of additional substituents may prevent a molecule from adopting the conformation that would be most ideal for stereoelectronic reasons and that may be realized to full extent only when a single substituent is introduced. Thus, in (365) the larger increase in the ionization rate upon introduction of the first cyclopropane ring is ascribed[344a] to a larger degree of twisting of the molecular skeleton, which enables better interaction between the cyclopropane ring and the positive carbon atom in the carbocation formed. With two or three cyclopropane rings attached, the rigidity of the molecule has increased so much that twisting is decreased.

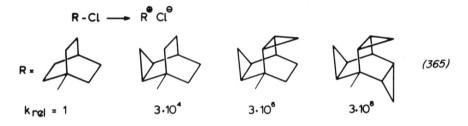

$$R\text{-Cl} \longrightarrow R^{\oplus}\ Cl^{\ominus}$$

R =

k_{rel} = 1 $3 \cdot 10^{4}$ $3 \cdot 10^{6}$ $3 \cdot 10^{8}$ (365)

However, *dedeuteration rates* of various deuterated haloforms[345] (Fig. 71) should show opposite trends under the influence of steric effects: Enhanced reactivity resulting from substitution of Cl by Br should increase even more on further substitution; additional bromine atoms should not only increase the substituent effect (presumably by stabilizing $^-CX_3$ by *d-orbital resonance* and/or by *polarization* of X) but also lead to increasing contributions of relief of steric strain in the ground state on anionization (see p. 253 ff). However, in actuality the effect of replacing chlorine by bromine decreases on further substitution (and increase in reactivity).

$$\begin{matrix} & |\bar{X}| & & & |\bar{X}| & \\ & | & & & | & \\ ^-|\underset{|}{\overset{|}{C}}-\bar{X}| & & \longleftrightarrow & & \underset{|}{\overset{|}{C}}=\bar{X}|^- & etc \\ & |\underline{X}| & & & |\underline{X}| & \end{matrix} \qquad (366)$$

It could be shown that in *protodetritiation of polymethylbenzenes* partial rate factors $f_{o,m,p}^{CH_3}$ decrease with increasing degree of substitution, even if steric factors remain approximately constant.[346] The last two examples show that the extent to which a substituent displays a stabilizing effect depends to some degree on the extent to which other substituents already have become active and that the "need" for

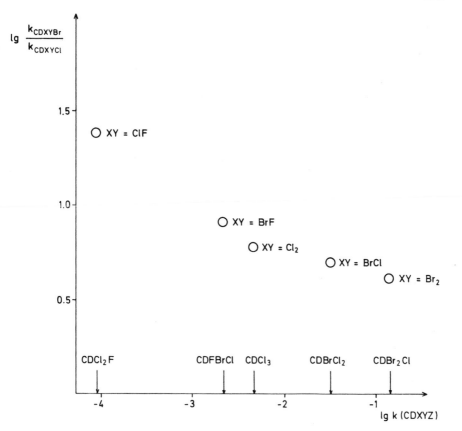

Fig. 71 Dependence of the effect of substitution of Cl by Br (log k_{CDXYBr}/k_{CDXYCl}) in reaction CDXYZ (Z = Cl or Br) + OH$^- \xrightarrow{k}$ CXYZ$^-$ + DOH on the reactivity (l mol^{-1} s^{-1}] of CDXYCl.

(additional) stabilization diminishes when the reactivity of the system increases.

One reason for the existence of *"demand-dependent"* free energy relationships *(367)* [which may be compared with *(368)*, representing a *linear* relationship]

$$\delta_X \, \Delta G^{\ddagger} = c \cdot \Delta G^{\ddagger} \cdot \delta_X \, \Delta G \qquad (367)$$

$$\delta_X \, \Delta G^{\ddagger} = c \cdot \Delta G_o^{\ddagger} \cdot \delta_X \, \Delta G \qquad (368)$$

where $\Delta G_o^{\ddagger}$ = free energy of activation of the parent compound of a reaction series

could be that the transition states of the more highly substituted (and therefore more reactive) members of a reaction series are more reactantlike (see pp. 327 and 352), with a correspondingly lower degree of substituent interaction. Another possibility is *saturation of resonance effects*: introduction of the first substituent can change the charge distribution in a given system in a way that allows further substituents to interact only to a lesser degree. A clear example in a system where steric inhibition of resonance is precluded and in which no ambiguity exists concerning the nature of transition states (since equilibria are considered) is shown in (*369*)[346a]: a phenylsulfonyl group at C-2 of fluorene lowers the pK (in Me$_2$SO) by 4.5 units. In 9-*p*-tolylthiofluorene the lowering is only 3.8 units, since in its anion the charge distribution acted on by the 2-phenylsulfonyl group is different from the one in the parent system.

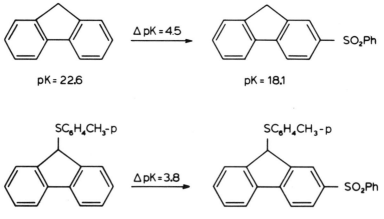

(*369*)

Substituent effects based on orbital overlap and formation of partial bonds cannot be cumulative if they result in changes in molecular geometry. For instance, calculations show[347] that the angle α in the stabilized homoallyl cation **258** is compressed by 35° with respect to the tetrahedral angle.

Thus it is reasonable that on ionization of symmetric tosylates of structure **259** only *one* of the two double bonds participates in the ionization. This can be seen from the fact that a second methyl group affects the acetolysis rate constant only by the *statistical* factor of 2 (*370*)[348]:

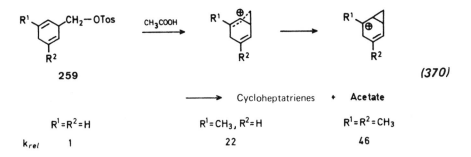

$$R^1 = R^2 = H \qquad\qquad R^1 = CH_3, R^2 = H \qquad\qquad R^1 = R^2 = CH_3$$

k_{rel} 1   22   46

The nmr spectrum of the 7-norbornadienyl cation shows that the pairs of protons at C-2,C-3 and C-5,C-6 are different, which means that the structure of the ion is not symmetric.

The values of the coupling constants lead to the conclusion that the C-7 bridge is tilted toward C-2,C-3 $(371)^{349}$:

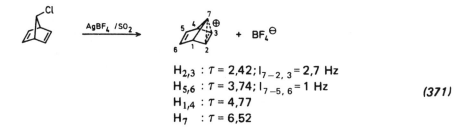

$$H_{2,3} : T = 2{,}42; I_{7-2,3} = 2{,}7 \text{ Hz}$$
$$H_{5,6} : T = 3{,}74; I_{7-5,6} = 1 \text{ Hz} \qquad\qquad (371)$$
$$H_{1,4} : T = 4{,}77$$
$$H_7 \ : T = 6{,}52$$

In accord with these results is the observation that the accelerating effect of the second double bond on the rate of ionization is much smaller than that of the first one (but larger than a purely statistical contribution)[350] (Fig. 72).

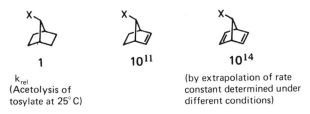

1   10^{11} 10^{14}

k_{rel}
(Acetolysis of   (by extrapolation of rate
tosylate at 25°C)   constant determined under
                        different conditions)

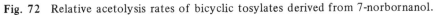

Fig. 72   Relative acetolysis rates of bicyclic tosylates derived from 7-norbornanol.

Similar situations are found in ionizations of related cyclopropyl-carbinyl systems.[351] Here, the remarkably low rate enhancement observed on going from the tricyclic system to the tetracyclic system may be due to adverse bishomoantiaromatic character of the corresponding carbocation A.[349a] (Fig. 73) As already mentioned on page

$$\text{1} \qquad\qquad \text{6·3·10}^8 \qquad\qquad \text{10}^{10} \qquad\qquad \text{A}$$

Fig. 73 Relative acetolysis rates of cyclopropylcarbinyl tosylates derived from 7-norbornanol.

212, on hydrolysis of *syn*-7-aryl-*anti*-7-norbornenyl-*p*-nitrobenzoates there is a strong interaction ($\rho = -5.27$) between the positive center at C-7 and the electron-donating substituents *p*-OCH$_3$ and *p*-N(CH$_3$)$_2$. This interaction wipes out completely the participation of the double bond; by contrast, with other substitutents strong interaction with the double bond results in only weak substituent effects in the aryl residue ($\rho = -2.30$). The same behavior was observed for 1-aryl-1-cyclopropyl-ethyl-*p*-nitrobenzoates.[351a]

An important application of the additivity principle of inductive-steric effects and of lack of additivity in bond-forming interactions is found in the work of Schleyer and collaborators on the participation of phenyl rings in β-position leading to phenonium ions.[352]

The rate constants k_t for the ionization of 1-aryl-2-propyl tosylates **260** may be written as a function of the rate constant k_0 of the unsubstituted 2-propyl tosylate:

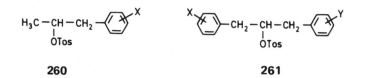

$$\text{260} \qquad\qquad\qquad\qquad \text{261}$$

$$k_t\ (\textbf{260}) = s \cdot k_0 + \Delta k_0 \qquad\qquad\qquad (372)$$

$$k_t\ (\textbf{260}) = k_0\ (s + \Delta) \qquad\qquad\qquad (373)$$

The coefficient s corrects for steric + inductive effects of the aryl substituent, Δ corrects for participation leading to the phenonium ion (according to the notation on p. 210: $sk_0 = k_s$; $\Delta \cdot k_0 = F \cdot k_\Delta$). Starting with the diaryl derivative **261** there are three possible transition states (**262-264**):

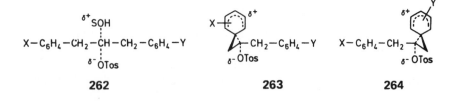

$$\text{262} \qquad\qquad\qquad\qquad \text{263} \qquad\qquad\qquad\qquad \text{264}$$

In **262** (nucleophilic participation of solvent SOH) inductive + steric effects of the two aryl rings operate simultaneously and according to the additivity principle this component of k_t is given by $k_o \cdot s_X \cdot s_Y$.

In **263** and **264** participation of *one* aryl ring occurs together with steric + inductive effects of the *other*. One thus has the components $k_o \cdot \Delta_X \cdot s_Y$ and $k_o \cdot \Delta_Y \cdot s_X$, and overall one has:

$$k_t \text{ (261)} = k_o (s_X s_Y + \Delta_X s_Y + \Delta_Y s_X) \qquad (374)$$

Using three different aryl groups (X or Y = p-H, p-OCH$_3$, p-NO$_2$) and measuring the rate constants for the six possible tosylates (three with X = Y and three with X $\neq$ Y) it was possible to compute the six unknowns. The contributions $F \cdot k_\Delta$ (= $\Delta \cdot k_o$) computed from the Δ values were the same as those obtained by independent means (p. 211). Thus, transition states leading to phenonium ions could be proved without reference to model compounds (which must be *assumed* to be unable to form phenonium ions; see Section 3.3.1.4).

CHAPTER **5**

Properties of Activated Complexes

5.1 ENERGY SURFACES AND REACTION PATHS

5.1.1 Calculation and Representation of the Course of a Reaction

Only the knowledge of the nature of the activated complex, of its enthalpy, geometry, and entropy, its electronic properties, and its polarity enables us to answer questions like:

1 Is a given reaction at all possible? That is, under a given set of conditions (type of reagent, temperature, solvent) is the value of $\Delta G^{\ddagger}$ such as to allow appreciable change within a realistic time span?

2 What is the susceptibility of the reaction toward changes in different parameters (structure of substrate and reagents, effect of substituents, of solvent)?

The inherent instability of activated complexes thus far has prevented their direct study*; the accumulated *empirical* knowledge (aside from thermodynamic and extrathermodynamic properties obtained from kinetic data) is *qualitative* and has been gathered *indirectly* (see Section 5.2).

Nonempirical, quantitative knowledge may be obtained from quantum mechanical calculations of the *potential energy hypersurfaces*, which would give the geometry and potential energy of transition states. In principle, using methods of statistical thermodynamics, one

---

*J. L. Kurz [*Acc. Chem. Res.* **5**, 1 (1972)] pointed out that although activated complexes cannot be isolated, *in principle* they could be observed. The time they spend in the transition state is about 10^{-13}s(h/kT, p. 229); therefore they could be detected spectroscopically if the period of the radiation used were shorter than this time span (ultraviolet and a large region of infrared). Only experimental difficulties associated with the extraction of signals produced by activated complexes from signals generated by other species present in overwhelming amounts prevent the direct observation of these complexes.

296

could then calculate the entropy $S^{\ddagger}$ and enthalpy $H^{\ddagger}$, and using quantum mechanics one could also determine electronic properties. Many such calculations have been performed in recent years.[353] They yield, in the form of contour diagrams (Figs. 74, 76, 77) or cinematographic representations (Figs. 78, 80) the required information with respect to favored (low energy) reaction paths and structures of the corresponding activated complexes.

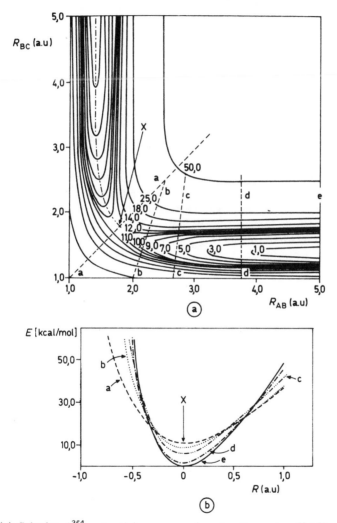

Fig. 74 (a) Calculated[354] potential energy surface for the system $H + H_2 \rightarrow H_A \cdots$ $H_B \cdots H_C \rightarrow H_2 + H$. The contour lines give configurations of equal energy (in kcal/mol relative to the initial states $H + H_2$). $R_{AB(BC)}$ are distances (in atomic units, 1 a.u. = 0.529 Å) between hydrogen atoms H_A and H_B (H_B and H_C). The transition state is at point X. (b) Cross sections a-e through the right-hand valley in Fig. 74a.

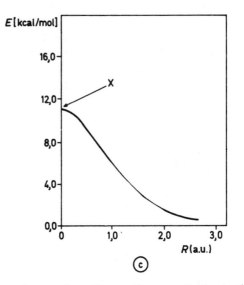

Fig. 74 (c) Energy change along the reaction coordinate starting from transition state X (see dashed line in the left-hand valley in Fig. 74a). R (a.u.) gives the deviation of the position of the departing H-atom from its position in the transition state (R = 0). [From I. Shavitt, R. M. Stevens, F. L. Minn, and M. Karplus, *J. Chem. Phys.*, **48**, 2700 (1968) (ref. 354).]

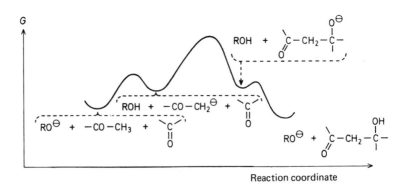

Fig. 75 Cross section along the reaction coordinate from the conceivable free energy surface of an aldol condensation.

298

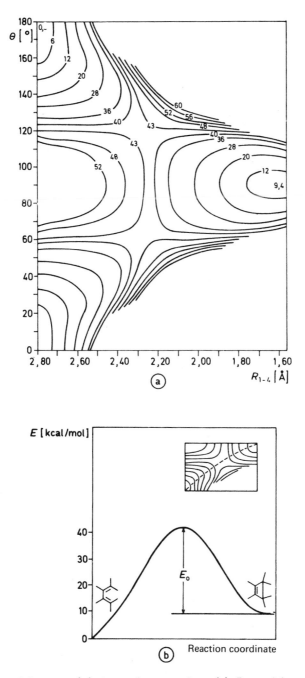

Fig. 76 Conrotatory cyclobutene ring opening. (*a*) Potential energy surface showing lowest possible energies for given values of R_{1-4} and θ. (*b*) Cross section along reaction coordinate. [From A. Rastelli, A. S. Pozzoli, and G. Del Re, *J. Chem. Soc., Perkin Trans. 2*, **1972**, 1571 (ref. 356).]

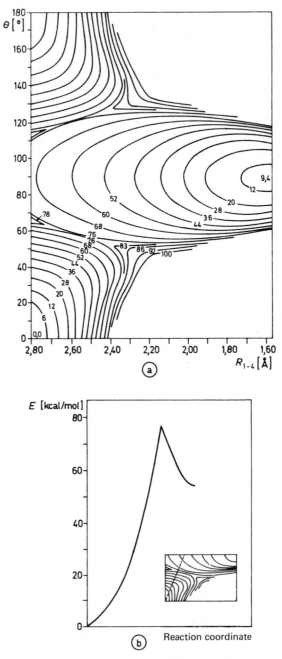

Fig. 77 Disrotatory cyclobutene ring opening. (*a*) Potential energy surface showing lowest possible energies for given values R_{1-4} and θ. (*b*) Cross section along reaction coordinate. [From A. Rastelli, A. J. Pozzoli, and G. Del Re, *J. Chem. Soc., Perkin Trans. 2,* **1972**, 1571 (ref. 356).]

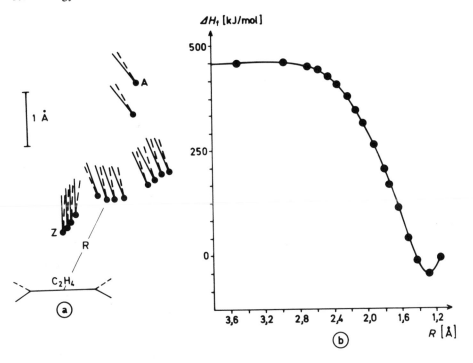

Fig. 78 Cinematographic representation (*a*) and enthalpy changes (*b*) for the calculated mode of addition of singlet methylene to ethylene ($CH_2 + C_2H_4 \rightarrow$ *cyclo*-C_3H_6). The process starts with the highest point A representing the approaching particle CH_2. The last point in the series (Z) corresponds to the position of CH_2 in the cyclopropane formed (solid and dashed lines represent bonds to the hydrogen atoms of CH_2 in front and in the back of the plane of the paper). Only motions in the plane of the paper were studied. Reaction coordinate R is the distance between the carbon atom and the middle of the C—C bond. [From N. Bodor, M. J. S. Dewar, and J. S. Wasson, *J. Amer. Chem. Soc.* **94**, 9100 (1972) (ref. 357).]

The simplest example is the calculation of the potential energy surface of the linear system $H_A \cdots H_B \cdots H_C$ for the reaction $H + H_2 \rightarrow H_2 + H$ (Fig. 74). (The calculations showed that bent systems $H \cdots\cdots H \cdot\cdot^{\cdot H}$ are of higher energy).

The calculation of this simplest energy surface, determined by only two geometric parameters started with a "basis set" of 15 atomic orbitals (for each atom one used the $1s$, $2s$, $2p_x$, $2p_y$, and $2p_z$ orbitals). The valley in the upper left-hand side (Fig. 74*a*) corresponds to the stable state $H_A - H_B + H_C$, the valley on the lower right [Fig. 74 (*b*), cross sections a-e] represents the stable state $H_B - H_C + H_A$. The dashed line represents the minimum energy path from one valley to the transition state (Fig. 74*c*; left valley, Fig. 74*a*) and is called *reaction coordinate.* It is characteristic that along this reaction coordinate a decrease in R_{BC} is coupled with increase in R_{AB}. The effect is largest at the *transition state* (X), where $-\Delta R_{BC} = +\Delta R_{AB}$ ($R_{AB} + R_{BC}$ = const). This corresponds to the asymmetric vibration $\vec{H}_C \cdots H_B \cdots \vec{H}_A$, leading to the collapse of the activated complex. Orthogonal to the reaction coordinate, $\Delta R_{BC} = \Delta R_{AB}$ ($R_{AB} - R_{BC} = 0$). This corresponds to the symmetric vibration $\vec{H}_C \cdots H_B \cdots \vec{H}_A$. With respect to such motions, which do not belong to the reaction coordinate, activated complexes represent an energy minimum, just like a normal molecule. This is in accord with their character as *saddle points* on the energy barrier one needs a minimum of 11 kcal/mol (from experimental data the value is 9.7 kcal/mol), and that in the linear activated complex of this symmetric reaction $R_{AB} = R_{BC} = 1.765$ a.u. (= 0.933 Å).

A section through the energy surface along the dashed reaction coordinate gives a diagram of the type shown in Fig. 54. Similar profiles for conceivable *free energy surfaces* are very popular as means for pictorial representations of reaction mechanisms. (Since they are arbitrary, no significance should be attributed to the slopes of lines connecting the extremes.) For instance, an aldol condensation (*375*) could be represented by the plot in Fig. 75.

$$RO^{\ominus} \; + \; -CO-CH_3 \;\; \rightleftharpoons \;\; ROH \; + \; -CO-CH_2^{\ominus} \qquad (375a)$$

$$-CO-CH_2^{\ominus} \; + \; \underset{O}{\overset{}{\underset{\parallel}{C}}} \;\; \rightleftharpoons \;\; -CO-CH_2-\underset{O^{\ominus}}{\overset{\mid}{\underset{\mid}{C}}}- \qquad (375b)$$

$$ROH \; + \; -CO-CH_2-\underset{O^{\ominus}}{\overset{\mid}{\underset{\mid}{C}}}- \;\; \rightleftharpoons \;\; RO^{\ominus} \; + \; -CO-CH_2-\underset{OH}{\overset{\mid}{\underset{\mid}{C}}}- \quad (375c)$$

It was necessary to use 350 computer hours to calculate 700 selected points on the 21-dimensional potential energy surface of cyclopropane and its unimolecular reactions.[355] For this reason in more complex systems only drastically simplified calculations can be used, taking into account only a few parameters most responsible for the reaction. This reduces drastically the dimensionality. Even such simplified calculations can yield valuable information with respect to the most favorable reaction path when several are possible. As an example we consider the two possible reaction paths in the thermal isomerization of cyclobutene to butadiene (Scheme 64).[356]

The parameters chosen as most significant for the course of the reaction are the distance R_{1-4} and the angle θ between the planes of the CH_2 groups and the plane of the four-membered ring ($\theta = 90°$ in cyclobutene, $\theta = 0°$ in butadiene, or 180° if rotation is in the opposite sense). The minimum energy of the system was calculated for each

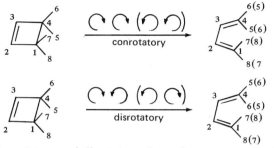

Scheme 64 Conrotatory and disrotatory thermal rearrangement of cyclobutene to 1,3-butadiene.

pair of parameters (each point on the contour diagram, Figs. 76a and 77a). The following simplifying assumptions were made: C-1-C-4 remain in a plane during the transformation, and the two CH_2 groups rotate to the same extent.

The diagram for the conrotatory reaction shows (for each of the two equally probable rotations) the transition state as saddle point (energy minimum with respect to variations orthogonal to the reaction coordinate) connecting the "cyclobutene valley" with the respective "butadiene valley." This character is absent in the diagram for the disrotatory reaction. The valleys of the stable compounds are separated by a sharp *discontinuity* (at high energy) and (by contrast with the conrotatory reaction) a favored reaction path over the "ridge" cannot be indicated. The conclusion (confirmed experimentally) is that conrotation is favored (see Section 5.2.4.2). A further illustration is the calculated reaction course for the addition of CH_2 to ethylene (Fig. 78).

Quite remarkable is an approach pioneered by Dunitz and collaborators[358a-358e] to determine *experimentally* the spatial properties of favorable reaction paths. Their working hypothesis is as follows: correlations describing the relationship between two or more independent structural parameters of a group of atoms in different surroundings (i.e., under the influence of different perturbations) represent at the same time the path of lowest potential energy for this group of atoms on the energy hypersurface (because the structural response of the atomic system to perturbations should follow local valleys on the energy hypersurface).

They found[358a] that in the system "tertiary amino group + carbonyl group" the relationship between the distance C ... N (d_1, Å) and the deviation (Δ, Å) of the $R^1R^2C=O$ group from planarity can be described by (376). Definitions of d_1 and Δ are shown in Fig. 79. Values for these parameters were obtained from the analysis of the crystal structures of appropriate compounds containing the two groups in different relative positions.

Fig. 79 Definition of d_1, d_2, and Δ. [From H. B. Bürgi, J. D. Dunitz, J. M. Lehn, and G. Wipf, *Tetrahedron* **30**, 1563 (1974) (ref. 358b).]

$$d_1 = -1.701 \log \Delta + 0.867 \text{ Å} \qquad (376a)$$

$$d_1 = -1.701 \log n + 1.479 \text{ Å} \qquad (376b)$$

where $n = \Delta/\Delta_{max}$ and $\Delta_{max} = 0.437$ Å (for a normal C—N distance of 1.479 Å). Similarly, the C—O distance (d_2, Å) was found to be

$$d_2 = -0.71 \log (2 - n) + 1.426 \text{ Å} \qquad (377)$$

Equations (*376*) and (*377*) should represent in a *general* sense the variation of the structural parameters along the reaction coordinate (i.e., along the lowest energy reaction path) for the addition of tertiary amines to carbonyl groups. A cinematographic representation of the positions of N,R^1,R^2, and O in stages A-F of such an addition is given in Fig. 80.

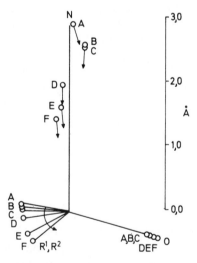

Fig. 80 Positions of N, R^1, R^2, and O in stages A-F of the most favorable reaction path for the addition of tertiary amines to the carbonyl group. The arrows represent the probable orientation of the unshared electron pair at nitrogen. [From H. B. Bürgi, J. D. Dunitz, and E. Shefter, *J. Amer. Chem. Soc.* **95**, 5065 (1973) (ref. 358a).]

Equation (*376a*) is very similar to the *calculated* lowest energy path for the addition of H⁻ to H₂C=O (*378*)[358b] (see Fig. 81).

$$d(\text{H}^- \ldots \text{C}) = -1.805 \log \Delta + 0.415 \text{ Å} \qquad (378)$$

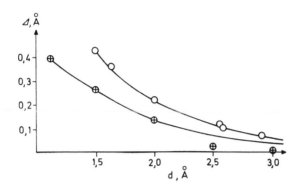

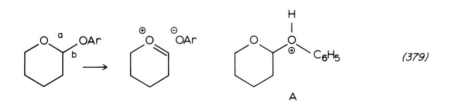

Fig. 81 Comparison between the experimentally determined most favorable reaction path for the addition of tertiary amines to C=O [circles (*376a*)] and the calculated reaction path for the addition of H⁻ to formaldehyde [crosses, (*378*)]. [From H. B. Bürgi, J. D. Dunitz, J. M. Lehn, and G. Wipf, *Tetrahedron* **30**, 1563 (1974) (ref. 358b).]

Whereas in its original version the method naturally does not yield information on energy, recent work indicates that energies of reaction paths may also be obtained. Thus, Jones and Kirby[358c] have not only found a linear relationship between the lengths of the two C—O bonds *a* and *b* of 2-phenoxytetrahydropyrans (*379*)

but also a correlation of these bond length alterations with the pK_a of the corresponding phenol ArOH (Fig. 81A). Since the pK_a of ArOH is itself linearly correlated with the free energy of activation for heterolysis of bond *b* in water ($\Delta G^\ddagger$ = 1.7 pK_a + 13.7), not only the geometric changes along the reaction path of (*379*) (and its microscopic reverse, addition of ArO⁻ to an oxycarbocation) become apparent, but

also the concomitant changes in energy. Compounds with a longer bond
b "have a lead in their journey" toward the transition state and the
remaining energy required to reach it ($\Delta G^{\ddagger}$) is least. Lengthening *b* by
1Å increases energy by about 1200 kJ/mol. For compound A the
correlation of Fig. 81A predicts *a* = 1.33Å and *b* = 1.52 Å. Since it is
known that for leaving groups better than phenol C—O cleavage sets in

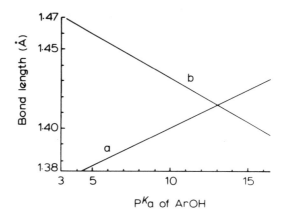

Fig. 81A Plot of acetal group bond lengths [*a* and *b* (*379*)] versus the pK_a of
ArOH. (From P. G. Jones and A. J. Kirby, *J. Chem. Soc., Chem. Commun.* **1979**,
288.)

before protonation is complete (general acid catalysis, see p. 176) the
two bond lengths obtained are maximum and minimum values, respectively, for the corresponding bond lengths (a and b, respectively) in the
transition states of the latter reactions. The intramolecular hydride
transfer reaction (*380*) has been studied[358d] for *n* = 1,2,3. Increasing *n*

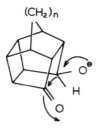

(*380*)

forces the reaction centers more closely together and this causes a decrease of the free energy of activation ($\geqslant$ 91, 80, and 72.5 kJ/mol).
As soon as accurate molecular structures are known for the reactants,
three points on *one* of the feasible reaction paths for H⁻ + C=O could
be mapped from the combined data.

It had been hypothesized that enzymes could achieve much of their catalytic ability by optimizing orientational relationships between active sites and that even a 10° misalignment of reactant groups relative to an ideal orientation may cause a massive decrease in rate. However, this notion of *"orbital steering"* was disproven when no difference in free energy of activation was found[358e] for the acid-catalyzed lactonizations of compounds **264a** and **264b**, which differ in the angle in which the hydroxy oxygen is juxtaposed to the carbonyl carbon atom but are very similar in all other respects (e.g., energy and O^1-C^2 distance).

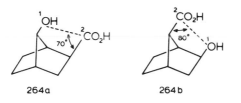

264a   264b

Variations in main structural characteristics [d(1-6)(Å) and $\triangle$ (Å)] within the series of annulenes **265-271** (Scheme 65) [change of C-1 and C-6 from the planar conformation ($\triangle \approx 0$) to the pyramidal one ($\triangle \approx 0.2$) on shrinking of d(1-6)] should correspond to the lowest energy reaction path of the disrotatory electrocyclic reaction (see Section 5.2.4.1) (*381*).[358f]

In symmetric reactions with transition states probably at "midway" and thus also symmetric (p. 321), the calculation of the most favorable reaction path can be limited to the computation of the minimum enthalpies of formation and geometries of the possible "midway structures"; examples are the two possibilities for a 1,5-hydrogen shift in 1,3-pentadiene: (*a*) suprafacial and (*b*) antarafacial, (Scheme 66) or substitution of $H^+(H^-)$ by $H^+(H^-)$ in methane [as model for "normal" $S_E 2$ ($S_N 2$) reactions][360], which can proceed either with inversion, (**274**) or retention, (**275**) (Scheme 67).

With rare exceptions (e.g., only when very similar systems are compared; see Table 82), none of the calculations can even remotely approximate the exact values of experimentally determined activation parameters.

Only answers in the sense of "better" or "worse" are obtainable. In general they agree with the experimental data. These results represent the progress offered by the transition state theory. This progress would not be apparent if one would only equate the activation parameters $\Delta H^{\ddagger}$ and $\Delta S^{\ddagger}$ with those of the Arrhenius equation or those of the collision theory of gas phase reactions. All three equations contain two empirical constants each, and differences seem to reside only in their analytical form and notations. However, none of the older theories can provide comparably concrete models for the stereoenergetic aspects of reactivity (*382-384*).

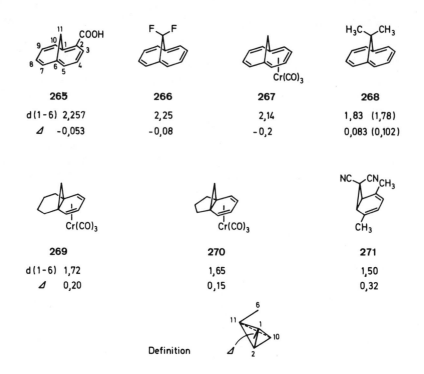

265	**266**	**267**	**268**
d(1-6) 2,257	2,25	2,14	1,83 (1,78)
⊿ -0,053	-0,08	-0,2	0,083 (0,102)

269	**270**	**271**
d(1-6) 1,72	1,65	1,50
⊿ 0,20	0,15	0,32

Definition

Scheme 65 Bicyclic and tricyclic annulenes as models for the various stages of the most favorable reaction path in transformation (*381*) (geometric parameters *d* and Δ in angstrom units)

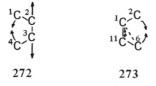

(e.g., **265**)   (e.g., **269**)

(381)

While in "normal" molecules *compression* of bond angles results in *stretching* of the C-2—C-3 bond between them (**272**) (presumably to relieve

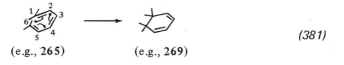

272   273

1,3-repulsions), it was observed that in several annulenes the reverse is true, namely, that bond *contraction* (C-1—C-11, **273**) takes place; this is indicative of an attractive interaction (dashed line, **273**) between C-1 and C-6 corresponding to the favorable reaction mode (*381*).

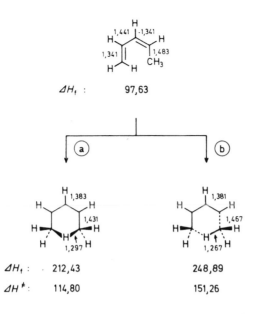

ΔH_f : 97,63

(a)	(b)

ΔH_f : 212,43 248,89

$\Delta H^{\neq}$: 114,80 151,26

Scheme 66 Calculated bond distances (Å) and enthalpies of formation ΔH_f (kJ/mol) of ground and transition states and activation enthalpies $\Delta H^{\ddagger}$ (kJ/mol) in suprafacial *(a)* and antarafacial *(b)* 1,5-hydrogen shifts in 1,3-pentadiene. [From R. C. Bingham and M. J. S. Dewar, *J. Amer. Chem. Soc.* **94**, 9107 (1972) (ref. 359).]

$\overrightarrow{H} - \underset{H}{\overset{H}{C}} - \overrightarrow{H}$ **274**   **275**

Structure	E_{rel}	
	$CH_5^{\oplus}$	$CH_5^{\ominus}$
274	14,67	0
275	0	137,99

Scheme 67 Calculated energy differences (kJ/mol) between structures **274** and **275** for CH_5^+ and CH_5^- (as models for transition states of bimolecular aliphatic electrophilic and nucleophilic substitutions).

TABLE 82   Calculated[361] and Experimental Activation
Energies of a Degenerate Thermal Cope Rearrangement

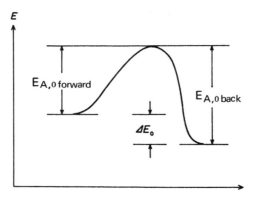

−X−	E_a (kJ/mol)	
---	Calc.	Exp.
−CH=CH−	47,3	49,4 53,6
−CH$_2$−	27,2	36,0
−	15,1	21,4

Eyring equation

$$k = \frac{kT}{h} \cdot e^{\Delta S^{\ddagger}/R} \cdot e^{-\Delta H^{\ddagger}/RT} \tag{382}$$

Arrhenius equation

$$k = A \cdot e^{-E_a/RT} \tag{383}$$

Collision theory

$$k = p \cdot Z \cdot e^{-E/RT} \tag{384}$$

where Z = collision number obtainable from the kinetic gas theory
       p = "steric factor"

5.1.2   The Principle of Microscopic Reversibility (pmr)

Pictures of energy surfaces (e.g., Figs. 76 and 82) illustrate the principle
of *microscopic reversibility*: if a given series of relative atomic con-

Fig. 82   Section through an energy surface along the reaction coordinate.

figurations represents the mechanism of the forward reaction, the same series of configurations must be involved in the mechanism of the back reaction. The pictures show that:

> The lowest potential energy path—and transition state—between reactant(s) and product(s) is the same for the forward and backward reactions.

The resulting relationship (385):

$$E_{A,0,\text{forward}} - E_{A,0,\text{backward}} = \Delta E_0 \qquad (385)$$

where $E_{A,0}$ = activation energies
ΔE_0 = difference in potential energies of products and reactants

corresponds to relationship (387), based on the thermodynamic-statistical formulation of pmr: *at equilibrium each molecular process proceeds at the same rate as its reverse*:

$$\log K = \log \frac{k_f}{k_b} \qquad (386)$$

$$\Delta G_f^{\ddagger} - \Delta G_b^{\ddagger} = \Delta G \qquad (387)$$

where subscripts f and b stand for forward and backward, respectively

For solvolyses of *exo* and *endo* isomers of 2-norbornyl esters (RX) (388) yielding the same product-forming cation [as shown by the iden-

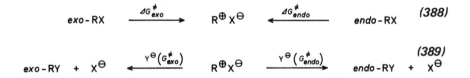

$$\text{exo-RX} \xrightarrow{\Delta G_{exo}^{\ddagger}} R^{\oplus} X^{\ominus} \xleftarrow{\Delta G_{endo}^{\ddagger}} \text{endo-RX} \qquad (388)$$

$$\text{exo-RY} + X^{\ominus} \xleftarrow{Y^{\ominus}\left(G_{exo}^{\ddagger}\right)} R^{\oplus} X^{\ominus} \xrightarrow{Y^{\ominus}\left(G_{endo}^{\ddagger}\right)} \text{endo-RY} + X^{\ominus} \qquad (389)$$

tity of product mixtures ($\%RY_{exo} : \%RY_{endo}$) obtained from both substances], Brown et al.[362] have shown that experimental product ratios can be calculated by means of (387) [as a difference (*endo* minus *exo*) and with $\Delta G_f^{\ddagger} = \Delta G^{\ddagger}$ (for RX → R$^+$X$^-$), $\Delta G_b^{\ddagger} = G^{\ddagger} - G_{R^+X^-}$, $\Delta G = G_{R^+X^-} - G_{RX}$] from the ratio of rates of ionization ($\rightarrow \Delta G_{exo}^{\ddagger} - \Delta G_{endo}^{\ddagger}$) and from ground state differences ($G_{endo\text{-}RX} - G_{exo\text{-}RX}$) [(389a), (389b), Table 83, and Fig. 83].

$$\Delta G_{endo}^{\ddagger} - \Delta G_{exo}^{\ddagger} - \left(G_{endo}^{\ddagger} - G_{exo}^{\ddagger}\right) = G_{exo\text{-}RX} - G_{endo\text{-}RX} \qquad (389a)$$

$$G_{endo}^{\ddagger} - G_{exo}^{\ddagger} = 2.3\, RT \log \frac{[exo\text{-product}]}{[endo\ \text{product}]} \qquad (389b)$$

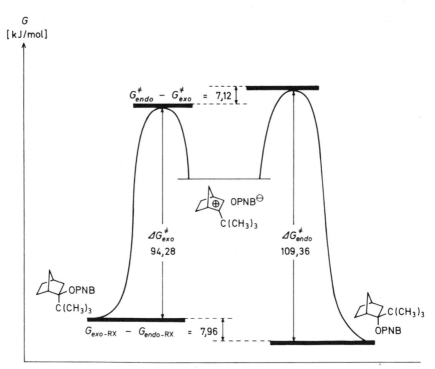

G
[kJ/mol]

$G^{\ddagger}_{endo} - G^{\ddagger}_{exo} = 7{,}12$

OPNB$^{\ominus}$
C(CH$_3$)$_3$

$\Delta G^{\ddagger}_{exo}$
94,28

$\Delta G^{\ddagger}_{endo}$
109,36

OPNB
C(CH$_3$)$_3$

C(CH$_3$)$_3$
OPNB

$G_{exo\text{-}RX} - G_{endo\text{-}RX} = 7{,}96$

Fig. 83 Free energy diagram (25°C) for the ionization of epimeric 2-*t*-butyl-2-norbornyl-*p*-nitrobenzoates [H$_2$O:acetone = 20:80; X (*388*) = OPNB].[362,363] [From E. N. Peters and H. C. Brown, *J. Amer. Chem. Soc.* **96**, 265 (1974) (ref. 362).]

The value $G^{\ddagger}_{endo} - G^{\ddagger}_{exo}$ for the reaction of the cation with water is thus practically the same as for its reaction with X⁻ [as described by (*387*), and (*389a*), derived from it]. According to pmr the transition state of the latter is identical to the transition state for ionization. Consequently, *exo:endo* product ratios are determined by the same properties of the transition state that control (together with $G_{exo\text{-}RX} - G_{endo\text{-}RX}$) the rates of ionization of the two epimers. This means that stereoselectivities in ionization and in product formation are *not independently* indicative of the potentially nonclassical character of the 2-norbornyl cation.

Comparison of data from the second and fourth columns in Table 83, suggests that steric effects responsible for the increase in ground state free energy of *exo*-RX are also present in the *exo* transition state and lead to a (somewhat) larger increase in $G^{\ddagger}_{exo}$. The *inherent* favoring of the *exo* transition state, so strongly visible in the unsubstituted

TABLE 83 Free Energies of Hydrolysis of *exo/endo* Isomers of 2-Norbornyl-*p*-Nitrobenzoates and Experimental and Calculated *exo/endo* Ratios of Products [kJ/mol, 25°C, see (*389a, 389b*)]

2-Norbornyl Group	Experimental			Calculated	
	$\Delta G^{\ddagger}_{endo} - \Delta G^{\ddagger}_{exo}$	$G_{exo\text{-}RX} - G_{endo\text{-}RX}$	$\% RY_{exo} : \% RY_{endo}$	$G^{\ddagger}_{endo} - G^{\ddagger}_{exo}$	$\% RX_{exo} : \% RX_{endo}$
(norbornyl, H)	18.86	−5.45	5,000	24.30	17,900
(norbornyl, CH$_3$)	16.76	−0.84	999	17.60	1,200
(norbornyl, *t*-C$_4$H$_9$)	15.08	7.96	19	7.12	17.6

secondary system (24.30 kJ/mol) could therefore exist in tertiary systems as well (albeit overshadowed by steric effects). However, it seems certain[364] that tertiary systems lead to classical 2-norbornyl cations and the same could then be true for the secondary system, where the strong favoring of *exo* processes was considered to be an argument for the formation of the nonclassical 2-norbornyl cation.

For multistep reactions pmr states that:

In the backward reaction of a multistep transformation, intermediates of the forward reaction are encountered in reversed order.

(One should realize that this means that in reactions of carbocations with nucleophiles, as well as in heterolyses of tosylates and similar substrates, which represent to some extent their reverse, the intermediates can be ion pairs; that is, the reaction $R^+ + Nu \rightarrow R\text{--}Nu^+$ need not be as simple as written above; see also Scheme 74).

If there are several competing reaction paths (I, II) leading from reactant(s) to product(s), the extent to which they manifest themselves must be the same in both directions (Scheme 68).

For *symmetric reactions* (e.g., isotope exchange), one is inclined (on the basis of pmr) to postulate symmetric reaction paths (Fig. 84a), since both the forward and the backward reactions proceed from the same state (if one ignores the isotope effect).

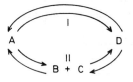

Scheme 68 Competition between two reaction paths (I, II) leading from A to D.

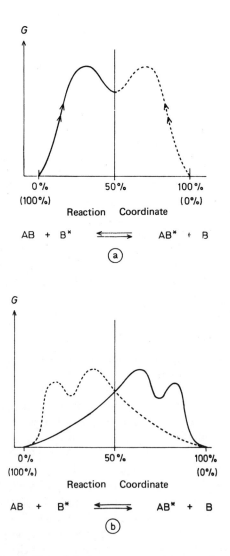

(a)

(b)

Fig. 84 Symmetric reactions (*a*) via *one* symmetric reaction path and (*b*) via *two* equivalent nonsymmetric reaction paths.

However, this is true only if a single reaction path is possible. If two reaction paths are possible, pmr allows any nonsymmetric course, provided the second path is the mirror image of the first with respect to a plane at the midpoint of the reaction coordinate (Fig. 84b). Since the tops of the highest activation barriers are equal for both paths, each is traversed by 50% of the reactants.[365] (Crossover from one path to the other at the "intersection" of the curves is not possible, since the reaction coordinates corresponding to the two curves traverse diastereomeric regions of the energy surface).

The free radical iodine isotope exchange in *cis*-diiodoethylene[366] (*390*):

$$I^* + \quad \substack{H \quad H \\ \diagdown \quad \diagup \\ \diagup \quad \diagdown \\ I \quad I} \quad \rightleftharpoons \quad I^* \substack{H \quad H \\ \diagdown \quad \diagup \\ \diagup \quad \diagdown \\ I \quad I} \cdot \quad \rightleftharpoons \quad \substack{H \quad H \\ \diagdown \quad \diagup \\ \diagup \quad \diagdown \\ I^* \quad I} \quad + \quad I \qquad (390)$$

(which is much faster than *cis-trans* isomerization) need not be as previously assumed a one-path reaction, which according to pmr must proceed via a planar (symmetric) triiodoethyl radical (**276**). (Of all possible comformers it is the only one possessing equivalent paths for the incoming I* and the outgoing I and does not offer any possibility for the direct formation of *trans*-diiodoethylene.) A reaction proceeding via diastereomeric tetrahedral radicals (**277a, 277b**) in which the incoming and outgoing I follow different paths, but which are also incapable of yielding the *trans* olefin does not violate pmr either (Fig. 85).

The pmr allows the elucidation of the mechanism of a reaction by studying the reverse reaction (e.g., when it is experimentally easier to perform). It has been shown that the decomposition of formocholine (Fig. 86) is *not* subject to general base catalysis; this allows the conclusion that the reverse reaction, namely, the addition of strongly basic amines to the carbonyl group (which is more difficult to study), does not occur under general acid catalysis.[367, /368/] A similar example is discussed in ref. 219a.

In the analysis of the mechanism of the Diels-Alder reaction (p. 435 and ff) the assignment of the enthalpy of formation of the transition state for the *formation* of the intermediate diradical A from two molecules of butadiene (*391a*) was based on the experimentally accessible enthalpy of formation of the transition state for the *decomposition* of a very similar diradical A' into two molecules of butadiene (*391b*).

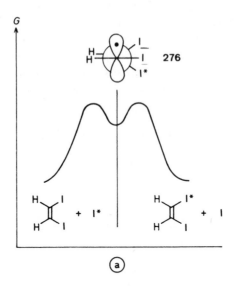

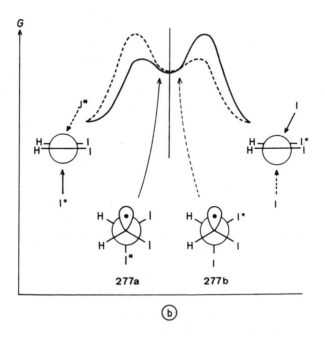

Fig. 85 Principle of microscopic reversibility in the iodine isotope exchange of *cis*-diiodoethylene. (*a*) One-path reaction involving a planar structure for the intermediate triiodoethyl radical. (*b*) Two path reaction via diastereomeric radicals.

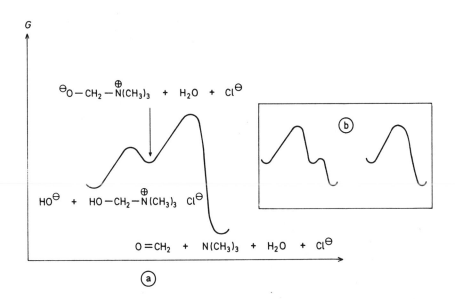

Fig. 86 Decomposition of formocholine chloride. (*a*) Specific base catalysis. (*b*) Possible free energy profiles for the *not observed* general base catalysis (composition of minima as in *a*).

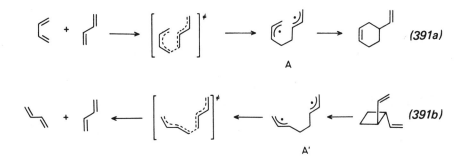

Rate constants for the dissociation of saturated hydrocarbons into radicals, for example, (*392*),

$$i\text{-}C_4H_{10} \underset{k_{\text{recombination}}}{\overset{k_{\text{dissociation}}}{\rightleftarrows}} CH_3\cdot + i\text{-}C_3H_7\cdot \qquad (392)$$

are difficult to extract from the complex kinetic data for the pyrolysis of paraffins. However, they can be easily calculated from the relationship $k_{\text{diss}} = K k_{\text{recomb}}$, since rate constants for recombination of simple alkyl radicals are known (Table 84) and enthalpies and entropies for hydrocarbons and their corresponding radicals are easily computed from increments available in tables,[43c] which allows the calculation of K.

The rate constant k_{AB} for recombination of nonidentical radicals A and B is given by (393):

$$k_{AB} = 2(k_{AA} \cdot k_{BB})^{\frac{1}{2}} \qquad (393)$$

where $k_{AA(BB)}$ = rate constant for recombination of two radicals A(B)

TABLE 84   Rate Constants for
Recombination of Identical Radicals[369]

Radicals	$\log k$ (l/mol s)
2 $CH_3\cdot$	10.34
2 $C_2H_5\cdot$	10.4
2 $C_3H_7\cdot$	10.4
2 $C_4H_9\cdot$	10.4
2 $i\text{-}C_3H_7\cdot$	9.9
2 $sec\text{-}C_4H_9\cdot$	9.9
2 $t\text{-}C_4H_9\cdot$	9.5
2 $C_3H_5\cdot$	9.0

5.2   QUALITATIVE STATEMENTS ABOUT ACTIVATED COMPLEXES

5.2.1   Molecules as Models; of Properties of Activated Complexes by Interpolation

Even when the complexity of the system prevents any calculative treatment (e.g., in reactions in condensed phases where intermolecular interactions are important), transition state theory can provide important qualitative statements with respect to changes in structure and energy along the reaction coordinate. Its two basic assumptions are essential in this respect: (a) the *similarity* of activated complexes and molecules makes it possible to use for activated complexes models and considerations proved to be successful in the interpretation of properties of normal molecules (e.g., mechanisms for substituent and solvent effects, electron delocalization, analysis of thermodynamic cycles); (b) the *continuity* of structural and energy changes in the system reactant(s)-activated complex-product(s) enables us to consider reactant(s) and product(s) as extreme models for the activated complex and to derive some properties of the activated complex from those of the models by interpolation. It is obvious that only known reactions can be analyzed in this fashion. One should not downgrade such a procedure because of its "a posteriori character," however, because this analysis makes it possible to *predict* the effect of *changes* in a given system on its reactivity.

For such analyses the following requirements have to be met:

1 An activated complex may incorporate only (interpolated) properties of a single reactant(s)-product(s) pair. Situations like the one depicted in Fig. 87 are considered to be unrealistic.[370]

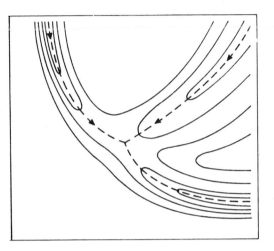

Fig. 87 Schematic representation of an energy surface in which two reaction coordinates (valleys) lead to the same transition state. [From M. R. Wright and P. G. Wright, *J. Phys. Chem.* **74**, 4394 (1970).]

2 The analysis must consider the properties of the two energy minima *directly* surrounding the rate determining transition state (situation a, Fig. 88). (Criteria defining this situation are discussed in Section 5.2.6.) For many reactions (e.g., ionization of RX) where the intermediate is very short-lived, this is impossible because the isolable end products result from subsequent transformations of the true primary product (situation b, Fig. 88). Conversely, if one postulates the involvement of an unstable, nonisolable intermediate in the rate-determining step, the properties considered for this intermediate must be to some extent measurable properties of the transition state ($H^{\ddagger}$, $S^{\ddagger}$, substituent and solvent dependence). This is the principle that allows to diagnose, for example, the existence of polar intermediates from the effect of solvent polarity on reaction rate.

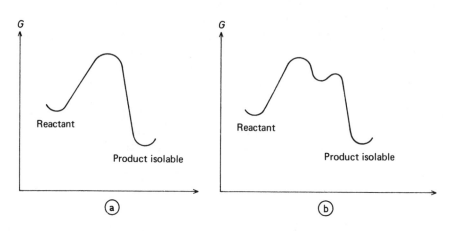

Fig. 88 Derivation of properties of transition states by interpolation from properties of reactant(s) and isolable product(s) is possible only in a one-step transformation (*a*).

An obvious way to envisage the transition state in terms of its *topology* is as a hybrid encompassing the properties of both the reactant(s) and product(s). In reactions proved to proceed in one step the topology of the transition state *must* lie "*between*" those of the reactant(s) and the product(s). How close it is to one of these two extremes cannot be ascertained without further assumptions (discussed in

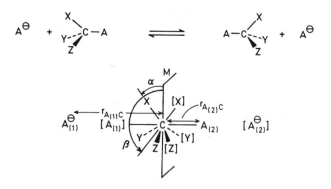

Fig. 89 Symmetry plane (M) and geometric parameters of a narcissistic S_N2 reaction. Positions of atoms after the reaction are in square brackets. Distances $r_{A(1)C}$ and $r_{A(2)C}$ for the *antisymmetric coordinate* $r_{A(2)C}$-$r_{A(1)C}$ are given for the initial state together with the angles (α between M and C–X and β between M and plane CYZ). The antisymmetric coordinate α-β can be constructed from the two angles. Antisymmetric coordinates have the value x *before* the reaction and $-x$ *after* the reaction.

Section 5.2.2.1.). A special case consists of "narcissistic reactions."[371] These are reactions in which reactant(s) and product(s) are mirror images with respect to a fixed plane of symmetry and where at least one anti-symmetric (with respect to this plane of symmetry) combination of coordinates of atoms participating in the reaction changes sign. The pertinent plane of symmetry can be made visible by superimposing reactants and products (Figs. 89 and 90).

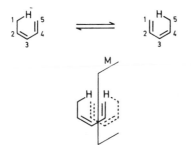

Antisymmetric Coordinates: $r_{1-H} - r_{5-H}$ and $r_{12} - r_{23}$ etc.

Fig. 90   Symmetry plane (M) and examples of antisymmetric coordinates ($r_{1-H} - r_{5-H}$, $r_{12} - r_{23}$, etc.) in a narcissistic 1,5-hydrogen shift.

When changes in bonding are highly synchronized all antisymmetric coordinates are strongly coupled as well; that is, they change in a nearly synchronous manner and it is highly probable that they traverse the value zero (the mirror plane) signaling the transition between "reactant side" to "product side" at the same time (see Figs. 91 and 92). Such narcissistic reactions have a symmetric transition state with respect to the mirror plane.

If however, one of the antisymmetric coordinates [e.g., $r_{12} - r_{13}$ ($r_{56} - r_{45}$)] in the narcissistic Cope rearrangement (Fig. 93) changes *faster* than the other [e.g., $r_{56} - r_{45}$ ($r_{12} - r_{23}$)] over large stretches of the reaction path, the symmetrical transition state can be bypassed even in a narcissistic reaction. In such cases two enantiometic paths are each traversed by 50% of the reactants (see also p. 315). In these instances the topology of the transition states (which are enantiomeric) is not known.

The symmetry postulate for synchronous narcissistic reactions simplifies considerably the calculation of their transition states; since the main features of the structure of the activated complex are given (all antisymmetric coordinates = 0) one can avoid a protracted searching of the energy surface for the saddle point and its definitive structure can be determined by energy minimization procedures just like in normal molecules.

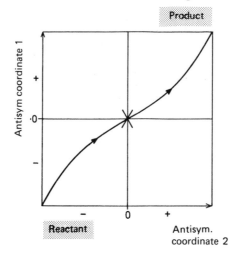

Fig. 91 Strongly concerted path of a narcissistic reaction. Symmetric transition state (x). [From L. Salem, *Acc. Chem. Res.* **4**, 322 (1971) (ref. 371).]

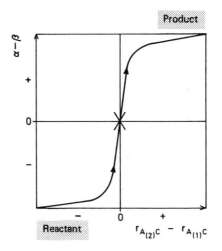

Fig. 92 Calculated concerted path of a narcissistic S_N2 reaction; actually the change in distance precedes the change in angle, but the transition state (x) is symmetric [see, e.g., A. Dedieu and A. Veillard, *J. Amer. Chem. Soc.* **94**, 6730 (1972). For the meaning of parameters, see Fig. 89.]

Using group theory McIver[371a] arrived at the conclusion that in multicenter reactions the probability of the lowest energy transition state being symmetric diminishes with increasing distance between the centers directly involved in bond making and/or bond breaking. (For

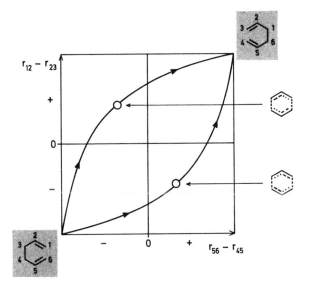

Fig. 93 Possible nonsynchronized course of a narcissistic Cope rearrangement with two enantiomeric reaction paths.

coupled changes of geometric coordinates on traversing the reaction coordinates, see also p. 303 and ff.)

To set *lower limits* for activation entropies in narcissistic gas phase reactions presumed to be concerted (*396*) and (*397*), Frey and co-workers[372] chose cyclic hydrocarbons as models for symmetric transition states. (Since partial bonds in transition states are replaced by full bonds in the models, these models correspond to more "compact" transition states, with less mobility and correspondingly lower entropy.)

Cope rearrangement

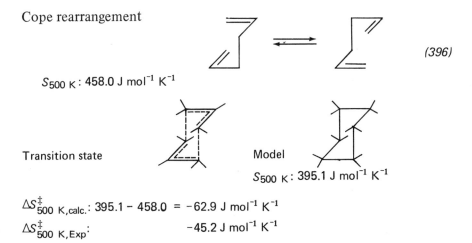

(396)

$S_{500 \text{ K}}$: 458.0 J mol^{-1} K^{-1}

Transition state   Model

$S_{500 \text{ K}}$: 395.1 J mol^{-1} K^{-1}

$\Delta S^{\ddagger}_{500 \text{ K,calc.}}$: 395.1 − 458.0 = −62.9 J mol^{-1} K^{-1}

$\Delta S^{\ddagger}_{500 \text{ K,Exp}}$:   −45.2 J mol^{-1} K^{-1}

1,5-Hydrogen shift

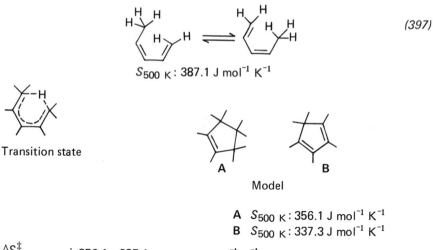

$S_{500 \text{ K}}$: 387.1 J mol^{-1} K^{-1}

Transition state

Model

A $S_{500 \text{ K}}$: 356.1 J mol^{-1} K^{-1}
B $S_{500 \text{ K}}$: 337.3 J mol^{-1} K^{-1}

$\Delta S^{\ddagger}_{500 \text{ K,Calc.}}$: 356.1 − 387.1 = −31.0 J mol^{-1} K^{-1}

337.3 − 387.1 = −49.8 J mol^{-1} K^{-1}

$\Delta S^{\ddagger}_{500 \text{ K,Exp}}$: = −45.2 J mol^{-1} K^{-1}

Steric influences of alkyl groups (R) on relative rates (respectively $\Delta\Delta G^{\ddagger}$) in acid-catalyzed hydrolyses of esters (*398*), (*399*) could be mimicked by the hypothetical transformations of hydrocarbons (*400*) and (*401*)[373]:

$$H_3C-COOC_2H_5 + H_2O \xrightarrow[\text{Model reaction}]{\Delta G^{\ddagger}_0} H_3C-\underset{\underset{\displaystyle OH}{|}}{\overset{\overset{\displaystyle OH}{|}}{C}}-OC_2H_5 \longrightarrow H_3C-COOH + C_2H_5OH \tag{398}$$

$$R-COOC_2H_5 + H_2O \xrightarrow{\Delta G^{\ddagger}_R} R-\underset{\underset{\displaystyle OH}{|}}{\overset{\overset{\displaystyle OH}{|}}{C}}-OC_2H_5 \longrightarrow R-COOH + C_2H_5OH \tag{399}$$

$$H_3C-CH(CH_3)_2 + CH_4 \xrightarrow[\text{Model reaction}]{\Delta G_0} H_3C-C(CH_3)_3 + H_2 \tag{400}$$

$$R-CH(CH_3)_2 + CH_4 \xrightarrow{\Delta G_R} R-C(CH_3)_3 + H_2 \tag{401}$$

$$-\Delta\Delta G^{\ddagger} = \Delta G^{\ddagger}_0 - \Delta G^{\ddagger}_R \approx \Delta G_0 - \Delta G_R = -\Delta\Delta G \tag{402}$$

Equations (*403*) and (*404*) show how even such very different molecules and reactions can indeed be used as models for reactant and transition state effects. In (*403*) and (*404*), *a* represents the effect on the free energy of replacing oxygen by methyl in the ground states and b represents the free energy difference between the product of the hydrocarbon transformation and the transition state for ester hydrolysis. The steric effect of replacing methyl by R should be different for ester and transition state.

$$\Delta G_0 = [(G_0^{\ddagger})_{ester} + b] - [(G_0)_{ester} + a] \qquad (403)$$

$$\Delta G_R = [(G_0^{\ddagger})_{ester} + b + \text{ster. effect}] - [(G_0)_{ester} + a + (\text{ster. effect})']$$
$$CH_3 \rightarrow R \qquad\qquad\qquad CH_3 \rightarrow R$$
$$(404)$$

When a and b are constant for all R's considered, subtraction of (403) and (404) gives (405):

$$\Delta\Delta G = \left(\frac{\text{ster. effect}}{CH_3 \rightarrow R}\right)_{\text{transition states}} - \left(\frac{\text{ster. effect}}{CH_3 \rightarrow R}\right)'_{\text{ester}} = \Delta\Delta G^{\ddagger}$$
$$(405)$$

One can conclude that in both systems reactivity is determined solely by steric effects and that the influence of R on the steric effect is so strong that replacement of oxygen ligands by methyl groups does not introduce any changes (a and b constant).

The substituent dependence of the free energy of transition states can even assume the form of a *linear combination* of the corresponding properties of reactants and products if both the forward and back reactions can be described by linear free energy relationships. An example is reaction (406, I $\rightleftarrows$ II)[374] for which one can write (407) and (408):

$$-1.23\,\sigma_X = \frac{-G_{(X)}^{\ddagger} + G_{(H)}^{\ddagger} - G_{(H)}^{I} + G_{(X)}^{I}}{2.3\,RT} \qquad (407)$$

$$0.63\,\sigma_X = \frac{-G_{(X)}^{\ddagger} + G_{(H)}^{\ddagger} - G_{(H)}^{II} + G_{(X)}^{II}}{2.3\,RT} \qquad (408)$$

Combining (407) and (408) and rearranging terms one obtains (409), showing the linear fashion in which the properties $G_{(X)} - G_{(H)}$ of product and reactant are blended to give $G_{(X)} - G_{(H)}$ of the transition state.

$$G^{\ddagger}_{(X)} - G^{\ddagger}_{(H)} = \alpha(G^{II}_{(X)} - G^{II}_{(H)}) + (1 - \alpha)(G^{I}_{(X)} - G^{I}_{(H)})$$

(409)

$$\alpha = \frac{1.23}{1.86} = \frac{\rho_f}{\rho_f - \rho_b}$$

Equation (409) is also valid for reactions obeying the Brönsted relationships (410),[375] in which case X and H represent variations of one of the two acid-base pairs and I and II, the left and right-hand sides, respectively, of (410). In this instance α is the Brönsted α for the corresponding reaction. [When reaction (410) obeys both the Brönsted and the Hammett equations for forward and backward reactions, $\alpha = \rho_f$: $(\rho_f - \rho_b)$.]

$$A^- + HB \underset{k_{-1}}{\overset{k_1}{\rightleftharpoons}} AH + B^-$$

(410)

The Brönsted coefficient α ($0 \leqslant \alpha \leqslant 1$) is then a direct measure of the extent to which the property of the activated complex, "free energy response to the structural change $H \rightarrow X$" resembles the same property in products (or reactants).

A farther reaching interpretation considers α not only as a measure for the change in that *one* particular property, but also as a measure of *"the position of the transition state along the reaction coordinate"* or for the relative "product-(reactant-)character" of the activated complex [100% (0%) for $\alpha = 1$; 0% (100%) for $\alpha = 0$]. (See also p. 358.) (A discussion of values of $\alpha(\beta)$ outside the limits $0 < \alpha(\beta) < 1$ (see p. 169) is given on p. 366).

An important consideration with respect to the values of α was first stated by Leffler.[375] This leads us directly to the topic of the next Section:

Since on traveling along the reaction coordinate, at the transition state the system traverses a free energy *maximum*, when two bases A^- and B^- compete for a proton, protonation is more advanced for the weaker base and not the stronger one, as is the case for stable states, which represent free energy *minima*.

If in our example A^- represents the weaker base, according to the conclusion above and the interpretation of α as the extent of progress along the reaction coordinate, α must have a value between 0.5 and 1.

5.2.2 Position of the Transition State on the Reaction Coordinate

5.2.2.1 The Hammond Postulate

The relationship between the properties of the transition state and those of reactant(s) or product(s), or in general, the position of the transition state on the reaction coordinate, either on the reactant or on the product side, may be expressed by a few simple statements. The properties considered are potential energy, structure (e.g., bond length, hybridization, electron delocalization, or charge distribution) or the dependence of the free energy on substituent, solvent, and reaction partner.

The simplest and most qualitative statement was proposed by G. S. Hammond[376] and is widely known as the "Hammond postulate." The basic idea is the *correspondence between energy and structure*: within limited regions of the energy surface *similar energies correspond to similar structures*. The Hammond postulate states that:

> When a system of atoms starting from its transition state (energy minimum) forms an energy minimum (transition state) of very similar energy, the structures of these two states should also be very similar.

The Hammond postulate may be applied to highly exothermal and highly endothermal elementary reactions. If *highly exothermal reactions* are fast,* their activation energies are low and the structures of reactant and transition state should be very similar. The reactant can serve as model for the transition state.

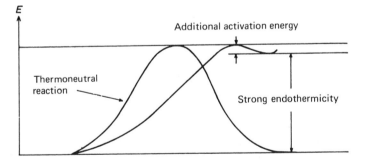

Fig. 94 The Hammond postulate. A strongly endothermal reaction proceeding (under similar conditions) at a rate similar to that of a neutral (or exothermal) reaction can have only a small excess activation energy.

*The Hammond postulate may not be applied to slow exothermal reactions. The Dewar benzene → benzene rearrangement[376a] is illustrative of such cases.

When highly endothermal reactions proceed under comparable conditions at *rates similar to those of thermoneutral* (or exothermal) processes they may possess only very low activation energies in excess of their endothermicity (Fig. 94).

In such instances, representing the microscopic reverse of the exothermal reactions mentioned above, the *product* may serve as model for the transition state.

In the region between the two extremes (e.g., thermoneutral reactions) conventional chemical species do not bear any analogy to transition states.

5.2.2.2 The BEP Principle

In its conclusions Hammond's postulate describes extremes of an analysis proposed by Bell[377] and Evans and Polanyi[378] (the BEP principle) dealing with relative reaction rates within series of analogous one-step reactions involving both bond making and bond breaking (substitutions, eliminations, rearrangements, and their reverse); in these series individual members differ mainly in their heat of reaction. The actual one-step reactions [e.g., (411)]

$$A + BC \rightarrow AB + C \tag{411}$$

are divided into hypothetical steps 1 and 2:

1 Dissociation of the breaking bond (without participation of the reaction partner) (412):

$$A + BC \rightarrow A + B + C \tag{412}$$

2 Recombination of one of the fragments with the reaction partner to reach the final state (413):

$$A + B + C \rightarrow AB + C \tag{413}$$

The solid curves in Fig. 95 show the energy change for both processes along a reaction coordinate stretching from 0 (no reaction) to 1 (complete transformation):

At its beginning the broken line representing the energy change in the actual one-step reaction (short dashes, Fig. 95) resembles closely the line of the hypothetical step 1, since the exothermal formation of the A—B bond can start only after a considerable endothermal distortion of the B—C bond. Having reached point X, where the systems A + B···C and A···B + C are of equal energy, the system does not follow

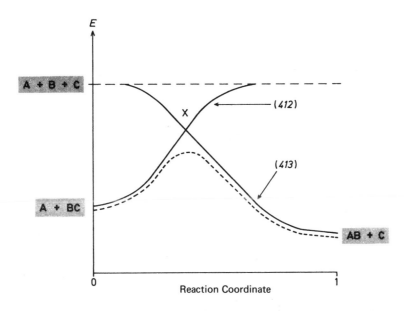

Fig. 95 The BEP principle. The one-step reaction A + BC → AB + C is dissected into two discrete steps: (1) BC → B + C and (2) A + B → AB.

path 1 leading to an even higher energy state A + B + C, but rather path 2 leading exothermally to the final stable state AB + C. Consequently, *X may be considered to be the transition state*, and the energy difference between X and A + BC, the *activation energy* of the one-step reaction.

If one applies the same analysis to a series of analogous reactions, where for instance reactants A_1, A_2, ..., A_i interact with BC to give A_1B, A_2B, ..., A_iB, the ascending curves (Fig. 95) will be the same for all systems, whereas the particularity of a given system will be expressed by the shape of the descending curve. If the shapes of descending curves are similar for the set of reactions and if in the vicinity of the crossover point the curves are close to linear, increasing exothermicity will shift X toward the reactant side and the activation energy will decrease (Fig. 96). For the more reactive substrate BC' curve 1 becomes more horizontal, and the effects are weaker. One may formulate two conclusions:

Within a *series of analogous reactions*

1 Increased exothermicity makes the transition state more reactant-like and the activation energy is lower.

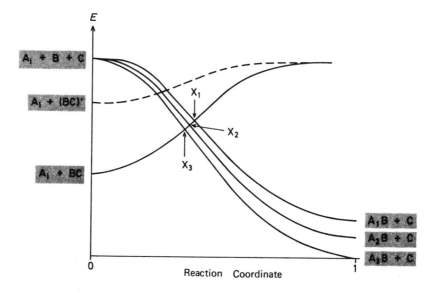

Fig. 96 Application of the BEP principle to a series of similar reactions in which reagent A is varied. The more reactive substrate (BC)' has the lower selectivity.

2 Increased reactivity of BC makes activation energies less responsive to variations of A, resulting in lower selectivity.

Finally it must be pointed out that according to the BEP analysis a *concerted path* (Fig. 95) requires a *lower* activation energy than a stepwise one through discrete reactions (*412*) and (*413*).

5.2.2.3 Use of Qualitative (Free) Energy Surfaces

It is possible to make more detailed statements with respect to structures of transition states, in particular regarding the *relative extent* to which changes in bonding occur during a transformation. One analyzes the properties of transition states of one-step processes in which *two* bonds are involved by means of the energetics of *two stepwise alternative reactions* where one of the bond changes is completed in the first step without changing the other bond involved in the overall process. In the case of (*414*)* the alternative stepwise reaction paths (*415*) and (*416*) are considered.

---

*As written above (*414*) is most suggestive of an $S_N 2$ reaction. If H is written instead of C it shows a proton transfer.[379,379a] Generally, (*414*) symbolizes all reactions (e.g., an E2 reaction)[378a] that can be dissected to yield two stepwise alternatives (e.g., E1 and E1cB) with intermediates whose degree of formation is characterized by parameters a and b, respectively (see below).

$$Nu^{\ominus} + C-L \longrightarrow Nu-C + L^{\ominus} \qquad (414)$$

$$Nu^{\ominus} + C-L \longrightarrow Nu-C-L^{\ominus} \longrightarrow Nu-C + L^{\ominus} \quad (415)$$

$$Nu^{\ominus} + C-L \longrightarrow Nu^{\ominus} + C^{\oplus} + L^{\ominus} \longrightarrow Nu-C + L^{\ominus} \quad (416)$$

The energetics of this system are shown in Scheme 69. (It is convenient to start with two pairs of reaction partners, of which one will be finally transformed into products and the other—after a temporary transformation—will be reformed at the end of the reaction sequence).

From Scheme 69 the heat of reaction (414) is:

$$E_{1,1} = E_A + E_B - E_{AB} \qquad (417)$$

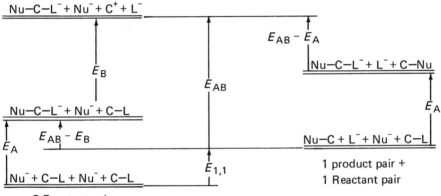

Scheme 69 Relationships between energies of reaction for individual steps in the stepwise reactions (415) and (416) and the energy of reaction of transformation (414):

E_A = energy of reaction for the addition step of (415)

E_B = energy of reaction for the heterolysis step of (416)

$-E_{AB}$ = energy of reaction for the reaction of the two intermediates:
Nu—C—L$^-$ + C$^+$ → Nu—C + C—L

$-(E_{AB} - E_B)$ = energy of reaction for the elimination of L$^-$ from Nu—C—L$^-$

$-(E_{AB} - E_A)$ = energy of reaction for recombination of C$^+$ and Nu$^-$

$E_{1,1}$ = energy of reaction for (414)

To obtain the energies of intermediate stages Critchlow[379] has made two simplifying assumptions:

1 In the formation of the two intermediates the energy should rise monotonously, as a linear function of the parameter representing the progress of the reaction [a for (415), b for (416), $0 \leqslant$ a, b $\leqslant 1$].

2 If the complete cleavage of L^- from $Nu-C-L^-$ ($b = 1$) sets free the amount of energy $E_{AB} - E_B$, then the cleavage of L^- from $Nu-C-L^-$ formed only to the extent a ($Nu..^a..C-L^-$) should release the amount $aE_{AB} - E_B$; partial cleavage of $Nu..^a..C-L^-$ to the extent b should release the amount $b(aE_{AB} - E_B)$ (Scheme 70).

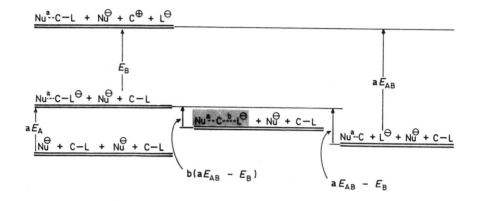

Scheme 70 Determination of the energy of the system $Nu ..^a.... C..^b.... L^-$ (relative to the initial state $Nu^- + C-L$). The parameter a ($0 \leqslant a \leqslant 1$) shows the extent of formation of the $Nu-C$ bond, parameter b ($0 \leqslant b \leqslant 1$) that of heterolysis of the $C-L$ bond.

Scheme 70 shows that the energy $E_{a,b}$ for points $Nu..^a.C..^b.L^-$ on the energy surface of the dual parameter reaction system is given by:

$$E_{a,b} = aE_A + bE_B - abE_{AB} \qquad (418)$$

$E_{1,1}$, E_A, and E_B are special cases of this general relationship. A numeric example is given in Fig. 97. Again, for the values of E_A, E_B, and E_{AB} used the concerted reaction path (over the saddle) is lower in energy than those of the two stepwise alternatives (along the sides of the square).

The coordinates ($a^\ddagger$, $b^\ddagger$) of the transition state (saddle point) are given by condition (419):

$$\left(\frac{\delta E_{ab}}{\delta_a}\right)_b = \left(\frac{\delta E_{ab}}{\delta_b}\right)_a = 0 \qquad (419)$$

$$a^\ddagger = \frac{E_B}{E_{AB}} \qquad (420a)$$

$$b^\ddagger = \frac{E_A}{E_{AB}} \qquad (420b)$$

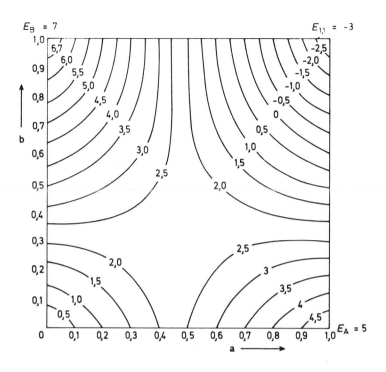

Fig. 97   Energy surface (in arbitrary units) $E_{a,b} = aE_A + bE_B - abE_{AB}$ for values $E_A = 5, E_B = 7, E_{AB} = 15$. The two parameters a and b $(0 \leqslant a, b \leqslant 1)$ indicate the extent to which the two bonds change as the reaction proceeds.

In the case shown in Fig. 97:

$$a^{\ddagger} = \frac{7}{15} \quad \text{and} \quad b^{\ddagger} = \frac{5}{15}$$

That is, *the energetically less demanding aspect of the total process is the most advanced in the transition state*. This result may be surprising in view of Hammond's postulate, but the latter cannot be applied to the individual steps of concerted processes.

The activation energy is given by

$$E^{\ddagger} = \frac{E_A E_B}{E_{AB}} \tag{421}$$

Since in practically all cases E_A, E_B, and E_{AB} are unknown, only qualitative applications of (420) and (421) are possible. (Most accessible are reactions where both partial processes involve proton transfer. Then E_A, E_B, and E_{AB} correspond to pK differences.[379]) It must also be pointed out that the analysis above considers only changes in bonding in the *encounter complex* with *optimally positioned* reaction partners.

The actual activation energy contains in addition the heat of formation E_i of the encounter complex. For members of a homogeneous series the values of E_i could be roughly equal.

Changes in reactivity on varying the system are due to the effect of these variations on E_A, E_B, and E_{AB}. Variations at the *periphery* of a concerted reaction system (e.g., changing the acid catalyst in the general acid-catalyzed hydrolysis of acetals, or the base in E2 reactions) have (as a first approximation) no effect on the other two characteristic energies, E_{AB} and E_B, so that for an E2 reaction satisfying the Brönsted equation one can write:

$$\beta = \frac{d \lg k_{(B^\ominus + HCCX)}}{d\, pK_{BH}} = \frac{d(-E_i - E^{\neq})}{2,3\, RT \cdot d\, pK_{BH}} \tag{422}$$

With $\Delta G = E_A = 2.3RT\, (pK_{HCCX} - pK_{BH})$ and $dpK_{BH} = -dE_A/2.3RT$ one obtains

$$\beta = \left(\frac{\delta(E_i + E^{\neq})}{\delta E_A}\right)_{E_B, E_{AB}} \tag{423a}$$

$$= \beta_i + \frac{E_B}{E_{AB}} \tag{423b}$$

$$= \beta_i + a^{\neq} \tag{423c}$$

If the pK dependence of $E^{\ddagger}$ is larger than that of E_i β (and α) may be again regarded as a measure for the degree of proton transfer in the transition state.

When variations occur at the periphery of a system reacting via a concerted pathway, linear free energy relationships (e.g., the Brönsted equation) describing the corresponding partial process (e.g., proton transfer) characterize the total process as well (e.g., an E2 reaction). However, if changes occur at the center of the system both partial processes are affected, so that if there are linear free energy relationships both for E_A and E_B (e.g., of the Hammett type: $E_{A(B)} = \rho\sigma + C; C = $ const), then

$$E_x^{\neq} = \frac{(C_A - \varrho_A \sigma_x)(C_B + \varrho_B \sigma_x)}{E_{AB}} \tag{424}$$

This expression can be readily transformed into a parabolic free energy relationship for the rate constant of the total process:

$$\log \frac{k_x}{k_\mathrm{H}} = a\sigma_x + b\sigma_x^2 \qquad (425)$$

In the preceding example we considered the common situation that E_A and E_B can be expressed by Hammett equations of opposite sign. In such instances, both electron-donating and electron-withdrawing substituents have an accelerating effect (to the left and to the right of the minimum on the $\rho\sigma$ plot, respectively). [Since lowering of, e.g., E_A results in lowering of $b^{\ddagger}$ (420), favoring of one partial process attenuates the unfavorable effect of the same variation on the total process due to the second partial process. (See also Section 3.3.1.6)]

The drawbacks of (421)—the unavailability of E_A, E_B, and especially E_AB and its being limited to the transformation within the encounter complex—were circumvented in a recent study by Gajewski.[379b] He applied (418) to *free energies* [read (418) with ΔG instead of E] of gas phase reactions. Furthermore, a new adjustable parameter p was introduced by assigning the reaction product the coordinates $a = b = p$, while keeping the conventional values $a = b = 0$, $G = 0$ for the reactant state and $a = 1$, $b = 0$, and $a = 0$, $b = 1$ for the respective intermediates of the two alternative stepwise processes. For gas phase reactions [e.g., the Cope rearrangement, whose possible stepwise modes are depicted in (426)], free energies of reaction for the overall transformation (ΔG) as well as for the conversion of the reactant into the two possible diradical intermediates (ΔG_A, ΔG_B) can be estimated relatively easily (see pp. 38 and 430). ΔG_AB was expressed by p (427a) (427b):

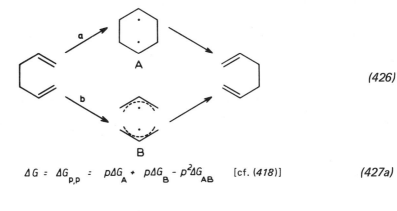

$$(426)$$

$$\Delta G = \Delta G_{p,p} = p\Delta G_\mathrm{A} + p\Delta G_\mathrm{B} - p^2\Delta G_\mathrm{AB} \qquad [\text{cf. } (418)] \qquad (427a)$$

$$\Delta G_\mathrm{AB} = \frac{\dfrac{-\Delta G}{p} + \Delta G_\mathrm{A} + \Delta G_\mathrm{B}}{p} \qquad (427b)$$

Substituting $(427b)$ into (421) [reading (421) with ΔG instead of E] yields

$$\Delta G^{\ddagger} = \frac{\Delta G_A \cdot \Delta G_B \cdot p}{\Delta G_A + \Delta G_B - \frac{\Delta G}{p}} \qquad (427c)$$

Remarkably, p appeared to be a constant with the value 1.5 for many Cope and Claisen rearrangements, thus allowing one to calculate substituent effects on these reactions through $(427c)$. Similarly, a constant value of $p = 1.6$ reproduced the experimental free energies of activation for the Diels-Alder reaction of various cyanosubstituted ethylenes with cyclopentadiene.

From $(427c)$ a crude indication can be obtained of when to expect a concerted or stepwise reaction, respectively. Using the values of p mentioned above and considering a reaction with $\Delta G = 0$, the condition $\Delta G_A = \Delta G_B$ leads to $\Delta G^{\ddagger} \approx 0.8 \, \Delta G_{A(B)}$; that is, the free energy of activation is lower for the concerted mode than for the stepwise ones. By contrast, for $\Delta G_A \gg \Delta G_B$ $\Delta G^{\ddagger} = 1.6 \, \Delta G_B$ is obtained. Now the stepwise path via the low energy intermediate is favored. Thus, a *thermodynamic* requirement for concerted behavior emerges: the two alternative intermediates must not differ greatly in free energy.[379c] [In addition, there can also be *quantum mechanical* requirements that must be fulfilled for concerted reactions to occur (see Section 5.2.4). For exergonic reactions ($\Delta G = -n \cdot \Delta G_A$; $n > 0$) concertedness is even indicated for $\Delta G_A \gg \Delta G_B$ if $n > p^2 - p$. For highly endergonic reactions $(427c)$ fails.]

5.2.2.4 The Harris and Kurz Analysis

Similar results were obtained by Harris and Kurz[380] (elaborating on a paper by Thornton[381]). They were considering the perturbation produced by a substituent on the potential energy of the transition state of an $S_N 2$ reaction

$$\overset{\alpha^-}{X} \ldots \ldots \overset{\beta^+}{CH_2 R} \ldots \ldots \overset{\gamma^-}{Y}$$
$$r_{XC} \qquad\qquad r_{CY}$$

($r_{XC(CY)}$:bond length; $\alpha(\beta, \gamma)$:partial charge).

Neglecting deformation vibrations and approximating X, Y, and CH_2R by mass points, the potential (V) of the unsubstituted system may be expressed by:

$$V = \frac{1}{2} k_{XC} \delta_{XC}^2 + \frac{1}{2} k_{CY} \delta_{CY}^2 + \beta \delta_{XC} \delta_{CY} \qquad (428)$$

Force constants k of both bonds and the coupling constant β are positive; the length $\delta_{XC}(\delta_{CY})$ is defined as $\delta_{XC} = (r_{XC}$ in any state*) $- (r_{XC}$ in the stationary state of the unsubstituted system).

The stationary state (of the unsubstituted system) $\delta_{XC} = \delta_{CY} = 0$ is an energy minimum in any case with respect to the symmetric stretching mode; however, with respect to the antisymmetric stretching mode this is true only when $k_{XC} + k_{CY} - 2\beta > 0$ (stable molecules). The alternative condition, $k_{XC} + k_{CY} - 2\beta < 0$ is indicative of a transition state, since the stationary state represents an energy maximum of the antisymmetric mode. [This can be easily shown by substituting $+a$ and $-a$ for δ_{XC} and δ_{CY}, respectively, in (428).]

If the introduction of a substituent results in a small perturbation, one can assume that its effect (P) on the potential is linear in δ_{XC} and δ_{CY} [(429)]:

$$P = m_{XC} \delta_{XC} + m_{CY} \delta_{CY} + C \qquad (429)$$

where $m_{XC(CY)}$ = proportionality factor
C = constant

For perturbation by an electron-attracting substituent at X: $m_{XC} < 0$, $m_{CY} > 0$, since primarily such a substituent reduces the negative polarization of X while at the same time increasing the positive polarization of the central carbon atom (but to a lesser extent). Consequently, stretching of the $X \cdots C$ bond is facilitated, whereas stretching of the $C \cdots Y$ bond becomes more difficult.

The potential energy of the perturbed transition state (V') is:

$$V' = \frac{1}{2} k_{XC} \delta_{XC}^2 + \frac{1}{2} k_{CY} \delta_{CY}^2 + \beta \delta_{XC} \delta_{CY} + m_{XC} \delta_{XC} + m_{CY} \delta_{CY} + C \qquad (430)$$

---

*This can be a nonstationary situation of the unsubstituted system, but also a stationary situation of a substituted system.

Partial differentiation with respect to δ_{XC} and δ_{CY} gives the coordinates δ'_{XC} and δ'_{CY} of the new extreme:

$$\delta'_{XC} = \frac{m_{CY}\beta - m_{XC}k_{CY}}{k_{XC}k_{CY} - \beta^2} \tag{431a}$$

$$\delta'_{CY} = \frac{m_{XC}\beta - m_{CY}k_{XC}}{k_{XC}k_{CY} - \beta^2} \tag{431b}$$

With appropriate values for m_{XC} and m_{CY} the formula states the following rule regarding the *direction* in which changes in substituents influence the structure of transition states ($\beta^2 > k_{XC} \cdot k_{CY}$):[380]

In any S_N2 transition state, an electron-withdrawing substituent in the nucleophile or leaving group will shorten (increase the order of) the reacting bond nearer to the substituent and lengthen (decrease the order of) the farther reacting bond.

Since an electron-withdrawing substituent in the nucleophile reduces the exothermicity of the substitution reaction, the rule for this case—stronger bonding of the nucleophile, weakening of the bond to the leaving group (i.e., shifting of the transition state on the reaction coordinate toward the product side)—agrees with the BEP principle. The same is true for the reverse situation: increased electronegativity of the leaving group increases the exothermicity, and the transition state is traversed earlier on the reaction coordinate. In the earlier analysis (Section 5.2.2.3) increased electronegativity of X results in an increase in E_A and in correspondingly higher values for parameter $b^{\ddagger}$ characterizing the extent to which bond C—Y is lengthened; however, in this simpler model bond $X \cdots C$ should remain unchanged.

Further statements (e.g., regarding the ratio $\delta'_{XC}:\delta'_{CY}$ and the more important ratio $\Delta n_{XC}:\Delta n_{CY}$) representing changes in bond orders* n are not possible, because they would require the knowledge of the ratios $m_{XC}:m_{CY}$ and $\beta:(k_{XC} \cdot k_{CY})^{1/2}$. These depend both on the

---

*The common methods for studying reacting bonds, like isotope effects (which depend not only on nuclear masses but also on force constants)[306] and substituent effects (which respond to formal charges) are primarily criteria for bond order and less for bond length. The relationship between bond lengths r_{ij}, bond orders n_{ij}, and force constants k_{ij} is: $r_{ij} = r_{ij(n_{ij}=1)} - 0.25 \ln n_{ij}$. Differentiation gives:

$$\frac{\Delta n_{XC}}{\Delta n_{CY}} = \left(\frac{n_{XC}}{n_{CY}} \cdot \frac{\delta_{XC}}{\delta_{CY}}\right) \quad ; \quad k_{ij} = n_{ij} k_{ij(n_{ij}=1)}$$

nature of X and Y and on the bonding in the unperturbed transition state.

Contrary to the prediction of effects of substituents at X and Y based exclusively on the assumption that the perturbation is small, predictions regarding the effects of substitution at the central carbon atom require special models for the estimation of m values.

5.2.3 Application of the Assumptions Regarding the Position of Transition States on the Reaction Coordinates

5.2.3.1 *Reaction Rates Affected by Relative Stabilities of Products*

The assumption that Hammond's postulate is valid for endothermic processes proceeding under normal conditions provides the basis for kinetic methods used to determine *bond dissociation energies* and *relative stabilities* of reactive intermediates. Because of the postulated small height of the "excess activation barrier" (Fig. 94) in such processes, activation enthalpies $\Delta H^{\ddagger}$ of unimolecular homolyses in the gas phase are equated to ΔH of reaction. Any stabilization of the *radicals* formed should manifest itself *fully* in the decrease in $\Delta H^{\ddagger}$. The examples given in Table 85 illustrate this point.

TABLE 85   Activation Enthalpies of Hydrocarbon Homolyses: On the Basis of Hammond's Postulate They Are Equated to Bond Dissociation Energies

Reaction	$\Delta H^{\ddagger} \approx \Delta H$ [kJ/mol]
$CH_4 \longrightarrow H_3C\cdot + H\cdot$	432
$\langle\rangle-CH_3 \longrightarrow \langle\rangle = CH_2 + H\cdot$	356
$H_3C-CH_3 \longrightarrow H_3C\cdot + \cdot CH_3$	361
$H_3C-CH_2-CH=CH_2 \longrightarrow H_3C\cdot + H_2C\dot=CH\dot=CH_2$	292

Lowering of $\Delta H^{\ddagger}$ by about 72.5 kJ/mol on substitution of H by C_6H_5 or $CH=CH_2$ corresponds to a decrease in bond dissociation energy as a result of the combined action of two factors:

1 Allylic or benzylic resonance (50-59 kJ/mol)[383] in products.
2 Formation of the stronger $C_{sp^2}-C_{sp^2}$ bonds in radicals stabilized by resonance.[384]

In homolysis of strained C—C bonds, in particular in *cyclopropane* and *cyclobutane*, comparison of heats of formation of the corresponding diradicals obtained by alternative methods with values obtained by summation of heats of formation of these strained rings and activation enthalpies of their homolysis showed that the heats of formation of the *transition states* for homolysis of cyclopropane and cyclobutane are, respectively, about 40 and 27 kJ/mol higher than those of the diradicals 385a. This situation is best understood if one considers the reverse reaction: whereas the highly exothermic recombination of, for example, two methyl radicals, does not require any activation energy because the only change is energy lowering due to bond formation, in the initial phase of the cyclization of diradicals to give strained ring compounds strain increases so fast that the increase in energy cannot be compensated by the energy lowering due to the still small extend of bond formation at this stage 385b. A similar situation pertains, when *bimolecular* recombination of radicals creates front-strain [reverse of (*287a*)]. See: C. Rüchardt et al., Angew. Chem. Int. Ed. *16*, 875 (1977).

The endothermic formation of ionic species [e.g., *carbocations* in solvolyses (*432*) or *carbanions* in base-catalyzed hydrogen isotope exchange reactions (*433*)] lends itself to less clear and rather qualitative interpretations:

$$\text{RX} \underset{(1-F)}{\overset{k}{\rightleftharpoons}} \text{R}^+\text{X}^- \xrightarrow{F} \text{product} \qquad (432)$$

$$\text{RD} + \text{B}^- \underset{(1-F)}{\overset{k}{\rightleftharpoons}} \text{R}^-\text{DB} \xrightarrow{F} \text{R}^-\text{HB} \longrightarrow \text{product} \qquad (433)$$

On one hand, the expression $2.3RT \log (k_{exp})_1/(k_{exp})_2$ is equal to $\Delta G_2^{\ddagger} - \Delta G_1^{\ddagger}$ for the crucial ionization step k only if the extent of internal return $(1 - F)$ is the same for both systems being compared; on the other hand, because of the importance of *solvation* and *association* effects, it is difficult to decide to what extent differences in stability calculated under the assumption $\Delta\Delta G^{\ddagger} = \Delta\Delta G$ for a given medium can be unequivocally ascribed to certain *structural factors*.

The experimentally determined value of the reaction enthalpy for the rearrangement *sec*-butyl cation → *t*-butyl cation in SO_2ClF at -80 to $-20°C$ (62 ± 2 kJ/mol) (Bittner et al.[386]) is remarkably close to the value determined by mass spectrometric methods for the same rearrangement in the gas phase. The conclusion was drawn that similar cations differ very little with respect to the extent of their *electrostatic* solvation and that the success of the cation theory as applied to aliphatic systems is probably due to the relatively small role of *specific* solvation effects in carbocationoid species.

The large contribution of solvation to the energetics of carbocation formation emerges from the analysis of the cyclic process for the hydrolysis of t-butyl chloride[387]:

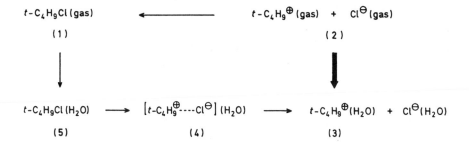

Scheme 71   Determination of the enthalpy of hydration of the t-butyl cation. State 4 is the transition state of the hydrolysis of t-butyl chloride.

The reaction enthalpy ΔH_{23} for the transformation of state (2) in Scheme 71 into state (3), that is, the enthalpy of hydration of state (3) is:

$$\Delta H_{23} = \Delta H_{21} + \Delta H_{15} + \Delta H_{54} + \Delta H_{43} \qquad (434)$$

According to Hammond's postulate $\Delta H_{43} = 0$ and using the following values for:

ΔH_{21} [enthalpy of formation of t-C_4H_9Cl
– enthalpy of formation t-butyl cation*
and chloride ion] $\qquad\qquad\qquad = -666\ \text{kJ/mol};$
ΔH_{15} [heat of solution of t-C_4H_9Cl in
H_2O – heat of vaporization of t-C_4H_9Cl] $= -17\ \text{kJ/mol}$
ΔH_{54} [activation enthalpy for the S_N1
hydrolysis of t-C_4H_9Cl] $\qquad\qquad = 97\ \text{kJ/mol}$

(434) gives $\Delta H_{23} \leqslant -586\ \text{kJ/mol}$

Subtracting the heat of hydration of chloride ion (–377 kJ/mol), the stabilizing hydration enthalpy of the t-butyl cation is:

$$\Delta H_{H_2O}\ (t\text{-}C_4H_9^+) \leqslant -209\ \text{kJ/mol}$$

*Obtainable from the heat of formation and the ionization potential of the t-butyl radical.

A similar analysis of the protonation of benzene gives for the hydration of the cyclohexadienyl cation an estimated value of

$$\Delta H_{H_2O} \left(\right) \leqslant -272 \text{ kJ/mol}$$

Along with quantum mechanical delocalization, solvation contributes significantly to the stabilization of this ion (and to the relative ease of electrophilic aromatic substitutions). In the gas phase charge distribution in the ion corresponds to maximum delocalization. In a polar medium it should be a compromise between stabilization by delocalization and stabilization by solvation. However, the latter is strongest when the positive charge is fully localized. It follows that charge distribution in the solvated ion should be less diffuse than in the gas phase. Application of quantum mechanical calculations of charge distribution and stability of such ionic species to reactions in condensed phases must take into account this point.

To determine relative stabilities of carbanions by means of kinetics of the base-catalyzed hydrogen exchange one should choose conditions for which the Hammond postulate is very likely to be obeyed. By using bases such as lithium cyclohexylamide or potassium t-butoxide, which are much weaker than the carbanions formed, one ensures that the reaction will be highly endothermic.

Conversely, when relative stabilities of (primary) products formed in an endothermic reaction are known from independent sources (e.g., see Section 3.1.1.4, reactivity indexes), the Hammond postulate allows the prediction of *relative reactivities*. The rules of aromatic substitution can be derived by applying the Hammond postulate: isomers derived from cyclohexadienyl intermediates most stabilized by the substituent at hand will be formed at the fastest rate.[387a]

The same can be done applying the BEP principle when the reaction is no longer strongly endothermic. Examples of enhanced reactivity as a consequence of increased product stability may be found in the aliphatic Claisen rearrangement.* The original version of this overall exothermic, synthetically important reaction (for stereoselective attachment of oxoethyl building blocks A to allyl building blocks B to give

---

*Comparison of activation energies of the Cope rearrangement of acyclic 1,5-hexadienes (E_A $ca.$ 145 kJ/mol) and the Claisen rearrangement of allylvinyl ethers (E_A $ca.$ 125 kJ/mol) shows the effect of the higher stability of the carbonyl products on the height of the activation barrier. Similar relationships are found in electrocyclic reactions. The activation enthalpy for the electrocyclic ring opening of oxetene is 100 kJ/mol,[387b] whereas that of cyclobutene is 136 kJ/mol. However both examples compare reactions in which atoms directly involved in bond reorganization are different. Consequently, it is impossible to decide to what extent the height of the activation barrier is affected by fundamental differences in properties of the transition states, in addition to the effect of product stability.

γ,δ-unsaturated carbonyl structures **C**, see p. 000) requires heating of the allylvinyl ether **D** (R$'$ = H or alkyl) at high temperatures for several hours (*435*).

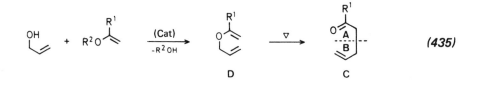

D   C   *(435)*

Although a direct comparison of reaction rates is impossible because the molecules are too different, data compiled in Table 86 clearly show that increasing stabilization of C by amide, ester, and carboxylate resonance (R$'$ = NR$_2$, OR, O⁻) results in a decrease in free energy of activation and allows the performance of rearrangements of sensitive substrates under mild conditions. (The first hypothetical step in a BEP analysis, endothermic homolysis of the C—O bond in **D**, is less affected by changes in R$'$ than the second, exothermic recombination of the two hypothetical radicals to give **C**, in which there is a stabilizing interaction between R$'$ and C=O).

Strong acceleration by the O⁻ substituent has also been found for the following types of reaction (*335a-335c*) see also ref. 387h):

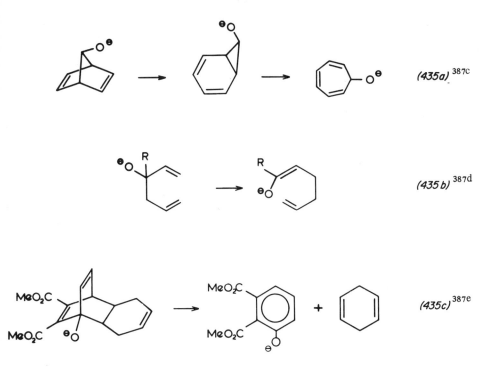

TABLE 86 Reaction Conditions in Aliphatic Claisen Rearrangements

R^1	Temperature (°C)	Reaction Time	Remarks	Reference
H, alkyl	190	15 min–4 h	—	A. W. Burgstahler and I. C. Norden, *J. Amer. Chem. Soc.* **83**, 198 (1961)
Alkyl	140	10 h	a	G. Saucy and R. Marbet, *Helv. Chim. Acta* **50**, 2091 (1967); K. Sakai, J. Ide, O. Oda, and N. Nakamura, *Tetrahedron Lett.* **1972**, 1287
$N(CH_3)_2$	140	20 h	These conditions refer to the dearomatizing (!) rearrangement of the benzyl analogue[a]	A. E. Wick, D. Felix, K. Steen, and A. Eschenmoser, *Helv. Chim. Acta* **47**, 2425 (1964)
OR	140	1 h	a	W. S. Johnson et al. *J. Amer. Chem. Soc.* **92**, 741 (1970)
O^-	25–70	10 min–3 h	b	R. E. Ireland and R. H. Mueller, *J. Amer. Chem. Soc.* **94**, 5897 (1972)

[a,b] **D** (*435*) rearranges *in situ* in the presence of the catalyst used (RCOOH, $Hg(OCOCH_3)_2$ (*a*) or Li-diisopropyl amide (*b*).

However, in these cases facilitation of the hypothetical first step [for (435b) concertedness has been proved] by O^{-387f} is also important besides the increase of exothermicity due to the transformation of an alkoxide into a stabler enolate or phenolate as in (435b) and (435c). A particularly striking example of type (435b) is the short route to nor-steroids shown below $(435d)^{387g}$:

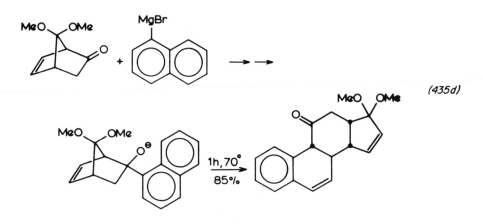

(435d)

The Cope rearrangement is also strongly accelerated when the O^- substituent in (435b) is replaced by C^+.[3871]

A further consequence of the effect of the product on reaction rates is the requirement of "*fair competition*," when comparisons of inherent stabilities of transition states of competing reaction paths are to be based on rate ratios: both paths should start from the same starting material and lead to the same final product. Scheme 72[388] describes such a situation, namely, the intramolecular cycloaddition in 5-allyl-cyclohexadiene-1,3.

The reaction product may arise either from a 2 + 2 or a 4 + 2 (Diels-Alder) cycloaddition. These reaction paths (as well as a third possibility—stepwise reaction via the diradical A) can be differentiated by deuterium labeling at the starred position. Analysis of the deuterium content at the corresponding positions in the product shows that

$$\left(\frac{k_{4+2}}{k_{2+2}}\right)_{184°} = 100 \qquad (436)$$

$$G^{\ddagger}_{4+2} - G^{\ddagger}_{2+2} = -17 \text{ kJ/mol} \qquad (437)$$

which means that the transition state of the 4 + 2 (Diels-Alder) cyclo-addition is at least 17 kJ/mol more favorable than the transition states of the two alternative reaction paths.

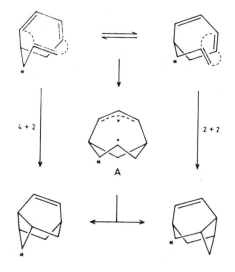

Scheme 72 Possible modes of formation of tricyclo[3.3.1.0²,⁷]non-3-ene from 5-allylcyclohexadiene-1,3. The reaction paths can be differentiated by labeling (*) of the starting material and determination of the position of the label in the product.

In common situations [e.g., (438)] :

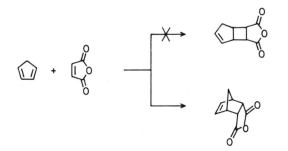

(438)

the nonoccurrence of the 2 + 2 addition is due not only to electronic factors (see p. 385) but also to the handicap of a less favorable product. Conversely, in interpreting the larger accelerating effect of a properly oriented cyclopropane ring [(in 279)* (440)] versus that of a double bond [(in 278); (439)] in solvolysis reactions one must take into account that the ionization 279 → 279⁺ is favored not only by potentially more favorable electronic factors (formation of a trishomocyclo-propenyl cation 279⁺ instead of a bishomocyclopropenyl cation 278⁺)

*One must point out the strong drop in reactivity when the electron cloud of the C—C bond in cyclopropane is turned away from the electrophilic center, as in iso-mer 280.

but also by the release of a larger amount of strain. All strained C—C
bonds in the original bicyclo[2.2.1]heptane skeleton are still present
in ion 278⁺, whereas in ion 279⁺ both the cyclopropane- and bicyclo-
[2.2.1]heptane strain present in the starting material 279 are
relieved.[390]

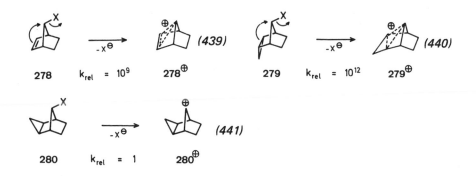

278	k_{rel} = 10^9	278⁺ (439)	279	k_{rel} = 10^{12}	279⁺ (440)

| 280 | k_{rel} = 1 | 280⁺ (441) |

The competition of *cis* and *trans* elimination of vinyl chloride to yield
acetylene might be considered another case of fair competition. However, in BEP analysis it is essential to look at the very first state in
which products have *just* become independent of each other. The
product state of *cis* elimination is then found to be potentially lower in
energy than that of *trans* elimation because of hydrogen bonding
between the chloride anion and the protonated base, which is, of
course, not possible on *trans* elimination.[467b]

The characteristic circumstance present in situations analyzed above
(i.e., the correspondence between the order of relative energies of
transition states and the order of relative energies of products) is sometimes called *"chemical noncrossing rule"* (i.e., of energy profiles along
the reaction coordinate[395a]). Deviations from this rule are observed in
transformations of highly reactive systems.

5.2.3.2 Highly Reactive Systems

According to Hammond's postulate, transition states of highly reactive
systems *must resemble the reactants*. Stabilizing or destabilizing interactions in products should be felt only to a very small extent in the
transition states; consequently, it may be assumed that (relative) stabilities of possible products will *not* manifest themselves in reaction rates
(or selectivities). The less stable of two possible products may be
formed faster if the *reactantlike* transition state leading to it is favored.
The decisive effects are primarily *electrostatic, steric and FMO* (p. 390)
interactions. Since reactantlike transition states are traversed at relatively large separations of reaction centers, energies involved in *bond
making* and *bond breaking* play only a subordinate role.

The increased tendency to form less stable enol derivatives (p. 156) observed with enolates whose reactivity is enhanced by use of dipolar aprotic solvents (p. 187) may be ascribed to the increased reactantlike character of the transition states.

What becomes important is the electrostatic interaction between the centers of opposite charge in the reaction partners. The transition state in which the positive pole of the attacking electrophile is turned toward the oxygen atom becomes increasingly favorable (Scheme 73).*

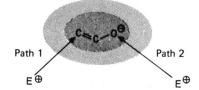

Path 1   Path 2

E $\oplus$ E $\oplus$

Scheme 73 The ambient reactivity of enolates with electrophiles (E$^+$). **Product-like transition states** (dark region): Path 1 is more favorable than path 2 because of the strength of the C=O bond being formed. **Reactantlike transition states** (light region): path 2 is more favorable than path 1 because the electrostatic energy of the transition state of path 2 is lower (the distance between E$^+$ and O$^-$ is smaller). (Of the two most important resonance structures of the enolate ion, only the dominant one is shown.)

The following examples [(442)-(444), and Table 87] show that reactions of delocalized carbanions with electrophiles occurring predominantly at or close to the *center of the negative charge* are the rule, even if this leads to the least stable of the possible products (see also refs. 191c, 390a and 467b):

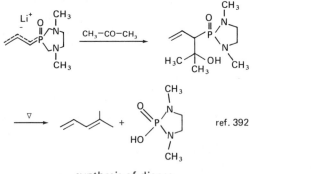

(442)

ref. 392

· synthesis of dienes

*According to Gomper and Wagner[191c] the effect of the solvent on the ratio of *O-* to *C-alkylation* is due not only to the shifting of the transition state on the reaction coordinate as a result of increased reactivity, but also to the increased polarity of the enolate ion caused by the dipolar aprotic solvent (i.e., to the difference in effective partial charges on the carbon and oxygen atoms).

TABLE 87   Carbon-13 Magnetic Resonance (CMR) Shifts[172a] and Reactions of Pentadienyl Anions[a, 394]

Anion	% Reaction at Central Carbon Atom with		
	H_2O	CH_3I	Ethylene Oxide
138,7 / 73,2 65,0	10	35	65
146,9 / 92,1 / 64,2 / 10,2			
92,1 / 141,8 / 83,6			77
30,8 31,4 / 91,7 / 127,9 / 78,1	60	100	
98,9 / 35,6 / 134,5 / 71,3	25		100
(eight-membered ring)	50	100	100

[a] As lithium salts. The numbers represent ^{13}C chemical shifts (in ppm) downfield from TMS. Values for even carbon atoms are in the region normal for *n*-alkenes. The stronger shielding of the odd carbon atoms shows their partial negative charge.

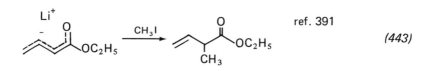

ref. 391

(443)

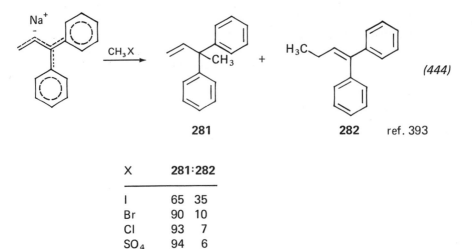

281   **282**   ref. 393

X	281:282	
I	65	35
Br	90	10
Cl	93	7
SO$_4$	94	6

A different view considers the selectivities of highly reactive systems to arise from the *principle of least nuclear motion*,[467b] which is dealt with in Section 5.2.5.

In accord with the assumption that in enolate reactions transition states are reactantlike is the observation that the reaction of the lithium enolate of 4-*t*-butylcyclohexanone with methyl iodide (*445a*) yields *cis* and *trans* products in almost equal amounts,[395] not predominantly the product that can be formed *directly* in the more favorable chair form, as is the case in less exothermal additions to half-chair cyclohexenes (see p. 26).

The example on pages 160 ff and 250 ff shows that in reactions of delocalized positive species with negatively charged reaction partners (B⁻) "*charge control*" plays a greater role than "*product stability control*" (the olefinic products are more stable than the isomeric cyclopropane derivatives); the latter prevails when the cations react with the corresponding (less reactive) conjugate acid (BH).

An example of charge control on protonation of a (nonalternant) reactive cyclopolyene (*445b*) was described by Haselbach.[395a]

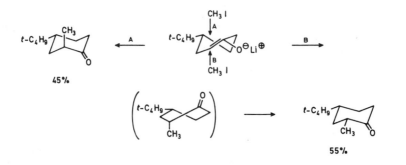

(*445a*)

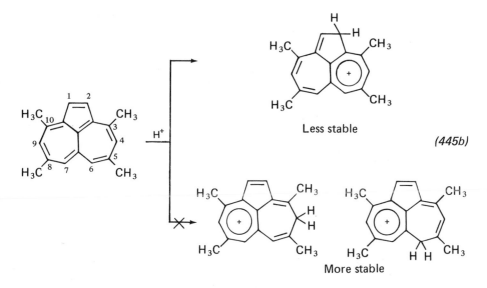

(445b)

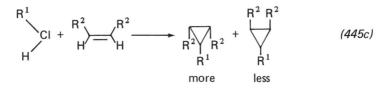

(445c)

Calculations show that in the starting material the electron density is greater at C-1(2) than at C-4(9) and C-6(7).

A further manifestation of the reactantlike character of transition states of highly exothermal addition reactions is the observation that when an unsymmetrically substituted carbene adds to a *cis* olefin (*445c*) the major product is the one resulting from a *sterically unfavorable* alignment of the initial reaction partners.

Possible reasons for this *steric attraction* between ligands facing each other are[395b]:

1 Favorable interactions between orbitals of these ligands in the transition state (see also secondary orbital interactions Section 5.2.4.6).
2 Van der Waals attractions in the transition state.
3 Attractions between dipoles arising in the transition state.
4 An interplay of factors 1-3.

When the distance between the reaction partners in the transition state becomes smaller, the interactions above are overshadowed by the strong repulsion between the doubly occupied molecular orbitals of the

individual reaction partners. As a result, the formation of the alternative transition state leading to sterically more favorable products
becomes increasingly important.

5.2.3.3 *Reacitivity-Selectivity Relationships*[395c, 395d]*: Interpretation of ρ, α (β) Values and of Primary Kinetic Isotope Effects*

The analysis of relationships between energy and structure of transition states and of energetics of reactions described in Sections 5.2.1-
5.2.2.4 may be considered as basis of the *reactivity-selectivity relationships* and of the "demand-dependent" substituent effects mentioned in Section 3.3.3.2. If, for instance, the lowering of the activation energy to be expected with increasing exothermicity affects each
of the competing reaction paths of a given reagent to the same extent,
$\Delta\Delta G^{\ddagger}$ will also be lowered correspondingly and the value of the competition constants will approach unity ($\Delta\Delta G^{\ddagger} = 0$).

The susceptibility factors (reaction constants) of linear free energy
relationships (α, β, ρ, s, m), reflecting the selectivity of a reagent or
substrate are often considered as measures for the *relative reactant-
product character*. Indeed at first glance certain *reactivity-selectivity
relationships* seem to confirm Hammond's postulate. For instance, the
decrease in absolute ρ values with increase in reactivity (and instability)
of the electrophilic reagent in electrophilic aromatic substitutions
(Table 88) may be considered to be indicative of decreased "positiva-
tion" of the transition states leading to the intermediate cyclohexadienyl cation and therefore of their increased reactantlike character.
However, to equate the percentage of the charge +1 "felt" by the substituent with percentage of product (reactant) contribution to the
corresponding transition state can be dangerous: for example, on comparing acetylation and ethylation it is conceivable that both transition
states may be found to have comparable positions on the two reaction
coordinates and that other factors may be responsible for differences in
absolute values of ρ.

The absolute values of ρ may be increased, for instance, by the
stronger electrophilicity of $AlCl_3$ compared to $GaBr_3$, or by the higher
electrophilicity of the CH_3CO group, if one considers the effects of the
entering groups on the cyclohexadienyl cation. Regarding the extent
to which charge localization determines the polar substituent effect one
must assign some unknown role to steric effects, solvent effects, and
association (related to R, M, and X) in the ion pair **A**.

                                                                          A

TABLE 88 ρ Values of Electrophilic Aromatic Substitutions[117]

Reaction	Conditions	ρ
Bromination	Br_2, HOAc–H_2O, 25°C	-12.1
Chlorination	Cl_2, HOAc, 25°C	-10.0
Acetylation	H_3CCOCl, $AlCl_3$, $C_2H_4Cl_2$, 25°C	-9.1
Bromination	HOBr, $HClO_4$, 50% dioxane, 25°C	-6.2
Ethylation	C_2H_5Br, $GaBr_3$, ArH, 25°C	-2.4

Transmission of charge effects is probably also influenced by the nature of R, M, and X. Finally, any interpretation of $|\rho|$ in terms of reactivity-selectivity relationships is very doubtful because *absolute* reaction rate constants and the temperature dependence of reaction constants ρ are mostly unknown.

Figure 53 shows reactivity-selectivity relationships in carbocations; qualitatively they are in agreement with our concepts about stability of carbocations. However, several factors may make these correlations fortuitous:

1 High values of log k may be due to high energies of ground states [see Table 72 and (440)], not to a special stability of carbocations.
2 The occurrence of several types of ion pairs may make competition constants of "carbocations" concentration dependent (see p. 89).
3 Diffusion control sets an upper limit on values of rate constants. On reaching this limit the more reactive of two competing nucleophiles may not achieve the higher value of its reaction rate appropriate to its reactivity; this will result in lower values for competition constants and lower selectivities. Data in Table 89 show that reactions of carbocations with strong nucleophiles may indeed proceed under diffusion control: whereas values of competition constants involving one very reactive and one less reactive nucleophile (k_{Cl^-}/k_{H_2O} and $k_{N_3^-}/k_{H_2O}$) are considerable, the values $k_{Cl^-}/k_{N_3^-}$ for 4,4′-dimethylbenzhydryl chloride, benzhydryl chloride, and t-butyl chloride are only 2.5, 0.7, and 1; this shows that the reactions of cations formed from these compounds with the two reactive nucleophiles may be diffusion controlled (Table 89).[396a]

Giese[397] described conditions under which the *temperature dependence* of competition constants would make it impossible to discuss reactivity-selectivity relationships because such temperature dependence could actually *reverse* the order of relative selectivities: if both competition reactions (1) and (2) of a variable substrate [e.g., A in (446)] obey the isokinetic relationship (330), and if both reactions

TABLE 89   Competition Constants for Reactions of Carbocations with Pairs of Nucleophiles Cl^-/H_2O (hydrolysis of RCl) and N_3^-/H_2O[396]

RCl	$k_{Cl\ominus}/k_{H_2O}$	$k_{N_3\ominus}/k_{H_2O}$	$k_{Cl\ominus}/k_{N_3\ominus}$
Triphenylmethylchloride	3100	280,000	0.01
4,4′–Dimethylbenzhydrylchloride	600	240	2.5
4–Methylbenzhydrylchloride	320		
Benzhydrylchloride	120	170	0.7
t–Butylchloride	4	4	1

$$\begin{array}{ll} & \xrightarrow{\;k_1\;} \quad \text{Reaction (1)} \\ & \\ & \xrightarrow{\;k_2\;} \quad \text{Reaction (2)} \end{array}$$

(446)

$$\delta_x \Delta H_1^{\ddagger} = \beta_1 \delta_x \Delta S_1^{\ddagger} \qquad (330\text{–}1)$$

$$\delta_x \Delta H_2^{\ddagger} = \beta_2 \delta_x \Delta S_2^{\ddagger} \qquad (330\text{–}2)$$

therefore show linear free energy relationships with the same variable parameters of A, variations in their activation data can be correlated as well. Dividing *(330-1)* by *(330-2)* and rearranging terms enables separation of variables yielding *(447a)* and *(447b)* ($\alpha_H = c/\beta_1$; $\alpha_S = c/\beta_2$):

$$\delta_x \Delta H_2^{\ddagger} = \alpha_H \delta_x \Delta H_1^{\ddagger} \qquad (447a)$$

$$\delta_x \Delta S_2^{\ddagger} = \alpha_S \delta_x \Delta S_1^{\ddagger} \qquad (447b)$$

The proportionality factors α_H and α_S reflect the relative sensitivities of reactions (1) and (2) toward δ_X. These can be different for enthalpy and entropy (see p. 281 ff).

Selectivity values [e.g., *(448)*] meeting in *one point* obey condition *(449)*

$$\lg \left(\frac{k_1}{k_2}\right)_x = (\rho_1 - \rho_2)\sigma_x \qquad (448)$$

$$\delta_x \lg \frac{k_1}{k_2} \equiv 0 \qquad (449)$$

Application of the Eyring equation to *(449)* gives *(450)*. From here one obtains the isoselective temperature T_{is}, where selectivity becomes *independent* of variations in A (*451*):

$$\delta_x \lg \frac{k_1}{k_2} \equiv 0 = \frac{\delta_x \Delta H_2^{\ddagger} - \delta_x \Delta H_1^{\ddagger}}{2,3\, R T_{is}} - \frac{\delta_x \Delta S_2^{\ddagger} - \delta_x \Delta S_1^{\ddagger}}{2,3\, R} \qquad (450)$$

$$T_{is} = \frac{(\alpha_H - 1)\, \delta_x \Delta H_1^{\ddagger}}{(\alpha_S - 1)\, \delta_x \Delta S_1^{\ddagger}} = \frac{\alpha_S (\alpha_H - 1)\, \delta_x \Delta H_2^{\ddagger}}{\alpha_H (\alpha_S - 1)\, \delta_x \Delta S_2^{\ddagger}} \qquad (451a)$$

$$T_{is} = \frac{(\alpha_H - 1)}{(\alpha_S - 1)}\, \beta_1 = \frac{\alpha_S (\alpha_H - 1)}{\alpha_H (\alpha_S - 1)}\, \beta_2 \qquad (451b)$$

The isoselective temperature T_{is} becomes identical to the isokinetic temperatures β_1 and β_2 only when $\alpha_S = \alpha_H$ and in most cases lies outside the measurable region: a reversal of selectivity orders due to changes in temperature does not occur in such cases. However, if the change in *activation entropy* is larger than the change in *activation enthalpy*, the isoselective point shifts to lower temperatures and may fall within the measurable region. An example of such a situation is the reaction of carbon radicals [which can be obtained even at $-20°C$ from organomercury salts and $NaBH_4$ (*452*)-(*454*)] with $BrCCl_3$ and CCl_4 [(*455*), Fig. 98].[398] Additions of dihalocarbenes (CF_2, CCl_2, CBr_2) to acyclic olefins also are largely governed by activation entropies.[398 a]

$$R-Hg-X \xrightarrow{\text{NaBH}_4} RHg-H \qquad (452)$$

$$R-Hg-H \longrightarrow R-Hg\cdot\ +\ H\cdot \qquad (453)$$

$$R-Hg\cdot \longrightarrow R\cdot\ +\ Hg \qquad (454)$$

$$R\cdot \quad \begin{array}{l} \xrightarrow{k_{Br}\,[Br-CCl_3]} RBr\ +\ \cdot CCl_3 \\ \xrightarrow{k_{Cl}\,[CCl_4]} RCl\ +\ \cdot CCl_3 \end{array} \qquad (455)$$

Figure 98 shows that the *isoselective relationship* (*451*) offers possibilities for the characterization of the reacting species: *for reactants (or reactive intermediates) of the same structural types T_{is} should be the same.*

Because the proportionality factors α_H and α_S depend on the isokinetic temperatures β_1 and β_2 of the two competing reactions (*456*)

$$\frac{\alpha_S}{\alpha_H} = \frac{\beta_1}{\beta_2} \qquad (456)$$

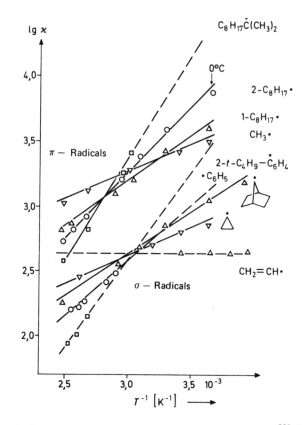

Fig. 98   Reversal of the selectivity order for some π and σ radicals[398] [$\kappa = k_{Br}/k_{Cl}$, see (455); π radicals contain the odd electron in a $2p$ orbital; σ orbitals contain the odd electron in an orbital with sp^2-character]. [From B. Giese, *Angew. Chem. Int. Ed.* **15**, 173 (1976) (ref. 398).]

the important situation, unequivocally yielding relative selectivities $\alpha_H = \alpha_S$, can occur only when $\beta_1 = \beta_2$. Consequently, *only very similar (competition) reactions may be discussed in terms of reactivity-selectivity*. [Even the difference between bromine and chlorine abstraction in the competing system (455) results in a reversal of the selectivity order between 40 and 80°C.]

    This condition of closest possible similarity between reactions to be compared [i.e., invariance—to the greatest extent possible—of essential properties of transition states within two (or several) reaction series] seems to be best realized in reaction types (457) and (458). In this instance the reaction center for all reaction series is the same, reactivities may be altered by changing substituents in one of the two phenyl rings, whereas substituent changes in the second ring make it possible to study eventual changes in selectivity caused by variations in reactivity.

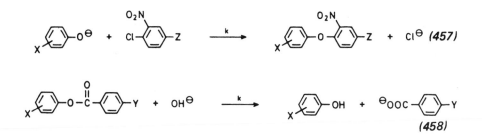

Tables 90 and 91 show that reaction series differing exclusively in their reactivity have practically the same ρ, in spite of substantial differences in reaction rates between the individual series!

TABLE 90 Rate Constants k ($1\ mol^{-1}\ s^{-1}$) and ρ Values for Two Reaction Series of Type (457)[399]

Z	H	p-CH$_3$	m-OCH$_3$	p-Cl	m-Cl	m-NO$_2$	p-NO$_2$	ρ
				X =				
NO$_2$	10 k	1,62	3,43	1,19	0,956	0,446	0,177	0,00845	−1,8
Cl	1,51·10^6 k	1,62	3,33	1,10	0,985	0,486			−1,8

TABLE 91 Rate Constants k ($1\ mol^{-1}\ min^{-1}$) and ρ Values for Some Reaction Series of Type (458)[400]

X	Y	p-N(CH$_3$)$_2$	p-CH$_3$	p-H	p-Cl	p-NO$_2$	$\rho_{Y(X=const.)}$
p-NO$_2$		0,897	20,95	48,40	109,6	1596	2,01
m-NO$_2$		0,5086	11,22	27,57	68,1	986	2,04
p-Cl			3,072	7,453	18,33	272,9	2,04
p-H		0,0725	1,41	3,345	10,50	117,7	1,98
p-CH$_3$		0,0478	0,836	1,993	5,83	89,7	2,03
$\rho_{X(Y=const.)}$		1,20	1,28	1,28	1,18	1,25	

$$[(457) : k_{Z=NO_2}/k_{Z=Cl} \approx 10^5, \qquad \rho = -1,8;$$
$$(458) : k_{Y=p-NO_2}/k_{Y=p-N(CH_3)_2} \approx 2\cdot10^3, \quad \rho = 1,2.]$$

From such findings one might conclude that reactivity and selectivity are *mutually independent* manifestations of special properties of a reacting system.[401,401a] The constant value of the selectivity log k_{Nu}/k_{H_2O} = N$_+$(Nu) for stable carbocations of different reactivities described on page 185 seems to confirm this.

Within individual reaction series an increase in reactivity does not result in reduced selectivity and curving of the correlation plot. In fact, invariance of ρ, α, etc. is the essential element of the Hammett equation, the Brönsted equation, etc. even in reaction series stretching over such a wide range of reactivities ($\Delta\Delta G^{\ddagger}$ up to 60 kJ/mol)[401,401a] that according to Hammond's postulate, or the BEP principle, one could expect considerable differences in positions of transition states on the reaction coordinates of individual members of the series.

Two alternative interpretations are possible:

1 The numerical value of ρ is *not* a measure of the changing position of the transition state along the reaction coordinate.

2 The position of the transition state along the reaction coordinate is the same for all members of a reaction series; that is, reactions obeying the Hammett equation (and other linear log-log correlations) are *not* covered by Hammond's postulate and the BEP principle.

Neither alternative allows the interpretation of lower $|\rho|$ values in electrophilic substitutions of reactive heterocycles (mentioned on p. 116, where these low values were attributed to deviations of substituent effects from additivity caused, probably, by decreased transmission coefficients) as a manifestation of a more reactantlike character (compared to benzene derivatives).[402]

An insight into the reasons for the discrepancy between the nonlinearity of log-log correlations expected on the basis of Hammond's postulate and the BEP principle and their actual extensive linear behavior may be gained from an analysis of Brönsted relationships (204)[403]:

The statements of Hammond's postulate (extended to free energy) namely:

1 In strongly endothermic reactions a change in free energy in the product state affects to the same extent the transition state leading to it.

2 In highly exothermic reactions variations in product stability have no effect on the reactantlike transition state leading to that product.

may be formulated as limiting values of α:

$$\text{(a)} \quad \Delta G = +|\Delta G_{max}| : \quad \alpha = \frac{d\Delta G^{\ddagger}}{d\Delta G} = 1 \qquad (459)$$

$$\text{(b)} \quad \Delta G = -|\Delta G_{max}| : \quad \alpha = \frac{d\Delta G^{\ddagger}}{d\Delta G} = 0 \qquad (460)$$

Assuming that between these two limits α is a linear function of ΔG (461) [this assumption is based on an analysis[403] of (409)]

$$\alpha = a\,\Delta G + b \qquad (461)$$

one obtains

$$\alpha = \frac{1}{2\,|\Delta G_{max}|} \cdot \Delta G + \frac{1}{2} \qquad (462)$$

Substitution of (462) in (409) gives after rearrangement of terms $(dG^{\ddagger} - dG_{react} = d\Delta G^{\ddagger})$ and integration (463):

$$dG^{\ddagger} = \alpha\, dG_{Prod.} + (1 - \alpha)\, dG_{react.} \qquad (409)$$

$$\Delta G^{\ddagger} = \frac{\Delta G^2}{4\,|\Delta G_{max}|} + \frac{1}{2}\,\Delta G + \Delta G_o^{\ddagger} \qquad (463)$$

where $\Delta G_0^{\ddagger}$ is the free activation energy for $\Delta G = 0$ and $\alpha = \frac{1}{2}$. It results from the two limiting conditions:

$$\Delta G^{\ddagger}_{\alpha = 1} = \Delta G_{\alpha = 1} = +|\Delta G_{max}| \qquad (459a)$$

$$\Delta G^{\ddagger}_{\alpha = 0} = 0, \; \Delta G_{\alpha = 0} = -|\Delta G_{max}| \qquad (460a)$$

$$\Delta G_o^{\ddagger} = \frac{1}{4}\,|\Delta G_{max}| \qquad (464)$$

Substitution of (464) in (462) and (463) gives

$$\alpha = \frac{\Delta G}{8\,\Delta G_o^{\ddagger}} + \frac{1}{2} \qquad (465)$$

$$\Delta G^{\ddagger} = \frac{\Delta G^2}{16\,\Delta G_o^{\ddagger}} + \Delta G_o^{\ddagger} + \frac{1}{2}\,\Delta G \qquad (466)$$

$$\Delta G^{\ddagger} = \Delta G_o^{\ddagger}\left(1 + \frac{\Delta G}{4\,\Delta G_o^{\ddagger}}\right)^2 \qquad (466a)$$

Formulas for β and $\Delta G^{\ddagger}$ for the reverse reaction are obtained by changing the sign of ΔG in (465)-$(466a)$.

A graphic representation of (465) and (466) is given in Fig. 99.

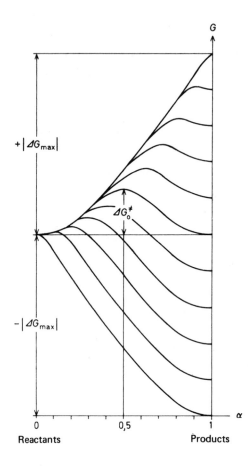

Fig. 99 Effect of ΔG on the position of the transition state on the reaction coordinate, α $(0 \leqslant \alpha \leqslant 1)$ and on $\Delta G^{\ddagger}$ $[(465)$ and $(466)]$.

An equation for H^{+}-transfer analogous to $(466a)$ was developed by Marcus[404] using a quantum mechanical approach. $\Delta G_{0}^{\ddagger}$ was termed as the *intrinsic barrier*. A relationship analogous to $(466a)$ evolves—as a special case—when E_{A} for H-transfer in the transition state $A \ldots H \ldots B$ is calculated as the energy at the intersection of the parabolas of the harmonic oscillators $AH(E_{AH} = k_{1} \cdot r^{2}/2)$ and $BH(E_{BH} = \Delta E + k_{2}(r - d)^{2}/2$; where r and d = distances, see Fig. 100, E = energy; k = force constant).[405]

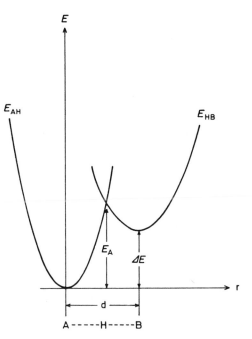

Fig. 100 Determination of the energy E_A of the transition state A $\cdots$ H $\cdots$ B as the energy of the intersection of vibrations of two harmonic oscillators AH and BH. [From G. W. Koeppl and A. J. Kresge, *J. Chem. Soc., Chem. Commun.* **1973**, 371 (ref. 405).]

For the special case when within a reaction series k_1, k_2, and d are constant and $k_1 = k_2$, (*467*) is replaced by a simpler relationship (*468*, $E_{A,0} = E_{A(\Delta E = 0)}$).

It follows from (*465*) that at *high intrinsic barriers* $\alpha \approx \beta \approx 0.5$ even for $\Delta G \neq 0$, though this is precisely true only for $\Delta G = 0$. In such cases the *curvature* of Brönsted plots *is small*. The same follows from (*464*): *at high intrinsic barriers* (slow reactions) *the reactivity range*, $2 |\Delta G_{max}|$ *for the transition from $\alpha = 0$ to $\alpha = 1$ is particularly broad. In considering narrow reactivity ranges, the relationship may appear to be linear.* These considerations may possibly apply to all linear log-log correlations.

$$E_A = \frac{k_1 d^2}{2 (1 - k_1/k_2)^2} \left[1 - \sqrt{\frac{k_1}{k_2} - \frac{2 (1 - k_1/k_2) \Delta E}{k_2 d^2}} \, \right]^2 \qquad (467)$$

$$E_A = E_{A,0} \left(1 + \frac{\Delta E}{4 E_{A,0}} \right)^2 \qquad (468)$$

Low intrinsic barriers (*ca.* 8.4 kJ/mol) and, consequently, *curvatures* in Brönsted plots are found for fast reactions such as proton transfer between oxygen and nitrogen atoms (e.g., Fig. 101).

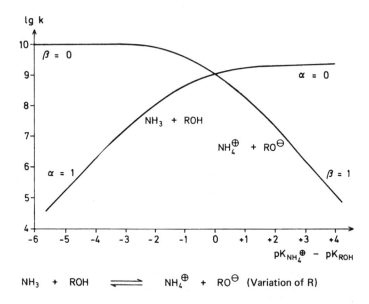

$$NH_3 \;+\; ROH \;\rightleftharpoons\; NH_4^{\oplus} \;+\; RO^{\ominus} \quad (\text{Variation of R})$$

Fig. 101 Brönsted plot for proton transfer (k in $1\ mol^{-1}\ s^{-1}$) between oxygen and nitrogen atoms. Between values -6 and 0 on the abscissa the reaction is endothermic from left to right; its reverse is exothermic. In the extreme left region of the diagram $\alpha = 1$, $\beta = 0$. In this region the reaction from right to left is diffusion controlled (as is the reaction from left to right in the extreme right region of the diagram). The different height of the two horizontal branches of the curves is due to differences in diffusion rates in neutral systems and systems containing opposite charges. Intersection of the two curves in such plots must always occur at $\Delta pK = 0$ ($K = 1$, $k = k_-$). [From M. Eigen, *Angew. Chem., Int. Ed.* **3**, 1 (1964) (ref. 406).]

The transition from $\alpha = 1$ to $\alpha = 0$ ($\beta = 0$ to $\beta = 1$) occurs here within an interval of about 6 pK units. On the other hand, Brönsted plots for protonation (deprotonation) of mesomeric hydrocarbons (Fig. 33) are almost linear over regions encompassing at least 10 pK units. The high intrinsic barriers of these processes (21-50.3 kJ/mol)[208a,407,467b] are probably related to deep-seated changes in structure (changes in hybridization, C—C bond length, substantial reorganization of the solvent shell), which on protonation (deprotonation) encompass the entire organic molecule, whereas proton exchange between oxygen and/ or nitrogen atoms is confined to the free electron pairs of the heteroatoms.

Ethylbenzene or cumene give, with n-amylsodium, ring-metallated compounds as the kinetically controlled products but benzylsodium derivatives as the thermodynamic products.[407a] This shows that there is a smaller intrinsic barrier to proton removal from sp^2-hybridized carbon, with no significant change in geometry (see also p. 125), than from the benzylic carbon, which is rehybridized from sp^3 to sp^2 upon anionization.

Of course, curvatures in Brönsted plots may be due to other causes as well, for example, changes in reaction mechanism or in the rate-determining step (see Section 3.2.3.2).

If the same assumptions as made for the free energy of transition states are considered to apply to the response of *structural* properties of the transition state to changes in reactant and product (see p. 326), one can substitute α in (465) by the structural change parameter $\lambda(\lambda_{\alpha=0} = 0, \lambda_{\alpha=1} = 1)$.[403] In this instance, within a reaction series amenable to linear correlations, processes with high intrinsic barriers do not show changes in the structure of transition states, in accord with the invariance of α.

Experimental values of α must still be interpreted with caution. For instance, *diffusion phenomena* should not be overlooked. Equations (465) and (466) consider only constants k_b and k_{-b} for the proton transfer in the encounter complex, whereas the *overall process* includes in addition the formation and decomposition of this complex.

$$A-H + B^- \underset{k_{-a}}{\overset{k_a}{\rightleftharpoons}} (A-H\text{--}B)^- \underset{k_{-b}}{\overset{k_b}{\rightleftharpoons}}$$

Encounter complex
reactants

$$(469)$$

$$(A\text{---}H-B)^- \underset{k_{-c}}{\overset{k_c}{\rightleftharpoons}} A^- + HB$$

Encounter complex products

$$k_{exp} = \frac{k_a k_b k_c}{[k_{-a}(k_{-b} + k_c) + k_b k_c]} \qquad (470)$$

$$K = \frac{k_a k_b k_c}{(k_{-a} k_{-b} k_{-c})} \qquad (471)$$

One actually obtains

$$\alpha_{exp} = \frac{d \log k_{exp}}{d \log K} \qquad (472)$$

Calculations on models show that α_{exp} is a good approximation for the "true α" (465) only in cases where internal return between the two encounter complexes is insignificant (when in the region $0 \leqslant \alpha_{exp} \leqslant 1$, $k_{-a} \gg k_b$; $k_c \gg k_{-b}$), when $\Delta G_0^{\ddagger}$ is high, and when over a broad range of the transition region ($\alpha_{exp} = 0 \rightarrow \alpha_{exp} = 1$) k_a and k_{-c} are larger than k_b and k_{-b}.[403]

For values of $\Delta G_0^{\ddagger} = 59$ kJ/mol, $k_a = 10^3$, $k_{-a} = 10^{11}$, $k_c = 10^6$ and $k_{-c} = 10^9$, α_{exp} for $\Delta G = \Delta pK = 0$ (i.e., $\Delta G_0^{\ddagger} = \Delta G^{\ddagger}$ and $k_b = k_{-b}$) turns out to be 0.38.[403]

If in the derivation of α from (467) ($\alpha = dE_A/d\Delta E$) one proceeds from the realistic assumption that when AH varies with ΔE, while B⁻ remains constant, k_1 and d vary also, one does not obtain the linear relationship (465) ($E_{A,0}$ instead of $\Delta G_0^{\ddagger}$, ΔE instead of ΔG, line A in Fig. 102, see also Fig. 100) but a curve B with $\alpha \simeq 0.6$ for $\Delta E = 0$.[405]

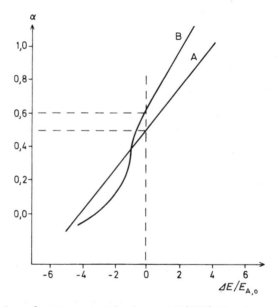

Fig. 102 Variation of α at constant k_1, k_2, and d (468), line A and variable k_1 and d, curve B. [From G. W. Koeppl and A. J. Kresge, *J. Chem. Soc., Chem. Commun.* **1973**, 371 (ref. 405).]

Constancy of Brönsted coefficients over large reactivity ranges (e.g., in reaction (472a)[401a]):

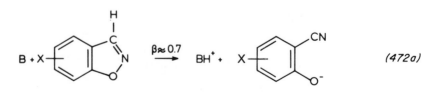

(472a)

can also be explained qualitatively[395d] by a model proposed by Jencks[379a] that makes use of two-dimensional free energy diagrams like Fig. 97. In Fig. 102a the two axes represent the breaking of the C—H

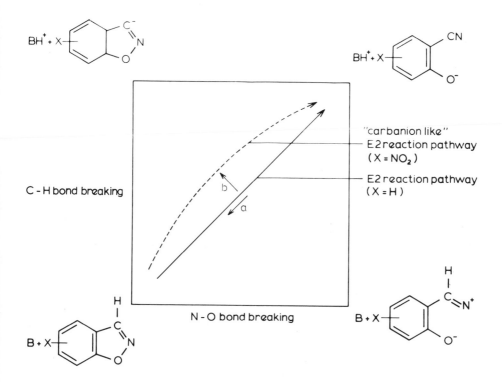

Fig. 102A Energy surface for the base-catalyzed elimination of substituted benzisoxazoles. [From A. Pross, *Adv. Phys. Org. Chem.* **14**, 69 (1977) (ref. 395d).]

and N—O bonds. An electron-attracting group X will not only stabilize the final state in the upper right-hand corner causing (according to the BEP principle) a shift of the transition state as indicated by arrow **a**, but also the hypothetical intermediate shown in the upper left-hand corner. The latter leads to a displacement of the transition state perpendicular to the original reaction coordinate (arrow **b**; this is sometimes called anti-Hammond behavior). (A shift of the transition state to the left is also seen to arise in Fig. 97 if in (*420a*) the value of E_B is lowered.) The result of **a** and **b** is that in the transition state of the more reactive system N—O breaking is less whereas the degree of proton transfer is nearly unchanged. (This represents a change of mechanism toward the E1cB extreme, a result that accords also with the Harris-Kurz-Thornton rule (see Section 5.2.2.4) if applied to an E2 reaction.)

Equations *(466)* and *(466a)* allow the interpretation of α and β values outside the region $0 \leqslant \alpha, \beta \leqslant 1$, for example, as in *(473)*.[408]

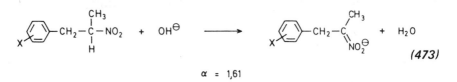

$$\alpha = 1{,}61$$

(473)

Differentiation of *(466a)* with respect to $d\Delta G$ gives *(474)*:

$$\alpha = 0{,}5 (1 + x) + (1 - x^2)\alpha_1$$

(474)

$$x = \frac{\Delta G}{4 \, \Delta G_o^{\ddagger}} \quad , \quad \alpha_1 = \frac{d \, \Delta G_o^{\ddagger}}{d \, \Delta G}$$

The first term in *(474)* is equal to the right-hand side of *(465)*. Transgression of the normal region of α and β is a consequence of the alteration in the height of the intrinsic barrier within a reaction series, expressed in the second term of *(474)*.

If one assumes that the intrinsic barrier of a reaction *(475)*

$$A - H + B^- \rightarrow BH + A^- \qquad\qquad (475)$$

represents the mean of the intrinsic barriers of reactions *(476)* and *(477)*

$$A - H + A^- \rightarrow A^- + H - A \qquad\qquad (476)$$

$$B - H + B^- \rightarrow B^- + H - B \qquad\qquad (477)$$

one may expect that changing base B, when dealing with an oxygen or nitrogen base, does not affect the mean value of $\Delta G_{0(475)}^{\ddagger}$ because the intrinsic barrier for proton transfer between such centers is low, so that $\alpha_1 = 0$ and α is described by *(465)*. However, changes in a CH acid AH will affect $\Delta G_{0(476)}^{\ddagger}$ and $\Delta G_{0(475)}^{\ddagger}$, since intrinsic barriers for proton transfer from and to such substrates are high. The term containing α_1 will contribute to α, which will then transgress its normal region and can no longer be used as criterion for the position of the transition state along the reaction coordinate.

An alternative interpretation[408a] is based on the formation of two *different anions*.

After the reversible formation of a (very weak) hydrogen-bridged complex, the rate (and $\Delta G^{\ddagger}$) determining step presumably leads to a pyramidal *nitrocarbanion*, which undergoes changes in geometry and in solvation to give the stable end product, the *nitronate ion*. Differences between the latter and the initial state give ΔG. Consequently, $\Delta G^{\ddagger}$ and ΔG do *not* refer to the *same* elementary process and system (*477a*) no longer falls under the regimen of equation (*409*); changes in structure can now influence $\Delta G^{\ddagger}$ and ΔG in totally different ways. Analogously, deviations from Brönsted correlation lines that are observed for charged catalysts (see Fig. 36) have been explained[408b] by electrostatic interactions between the catalyst and the cationic center developing on the substrate in the transition state of proton transfer. These interactions change $\Delta G^{\ddagger}$ but leave ΔG unaffected, since no positive charge is present in the initial state and in the final state the reaction partners are no longer in close proximity.

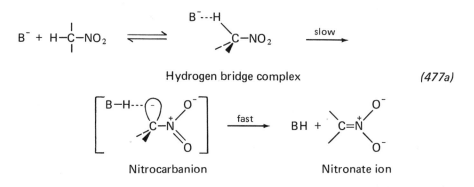

Hydrogen bridge complex   (*477a*)

Nitrocarbanion   Nitronate ion

Pross[409] has recently shown that constant selectivities observed for N_+ *correlations* can also be reconciled with Hammond's postulate. Ritchie's hypothesis that in these reactions $\Delta G^{\ddagger}$ is determined only by breaking open of the solvation shell of the nucleophile (p. 187), while solvation of the electrophile remains the same in the initial and transition states is considered doubtful, since it cannot explain the huge differences in absolute reaction rate constants of individual electrophiles. The probable mechanism of a cation-anion combination is given in Scheme 74: the solvated ions first form the ion pair 283 separated by solvent molecules, in which the anion is partially desolvated. This ion pair is transformed in the rate-determining step into the contact ion pair 284, where partial desolvation of the cation has also taken place. This is followed by the formation of RX.

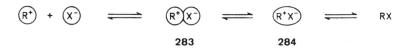

283   284

Scheme 74   Recombination of solvated ions R^+ and X^- via solvent separated ion pair (283) and contact ion pair (284).

The free activation energies for combinations of electrophile 1 (E_1) and nucleophiles 1 and 2 ($Nu_{1(2)}$) in solvents 1 and 2 ($S_{1(2)}$), $\Delta G_{S_1}^{\ddagger}(Nu_1)(E_1)$ and $\Delta G_{S_2}^{\ddagger}(Nu_2)(E_1)$ are given by:

$$\Delta G_{S_1}^{\neq}(Nu_1)(E_1) = \alpha_1 G_{S_1}(Nu_1) + \beta_1 G_{S_1}(E_1) \qquad (478a)$$

$$\Delta G_{S_2}^{\neq}(Nu_2)(E_1) = \alpha_2 G_{S_2}(Nu_2) + \beta_1' G_{S_2}(E_1) \qquad (478b)$$

$G_{S_{1(2)}}(Nu_{1(2)})$ is the free energy of solvation of the two nucleophiles in the two solvents, $G_{S_{1(2)}}(E_1)$ is the free energy of solvation of electrophile E_1. $\alpha_{1(2)}$ represents the extent of desolvation of the two nucleophiles in the two transition states, β_1 and β_1' represents the desolvation of E_1. α and β (β') should depend on the inherent reactivities of the nucleophiles ($0 \leqslant \alpha, \beta \leqslant 1$). The difference in free activation energies of E_1 with the two nucleophilic systems is given by subtraction of (478a) and (478b):

$$\Delta\Delta G^{\neq}(E_1) = \alpha_2 G_{S_2}(Nu_2) + \beta_1' G_{S_2}(E_1) - \alpha_1 G_{S_1}(Nu_1) - \beta_1 G_{S_1}(E_1) \qquad (479a)$$

Substituting E_2 for E_1 (changing the electrophile) gives (479b):

$$\Delta\Delta G^{\neq}(E_2) = \alpha_2 G_{S_2}(Nu_2) + \beta_2' G_{S_2}(E_2) - \alpha_1 G_{S_1}(Nu_1) - \beta_2 G_{S_1}(E_2) \qquad (479b)$$

The influence of nucleophiles can be eliminated by subtracting (479a) from (479b):

$$\Delta\Delta G^{\neq}(E_2) - \Delta\Delta G^{\neq}(E_1) = \beta_2' G_{S_2}(E_2) - \beta_2 G_{S_1}(E_2) - \beta_1' G_{S_2}(E_1) + \beta_1 G_{S_1}(E_1) \qquad (480)$$

The constant selectivity of the N_+ systems can be represented by (481):

$$\Delta\Delta G^{\neq}(E_2) - \Delta\Delta G^{\neq}(E_1) = 0 \qquad (481)$$

It follows that:

$$\beta_2' G_{S_2}(E_2) - \beta_2 G_{S_1}(E_2) = \beta_1' G_{S_2}(E_1) - \beta_1 G_{S_1}(E_1) \qquad (482)$$

that is, the differences in free enthalpies of desolvation of E_1 and E_2 in their transition states with $Nu_1 - S_1$, and $Nu_2 - S_2$ are equal. If, in addition, one introduces the approximation $\beta = \beta'$ (i.e., the extent of desolvation of the electrophile depends on it alone, not on the reaction partner), one obtains

$$\beta[G_{S_2}(E) - G_{S_1}(E)] = const \qquad (483)$$

If one considers that the differences in free energies of solvation in the two solvents S_1 and S_2 are proportional to the free enthalpies of solvation and if one assumes that the *more reactive electrophile is stronger solvated*, (483) can be formulated as follows:

Reactive and strongly solvated electrophiles undergo weak desolvation in their transition states $(\beta \approx 0)$, whereas weakly solvated electrophiles must be desolvated to a greater extent $(\beta \approx 1)$.

This statement is in complete agreement with Hammond's postulate: reactive electrophiles give rise to reactantlike transition states that remain strongly solvated. The constant selectivity in N_+ relationships is the result of two opposing factors: a reactive electrophile is stronger solvated, which should result in an *increase* in selectivity; however, its transition states are reactantlike and by this factor its selectivity should *decrease*. These relationships result because in the case at hand solvation energies do not change the order of inherent energies (and reactivities) [as is the case, e.g., in the solvation of halide anions by protic solvents (p. 189)]. They are presented again for a single solvent in Fig. 103.

A kind of selectivity quite opposite to the one treated so far (Hammond-BEP behavior) is encountered in *frontier orbital (FMO) controlled reactions* whose transition states are reactantlike. In these cases properties of the reactants determine selectivity and *the most reactive systems are also the most selective ones* while their transition states have moved further along the reaction coordinate than those of less reactive systems. Further amplification and examples are given on page 395. At this point it is only pointed out that cases may exist in which both product variation and reactant variation increase reactivity, in which case their respective influences on the transition state position may cancel. In such reaction series the transition state position (and selectivity) varies only slightly, although product and reactant energies vary widely (see Fig. 103A). Naturally, the possibility of both FMO and Hammond-BEP factors contributing to the nature of the transition state seems to be greatest for reactions whose transition state is about halfway along the reaction coordinate.[409a]

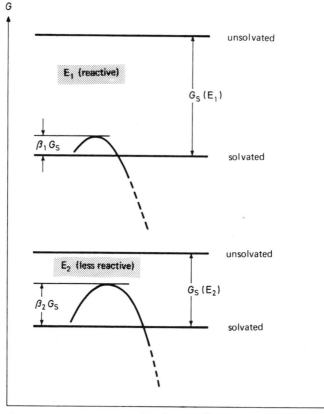

G

unsolvated

E₁ (reactive)

$G_S (E_1)$

$\beta_1 G_S$

solvated

unsolvated

E₂ (less reactive)

$G_S (E_2)$

$\beta_2 G_S$

solvated

Reaction coordinate

Fig. 103 Activation barriers in reactions of reactive (and therefore strongly solvated) electrophiles E_1 and less reactive (and therefore weakly solvated) electrophiles E_2 with nucleophiles. For clarity (constant) contributions of nucleophiles are omitted, Symbols explained in text. [From A. Pross, *J. Amer. Chem. Soc.* **98**, 776 (1976) (ref. 409).]

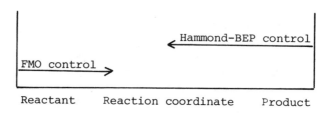

Hammond–BEP control

FMO control

Reactant   Reaction coordinate   Product

Fig. 103A Counteracting influences of Hammond-BEP and FMO control on transition state position. The arrows indicate the change in transition state position accompanying an increase in reactivity.

Based on the simplified zero-point energy interpretation of *kinetic isotope effects* (p. 260), a *maximum* of the primary kinetic isotope effect within a reaction series for $\Delta G = 0$ is considered as indicative of a "symmetric" transition state (where the forces affecting the atom to be transferred from one reactant to the other are in equilibrium). In light of the many factors that could influence the value of, for example,

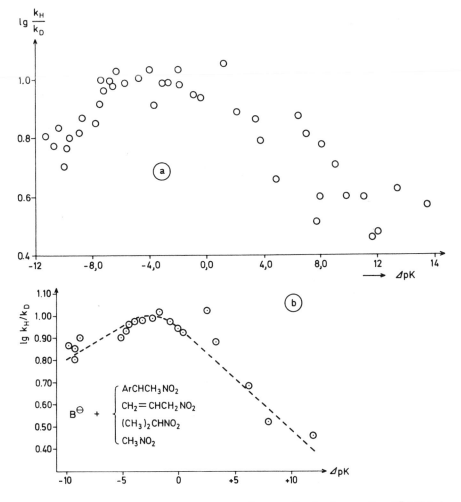

Fig. 104   Variation of the primary isotope effect k_H/k_D in the reaction $B^- + HA \rightarrow$ $BH + A^-$ (HA = CH acid) as a function of ΔG [$\Delta G = 2.3RT$ ($pK_{HA} - pK_{HB}$) = $2.3\,RT\Delta pK$]. (*a*) ΔpK Dependence of the primary kinetic isotope effect for unrestricted variation of the CH acid and base. [From F. G. Bordwell and W. J. Boyle, Jr., *J. Amer. Chem. Soc.*, **93**, 512 (1971) (ref. 414).] (*b*) ΔpK Dependence of the primary kinetic isotope effect when only nitroalkanes are considered. [From F. G. Bordwell and W. J. Boyle, Jr., *J. Amer. Chem. Soc.* **97**, 3447 (1975) (ref. 408a).]

k_H/k_D (p. 000, ff),[401,][412] such an interpretation may be questionable (with the possible exception of very similar reaction series), since in many instances variations of k_H/k_D with ΔG turn out to be relatively small. Data for reactions of various CH acids HA with various bases B⁻ are given in Fig. 104.[414]

The wide scattering of $\log(k_H/k_D)$ values shown in Fig. 104a becomes smaller when only a single class of compounds is considered (Fig. 104b).[408a] The relatively small variation of the (practically maximum, see Table 76) isotope effect in the region $-5 < \Delta pK < 5$ was interpreted in the same way as the anomalous values of Brönsted coefficients of nitroalkanes: k_H/k_D should refer to the formation of the nitrocarbanion from the hydrogen-bridged complex. The structure dependence of ΔG (assumption: $\Delta G \approx 0$) in *this* step of the reaction should be substantially smaller than the structure dependence of ΔpK in the overall reaction (477a).

Streitwieser et al.[209] found that triphenylmethane and 9-phenyl-fluorene show practically identical H/D effects on deprotonation with CH_3O^-, although the latter hydrocarbon is about 13 pK units more acidic. Only slight variations in k_H/k_D with reactivity were also found in Hofmann eliminations (when ethoxide was replaced by t-butoxide or upon addition of DMSO to the alcoholic media).[415]

5.2.3.4 Other (Semiquantitative) Indications of Structures of Transition States

Use of Rigid Model Compounds. The fact that the deamination of A-NH$_2$ (484) gives the corresponding alcohol just as in the case of "normal," more flexible alicyclic primary amines[416] shows that the calculated[164] strain difference:

$$(H_{R^+} - H_{RH})_A - (H_{R^+} - H_{RH})_B = 110\ kJ/mol$$

considered to be the main reason for the extreme inertness of A-OTos (compared to B-OTos) in *endothermic* ionizations has very little effect on the relative free energies of transition states in the *exothermic* heterolysis of diazonium ions; consequently one may consider that the latter transition states have a high degree of reactantlike character.

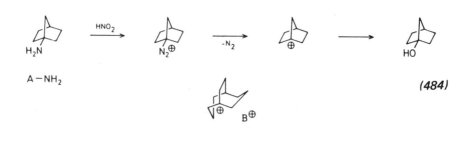

Lamaty and co-workers compared rate constants for axial and equatorial attack of 4-*t*-butylcyclohexanone by nucleophiles BH_4^-, SO_3^{2-}, and NH_2OH with rate constants for the addition of the same nucleophiles to adamantanone[417] (Scheme 75).

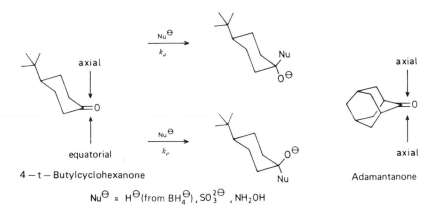

Scheme 75 Addition of nucleophiles to 4-*t*-butylcyclohexanone and adamantanone (see accompanying table).

Interpretation of Results.   If the transformation occurs via a transition state with a *high degree* of *reactantlike* character, the reagent attacking adamantanone "sees" the axial site of 4-*t*-butylcyclohexanone. In the limiting case $k = 2k_a$. This situation is approached on addition of SO_3^{2-}. In transition states with *high productlike character* the *oxygen atom* of adamantanone assumes a position that can be compared to that of the oxygen atom in the equatorial addition to 4-*t*-butylcyclohexanone. If in such a situation the steric effect of Nu^- can be disregarded (probably on addition of BH_4^-), in the limiting case $k = 2k_e$. This is approximately true for BH_4^-.

Table for Scheme 75

	4-*t*-Butylcyclohexanone				
Reagent	k_{total}	% Axial Product	$k_a{}^a$	k_e	k for Adamantanone
BH_4^-	10.7^b	80	8.55	2.15	5.14^b
SO_3^{2-}	0.7^c	10	0.07	0.63	0.16^c
NH_2OH	$464 \times 10^{4\ d}$	?	?	?	$44 \times 10^{4\ d}$

^{a} Calculated: % axial product: $100 = k_a : k_{total}$.
^{b} In l/mol min, H_2O/dioxane, 50/50, 25°C.
^{c} In l/mol s, H_2O, pH = 4, 25°C.
^{d} In l/mol min, H_2O/C_2H_5OH, 95/5, pH = 1, 25°C.

Based on the high value of k_{total} and of k for addition of NH_2OH, one may assume that in this reaction $k = 2k_a$ (reactantlike transition states); it follows that oxime formation in 4-t-butylcyclohexanone occurs to the extent of about 95% from the equatorial side.

The assignments—*reactantlike character* for the addition of SO_3^{2-}, *productlike character* for the addition of BH_4^-—are in agreement with values of other selectivity parameters for ketones, and in particular with the fact that substituents of aromatic ketones "feel" less negative charge in reactions with doubly negative anions SO_3^{2-} ($\rho = +1.27$) than in reactions with simple anions ($\rho = +2.33, +3.06$, Table 92)[418] (see however, comments on p. 352).

TABLE 92 Selectivity Parameters of Nucleophilic Reagents in Reactions with Cyclic and Aromatic Ketones[418]

Reagent	Selectivity Parameter			
	$k_{C_6}/k_{C_s}{}^a$	k_{C_6}/k_{C_7}	ρ	α
BH_4^-	46	158	+3.06	
CN^-	24	50	+2.33	0.74
NH_2OH	23	23	+0.32	0.56
SO_3^{2-}	15	12	+1.27	0.49

$^a k_{C_i}$ = rate constant of the reaction with $(CH_2)_{i-1}C{=}O$.

Enthalpies of Transfer of Transition States from One Solvent to Another.[419] Geometric and electronic changes occurring along the reaction coordinate should be paralleled by continuous changes in inter- action with solvent. Within a reaction series changes in *differences in interaction* of two sufficiently different solvents and transition states should be a *measure of different progress along the reaction coordinate.*

Experimental results and their interpretation are best visualized by plotting results for nine S_N2 reactions in solvents *methanol* and *di- methylformamide* (DMF) (Fig. 105).

The relations $P = \delta\Delta H_s^{\text{products}}$, $R = \delta\Delta H_s^{\text{reactants}}$ are differences in heats of solution of the corresponding compounds in methanol and DMF; they are equal to the enthalpy of transfer from methanol to DMF; $T = \delta H^t$ is the enthalpy of transfer of the transition states. It is given by

$$\delta H^t = \delta\Delta H_s^{\text{reactants}} + \delta\Delta H^{\ddagger} \qquad (485)$$

where $\delta\Delta H^{\ddagger}$ is the difference in activation enthalpies of a given reac- tion in the two solvents. If in comparing two reactions the shift of T

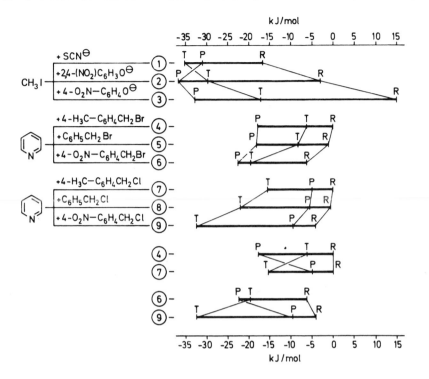

Fig. 105 Enthalpies of transfer from methanol to dimethylformamide for various states of nine S_N2 reactions: R, reactants; T, transition states; P, products.[419]

in Fig. 105 to the right (left) is larger than that of P and R, the reactant-(product)-like character of the transition state should have increased. One may draw the following conclusions:

1. Increase in basicity of the nucleophile increases the reactantlike character of the transition states (reactions 1-3, Fig. 105).
2. Withdrawing electrons at the central carbon atom increases the productlike character of the transition state (reactions 4-6 and 7-9; the corresponding prediction of the Harris-Kurz analysis was not discussed in Section 5.2.2.4).
3. Increase in basicity of the leaving group increases the productlike character of the transition state (cf. reactions 4 and 7, and 6 and 9).

All three conclusions are in accord with the results of the Harris-Kurz analysis.

Acid Dissociation Constants of Transition States.[420] If in a reaction both the rate constant k_S for the transformation of the substrate S and the rate constant for the transformation of the protonated substrate SH^+ are known, it is possible to compute the pK of the transition state from the two rate constants and the pK of the protonated substrate $[(486)-(491)]$.

$$S + X \xrightarrow{k_S} [S^{\neq}] \longrightarrow products \tag{486}$$

$$SH^+ + X \xrightarrow{k_{SH}} [SH^{\neq}]^+ \longrightarrow products \tag{487}$$

$$k_S = \kappa \frac{kT}{h} K_S^{\neq} \tag{488}$$

$$k_{SH} = \kappa \frac{kT}{h} K_{SH}^{\neq} \tag{489}$$

The two virtual equilibrium constants may be correlated via a cyclic process [Scheme 76, (490), (491)].

$$H^{\oplus} + S + X \overset{K_S^{\neq}}{\rightleftharpoons} S^{\neq} + H^{\oplus} \tag{490}$$

$$\Big\updownarrow K_{SH}^{\oplus} \qquad\qquad\qquad \Big\updownarrow \kappa^{\neq}$$

$$SH^{\oplus} + X \overset{K_{SH}^{\neq}}{\rightleftharpoons} [SH^{\neq}]^{\oplus} \tag{491}$$

Scheme 76   Thermodynamic cycle for the determination of pK values of transition states.

The $pK^{\ddagger}$ of the transition state for the hydrolysis of a methyl halide corresponds to the ionization:

$$\tag{492}$$

By means of empirical correlations between effects of *charge, dipole moments of substituents, distances,* and *relative orientation,* and pK values for "normal" acids [e.g., (493), (494),] or by using model compounds, one can try to reproduce experimental values for $pK^{\ddagger}$ by choosing an appropriate structural model for the transition state of the reaction under consideration.

$$\log \Delta pK = 1.231 - 0.212\, r_{\text{rms}} \tag{493}$$

pK changes induced by the introduction of a negative charge at an atom residing at an average distance r_{rms} (Å)[*] from the atom bearing the acidic proton

$$\lg \frac{\Delta pK}{\mu \cos \theta} = 1{,}250 - 0{,}383\, r \tag{494}$$

pK-variation produced by the dipole moment of a halide substituent

where μ = dipole moment of the corresponding methyl halide
θ = angle between the bond dipole and the line connecting the midpoint of the dipole and the basic site in the extended conformation of the substrate
r = length of this line (Å)

From rates of reaction of CH_3Cl with $H_2O(SH^+)$ and $OH^-(S)$ one finds $pK^{\ddagger} = 11.9$ For a strongly productlike structure of the transition state $(285, R = H)$[§]

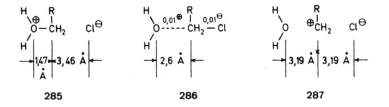

(493) gives for the change in pK value (ΔpK) in going from protonated methanol ($pK = -2.2$) to structure 285:

$$\Delta pK = pK^{\ddagger}_{285} - pK_{CH_3\overset{+}{O}H_2} = 1.5$$
$$pK^{\ddagger}_{285} = -0.7$$

[*]S. J. G. Kirkwood and F. H. Westheimer, *J. Chem. Phys.* **6**, 506 (1938), p. 510. In 285 r_{rms} is equal to the distance between $\overset{+}{O}$ and Cl^-.

[§]Distances between nonbonded atoms in models for transition states are equilibrium distances of noble gas clusters, Ne, Ar in 285, Ne, Ne in 286.

Thus it is clear that the activated complex is not similar to **285**. Although the strongly reactantlike model **286** (R = H) does approximate via (*494*) (effect of the dipole moment of CH_3Cl on the pK value of water) the experimental value for p$K^{\ddagger}$, the same is true for structure **287** (R = H), which corresponds to a transition state of an S_N2 reaction taking place via an ion pair. The latter possibility is a real one, according to Sneen:

> Ion pairs have been established as substrates for nucleophilic attack; covalent, saturated carbon has not.[421]

To check this matter, Kurz and Harris[422] studied S_N2 reactions of bromoacetic acid ($=SH^+$) and bromoacetate ion ($=S$) (**285, 286, 287**: R = COOH, COO$^-$, Br instead of Cl). In these reactions one determines the pK value of the carboxy proton in the transition state. Because of the positive charge at the central carbon atom, it should be lower for **287** than for **286**. The experimental results correspond to values estimated for **286**. These results show that the rate-determining processes in *hydrolyses* of alkyl halides and similar compounds must be connected with *changes in solvent configuration* and not with *changes in the inner structure* of the nucleophile-substrate pair (see, however, p. 450).

5.2.4   "Allowed" and "Forbidden" Transition States; The Role of Orbital Symmetry and the Woodward-Hoffmann Rules[423]

5.2.4.1   Pericyclic Reactions

We have mentioned repeatedly stereospecific reactions that can be interpreted as (intramolecular as well as intermolecular) cyclic shifts of "mobile" electron pairs. The (often astonishing) favoring of one of the many possible stereochemically different reaction paths in such *pericyclic* reactions can be ascribed, regardless of reactant-product thermodynamics, to special bonding relationships in the more favorable transition states.

Pericyclic Reaction Types

1   **Cycloadditions (Reverse: Cycloreversions):** Closure (opening) of a ring by formation (breaking) of two σ bonds between the ends of two chains with m ($m - 2$) and n ($n - 2$) conjugated mobile electrons.

2 **Electrocyclic Reactions:** Ring closure (ring opening) by formation (breaking) of a σ bond between the ends of a system of m ($m - 2$) conjugated mobile electrons.

3 **Sigmatropic Rearrangements of Order** [m, n]: Migration of a σ bond between the ends of two conjugated systems of m, respectively n mobile electrons arising through hypothetical homolysis of the appropriate σ bond.

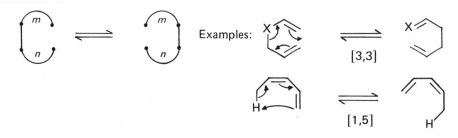

4 **Cheletropic Reactions:** Breaking (formation) of two σ bonds between the ends of a conjugated system with m ($m + 2$) mobile electrons and a single atom.

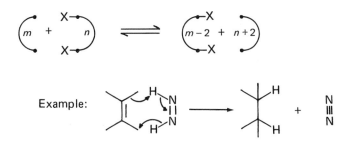

5 **Group Transfer Reactions:** Transfer of end ligands of a conjugated system to both ends of another system:

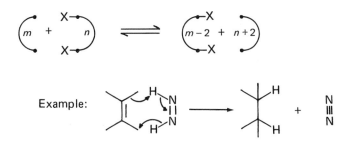

5.2.4.2 Conservation of Orbital Symmetry

The breakthrough in the interpretation of the favoring of certain stereo-chemical reaction paths in such reactions is due to Woodward and

Hoffmann: in general, reactions can take advantage of the energetically favorable concerted bond breaking and bond making (see p. 330 ff) and avoid energy-rich intermediates only if the occupied orbitals of the reactant are *continuously* transformed into those of the product. A necessary condition is that the *symmetry* of these orbitals remain *unchanged* throughout the overall transformation [reactant(s) → activated complex → product(s)].

This condition establishes the concerted stereochemistry-determining motions of atoms required for the achievement of the reaction path having the lowest energy.* This is the essence of the **Woodward-Hoffmann rules**, which are presented in a most general form in Section 5.2.4.5.

To illustrate the analysis we first consider the historically important example of the electrocyclic thermal transformation of cyclobutene into butadiene-1,3[424] (Fig. 106; see also Figs. 76 and 77): on transformation into butadiene both CR^1R^2 groups and the orbitals forming the σ bond to be broken must rotate 90°. There are two stereochemical alternatives, leading to different products. In the conrotatory opening (*495*) both groups rotate in the *same* sense, in the *disrotatory* opening (*496*) they rotate in *opposite* sense. The symmetry element present in every phase of the synchronous (strongly coupled, see p. 321) conrotatory opening (disregarding differences between R^1 and R^2) is a *twofold* axis; the symmetry element in the disrotatory opening is the mirror plane shown. Orbitals of reactant and product must be classified with respect to these two elements (see Fig. 106).

Correlation lines in Fig. 106 show that a continuous symmetry-conserving transition of ground state occupied orbitals of cyclobutene to ground state occupied orbitals of butadiene is possible only in the *conrotatory* process. In the disrotatory opening conservation of orbital symmetry would lead to the transformation of a ground state orbital of cyclobutene into an unoccupied higher orbital of butadiene. Experiment shows that the thermal opening of ordinary cyclobutenes is strictly conrotatory.

The correlation diagram for the *disrotatory* transformation is followed when the reverse reaction, the formation of a cyclobutene from butadiene occurs *photochemically*, involving the first excited singlet state of butadiene: conservation of orbital symmetry leads to the electronic ground state of cyclobutene. This reaction course is again confirmed by many examples. In a general sense, the fact that *photochemical pericyclic reactions* normally lead to results different from those obtained through the analogous thermal transformations may be ascribed to the participation of higher orbitals of different symmetries.

*With respect to orbital changes! However, one has to keep in mind that other interactions (e.g., steric ones) may contribute decisively to the total activation energy.

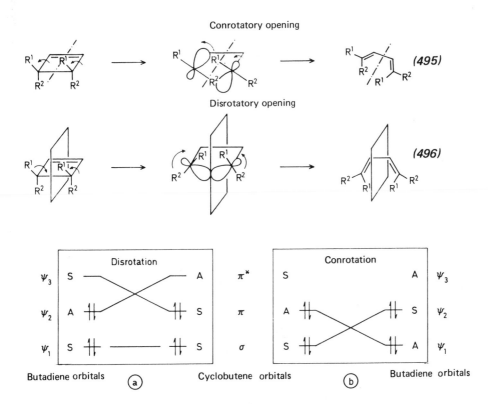

Fig. 106 Orbital symmetry correlations for the disrotatory (a) and conrotatory (b) cyclobutene-butadiene transformation. Taken into account are the breaking σ bond and the π and π^* orbitals of cyclobutene and the molecular orbitals ψ_1-ψ_3 of butadiene; S (A), orbital is symmetric (antisymmetric) with respect to the symmetry element of reaction (*495*) or (*496*).

However, many photochemical reactions proceed by unique mechanisms, often involving triplet states, so that predictions based on orbital symmetry should be made with caution.

The same analysis can be applied to electrocyclic reactions of other systems (including ionic ones, Table 98),[425] as well as to cheletropic reactions, group transfer reactions, and cycloadditions.[423] The stereochemical outcome in the latter depends on the direction from which the ends of the reacting systems may be attacked (Scheme 77). In *suprafacial attack* for π systems, the two new bonds are formed on the same side of the nodal plane formed by the atomic nuclei; in *antarafacial attack* the two new bonds are formed at opposite sides of the nodal plane. Since strained σ bonds may also function as components in cycloadditions, one must define the two stereochemical alternatives for this type of bonding:

Suprafacial The two ends of the bond react in the same sense on bond formation (both by retention or inversion).

Antarafacial The two ends react in opposite sense (Scheme 77).

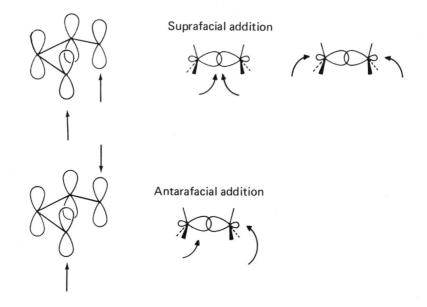

Suprafacial addition

Antarafacial addition

Scheme 77 Suprafacial and antarafacial additions to π systems and to σ bonds. The new σ bonds are formed from directions indicated by arrows.

Sigmatropic rearrangements may also occur suprafacially or antarafacially, depending on whether the relevant σ bond migrates along the

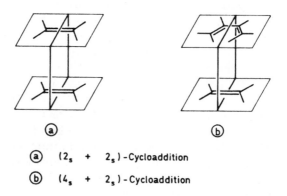

(a)   (b)

(a) $(2_s + 2_s)$ - Cycloaddition

(b) $(4_s + 2_s)$ - Cycloaddition

Fig. 107 Symmetry planes for doubly suprafacial cycloadditions.

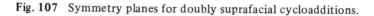

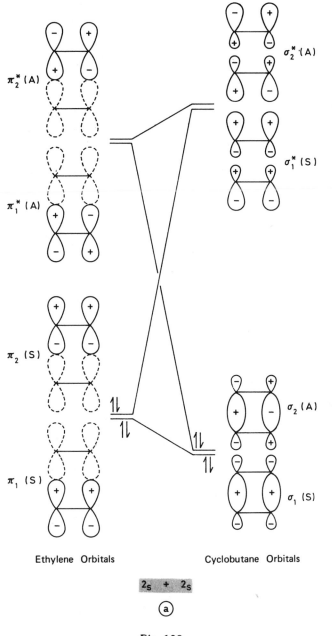

Ethylene Orbitals   Cyclobutane Orbitals

$2_s + 2_s$

(a)

Fig. 108

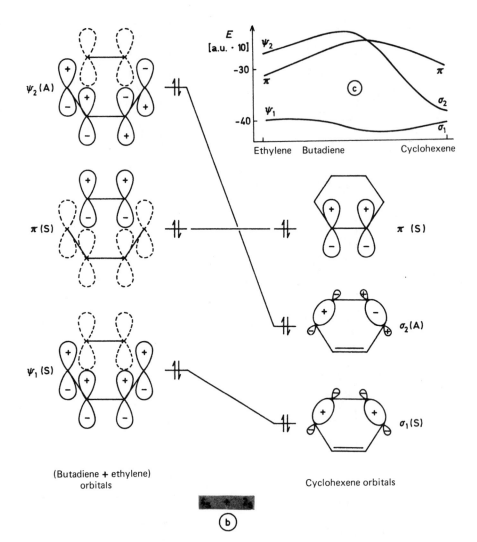

(Butadiene + ethylene) orbitals

Cyclohexene orbitals

Fig. 108 Correlation diagrams for thermal doubly suprafacial cycloadditions. Thermal dimerization of ethylene (*a*) is "orbital symmetry forbidden" because conservation of symmetry (and signs) of the reactant orbital π_2 leads to the excited product orbital σ_1^*. The qualitative correlation diagram for the "allowed" Diels-Alder reaction of 1,3-butadiene and ethylene (*b*) corresponds completely to the correlation diagram based on rigorous quantum mechanical calculations for this reaction (*c*).[477] The construction of the correlation diagrams is based on the quantum mechanical "noncrossing rule," which states that correlation lines between states of equal symmetry may not cross.

384

same side, or toward the opposite side of the π system. When individual carbon atoms are considered, the two new modes become the well-known modes of retention and inversion. The stereochemistry for the reaction of a given component is indicated by a subscript (s, a) next to the number of electrons characterizing that component. The Diels-Alder reaction is a $(4_s + 2_s)$-cycloaddition; thermally "allowed" migrations of carbon in sigmatropic rearrangements are of the order $[1_a, 3_s]$; that is, the migrating carbon undergoes inversion of configuration.

In constructing correlation diagrams for (2s + 2s)- and (4s + 2s)-cycloadditions (Fig. 107), the decisive symmetry element for reactants, products, and transition state is the mirror plane perpendicular to the planes of the molecules of the reaction partners.

The reactant states in orbital correlation diagrams for such cyclo-additions (Fig. 108) consist of *pseudomolecules* (encounter complexes) in which both reactants have come close enough for incipient inter-action. The molecular orbitals of such pseudomolecules are the MOs of the individual reactants, which may be considered to be MOs of the pseudomolecule fully localized in the region of the respective reactant. In the transition state and in the cyclic adduct the "same" orbitals extend in a different form, but with the same symmetry (and phase) over other regions of the species formed from the pseudomolecules.

The correlation diagram for the (2s + 2s) cycloaddition shows that this reaction is thermally "forbidden" because both occupied ground state orbitals of the pseudomolecule are symmetric, whereas one of the molecular orbitals formed by the two σ bonds of cyclobutane is anti-symmetric, so that there is no correlation between *all* ground state MOs with conservation of symmetry; this is not the case in thermally "allowed" (4s + 2s)-cycloadditions.

One of the most important results of orbital symmetry analysis by means of correlation diagrams is the recognition that a certain reaction type is not confined to a single classical example to which the analysis was initially applied; similar correlation diagrams can be constructed for *isoelectronic* systems, as well as for *electron richer* and *electron poorer* conjugated systems (also *ionic* ones).[425a]

In addition to Diels-Alder reactions, the number of ground state occupied symmetric (or antisymmetric) orbitals of the reactants is equal to the number of symmetric (antisymmetric) ground state occupied orbitals of the product in many other (s + s)-cycloadditions involving one reactant with $(4n + 2)$ π electrons and another with $4n$ π electrons; thus symmetry correlations are possible and the reactions are thermally allowed (Table 93). (Examples of such reactions are given in Table 99, below.)

TABLE 93 Number of Symmetric (s) and Antisymmetric (a) Molecular
Orbitals in (s + s)-Cycloadditions of Two Reactants with $4n + 2$ and
$4n$ π Electrons

Species and Bond	Number of Participating Electrons (type)	Occupied MOs		
		Number	of which s	of which a
Reactant 1	$4n + 2$ (π)	$2n + 1$	$n + 1$	n
Reactant 2	$4n$ (π)	$2n$	n	n
	sum for reactants:		$2n + 1$	$2n$
Product	sum for product:		$2n + 1$	$2n$
Partial structure from Reactant 1	$4n$ (π)	$2n$	n	n
Partial structure from Reactant 2	$4n - 2$ (π)	$2n-1$	n	$n-1$
New σ bonds	4 (σ)	2	1	1

The *1,3-dipolar addition*, exceedingly important for the synthesis of
five-membered ring heterocycles, turns out to be a close relative of the
Diels-Alder reaction; the 1,3-dipole assumes the role of the 4-electron
component[426] (Fig. 109).

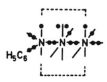

Fig. 109 The 4π electron system (delocalized within the plane indicated by dashed
lines) involved in the 1,3-dipolar addition of phenyl azide (only axes of $2p, sp^2$, and
sp orbitals are shown).

The ability of the two occupied π orbitals of *bicyclo[2.2.1]hepta-
diene-2,5* to interact through space and form molecular orbitals with
the same symmetry properties (with respect to the mirror plane
through C-1—C-7—C-4) as ψ_1 and ψ_2 of butadiene (Fig. 110) explains
the observation that this diene undergoes a number of *homo-Diels-
Alder* reactions (*496a*). The correlation diagram for these reactions
differs from that of "normal" $(4_s + 2_s)$-additions only in that the symmetric π orbital of the dienophile correlates with the symmetric orbital
of the newly formed *cyclopropane* bond of the product.

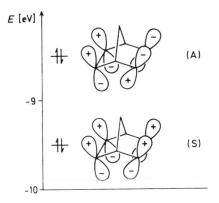

Fig. 110 Energies of the symmetric and antisymmetric combination of double bond orbitals in bicyclo[2.2.1]hepta-2,5-diene determined by photoelectron spectroscopy.[427]

(496a)

Qualitative orbital correlation diagrams can be constructed by means of the Hückel MO theory,[428] *without* considering the symmetries of reactants and products.* One starts by defining a basis set of orbitals for reactant and product in accord with the geometry of the reaction. The transition from one basis set to the other is taken into account by writing the *resonance integral as a function of the progress r along the reaction coordinate* ($0 \leqslant r \leqslant 1$). Using these expressions for the resonance integral and the common assumptions of the Hückel MO theory, one can write the secular determinant. Its solutions give the energies of the molecular orbitals as functions of the reaction coordinate r. Their graphic representation (Fig. 111) gives the correlation diagram. An example is the disrotatory opening of a cyclopropyl system to an allyl system (*497*):

---

*This treatment shows that the symmetry correlation is not a postulate in itself of quantum theory, but rather a consequence of the continuous character of the MO transformations along the reaction coordinate. Even systems without symmetry elements can be analyzed in this way.

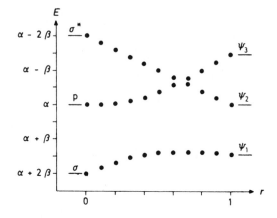

Fig. 111 Orbital correlation diagram for the disrotatory cyclopropyl → allyl rearrangement derived from Hückel's MO theory. [From J. C. Dalton and L. E. Friedrich, *J. Chem. Educ.* **52**, 721 (1975) (ref. 428).]

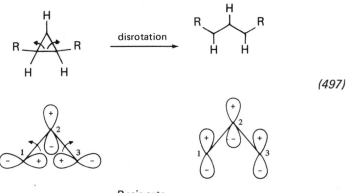

Basis sets

(497)

The simplest assumptions with respect to the variation of the resonance integral H_{ij} give (498):

$$H_{12} = H_{21} = H_{23} = H_{32} = \beta r$$
$$H_{13} = H_{31} = 2\beta(1-r)$$
$$0 \leqslant r \leqslant 1$$

(498)

For $r = 0$ (498) gives the resonance integral of the cyclopropyl system, for $r = 1$, that of the allyl system. In the intermediate region H_{ij} varies linearly with r. With the simplifying assumption that the Coulomb integrals α of the various carbon atoms are equal and constant and for $x = (\alpha - E)/\beta$, the secular determinant is:

$$\begin{vmatrix} x & r & 2(1-r) \\ r & x & r \\ 2(1-r) & r & x \end{vmatrix} = 0 \qquad (499)$$

Its solutions are:

$$x_1 = 2(1-r) , \quad x_{2,3} = r - 1 \pm \sqrt{3r^2 - 2r + 1} \qquad (500)$$

From these, the orbital energies are:

$$E_a = \alpha - 2(1-r)\beta$$

$$E_b = \alpha - (r - 1 + \sqrt{3r^2 - 2r + 1})\beta \qquad (501)$$

$$E_c = \alpha - (r - 1 - \sqrt{3r^2 - 2r + 1})\beta$$

Plotting these energies versus r gives the correlation diagram in Fig. 111. It shows that the σ orbital of the cyclopropyl system correlates with ψ_1 of the allyl system and that p and σ^* are transformed into ψ_3 and ψ_2, respectively. Consequently, the thermal *disrotatory* cyclopropyl-allyl rearrangement is allowed only for *cations*, since only in this instance does the ground state occupied orbital of the reactant yield the ground state occupied orbital of the product; the highest occupied molecular orbital (HOMO) of the cyclopropyl radical or the cyclopropyl anion would lead to an excited level as the HOMO of the corresponding allyl system. (The latter rearrangements are allowed if they proceed in a *conrotatory* way, as shown in the corresponding analysis of the back reaction, allyl → cyclopropyl; basis sets (502) and resonance integrals (503) must be used in this instance).

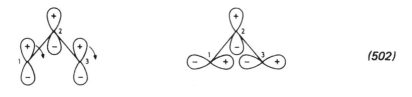

$$(502)$$

$$H_{12} = H_{21} = H_{23} = H_{32} = \beta(1-r)$$
$$H_{13} = H_{31} = -2\beta r \qquad (503)$$

In many instances the search for new variants of known reactions inspired by the Woodward-Hoffmann analysis proved to be highly successful (see Tables 98 and 99).

5.2.4.3 The Frontier Orbital (FMO) Analysis[428a]

The success of Woodward and Hoffmann prompted alternative interpretations of pericyclic reactions, which led to the same results. Fukui's frontier orbital method[429] states that mutual *perturbations* between orbitals of the reaction partners in the *initial phase* of a cycloaddition are the main elements governing the course (and stereochemistry) of the transformation. Basically, one should add up the effects of all possible combinations of orbitals of the reaction partners taken in pairs. However, since interaction of two doubly occupied orbitals does not result in a net gain or net loss in energy (Fig. 112a), only interactions between an empty orbital (Fig. 112b) or a singly occupied orbital (Fig. 112c) and an occupied orbital are essential.

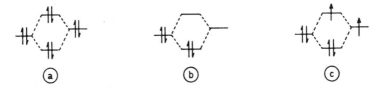

Fig. 112 Electron energies on interaction of two orbitals.

Since the intensity of the interaction is inversely proportional to the difference in energy between orbitals, it turns out that the dominant factors in such interactions are the highest doubly occupied (HOMO) and lowest unoccupied molecular orbitals (LUMO).

The stabilization energy (*SE*) of the systems shown in Figs. 113 and 114 is given by (504)[430] [corresponding to the covalent term in (189)].

$$SE = \frac{2\,(1 \cdot k + 1' \cdot k')^2 \cdot \beta^2}{\alpha(HOMO)_S - \alpha(LUMO)_R} + \frac{2(m \cdot n + m' \cdot n')^2 \cdot \beta^2}{\alpha(HOMO)_R - \alpha(LUMO)_S} \quad (504)$$

where α = orbital energy
          β = resonance integral
      $1 - k'$ = atomic orbital coefficient

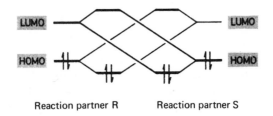

Reaction partner R   Reaction partner S

Fig. 113 Frontier orbital interactions of two reaction partners R and S.

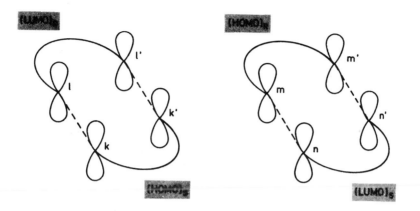

Fig. 114 Atomic orbital coefficients in HOMO-LUMO interactions of two cyclo-addition partners R and S; see (504).

Apart from values of orbital energy differences and of the resonance integral, the relative signs of products of atomic orbital coefficients are crucial. They are given by the phase properties (number of nodes) of the corresponding wave functions.

The signs of coefficients of terminal atomic orbitals in HOMOs (LUMOs) of systems with $4n$ ($4n + 2$) π electrons are opposite; in HOMOs (LUMOs) of systems with $4n + 2$ ($4n$) π electrons they are identical. Consequently, only for doubly suprafacial cycloadditions of type $(4n + 2) + 4m$ is SE appreciable. For combinations of type $(4n) + (4m)$ and $(4n + 2) + (4m + 2)$, $SE \approx 0$ because in both HOMO-LUMO interactions an orbital with coefficients of opposite sign must interact with an orbital with coefficients of equal sign. The products of atomic orbital coefficients then have opposite signs in both terms of (504) and (in hydrocarbons with symmetric distribution of electrons) they cancel each other. The thermal formation of cyclobutane from two molecules of ethylene is an example of such an unfruitful situation, whereas the Diels-Alder reaction (and other doubly suprafacial cycloadditions with $4n + 2$ participating mobile electrons) represents a favorable situation.

The *relative reactivity* and the *regioselectivity* of the *Diels-Alder* reaction and of *1,3-dipolar additions* can also be described by (504).[431,432] In a normal Diels-Alder reaction the interaction $HOMO_{diene}$-$LUMO_{dienophile}$ is so strong that the second term in (504), which takes into account the intraction $HOMO_{dienophile}$-$LUMO_{diene}$ can be neglected (see also Scheme 78). Introduction of electron-with-drawing substituents in the dienophile reduces the energies of its frontier orbitals, and if the diene remains unchanged this will result in an increase in the (negative) stabilization energy and in log k for the reaction under consideration. The same is true if the dienophile remains unchanged and the energy of the HOMO of the diene is increased by

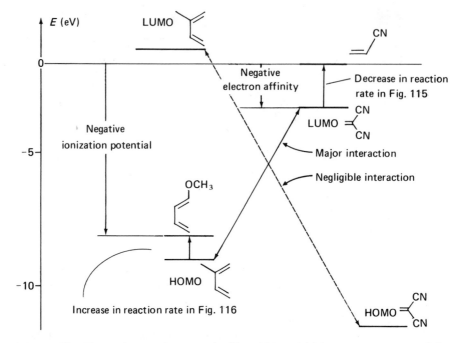

Scheme 78 Change in reaction rates in Figs. 115 and 116 as a consequence of the variation in the dominant frontier orbital interaction. Negative values of the ionization potential and of the electron affinity of a compound are taken as energies of its HOMO and LUMO, respectively.

the introduction of electron-donating substituents. Both situations are represented in Scheme 78 and Figs. 115 and 116. Obviously, in the reaction series shown in Fig. 115 and in the series of acyclic butadienes in Fig. 116 other factors (e.g., steric-conformational, or electrostatic interactions) must be constant (or show a linear dependence on the HOMO-LUMO difference); only in this instance will the expected linear dependence of $\log k$ and $[\alpha_{HOMO\text{-}diene} - \alpha_{LUMO\text{-}dienophile}]^{-1}$ *(504)* or the hyperbolic dependence of $\log k$ and $\alpha_{HOMO\text{-}diene} - \alpha_{LUMO\text{-}dienophile}$ be observed. [Since *(504)* is valid for weak perturbations in the beginning stage of a transformation, such correlations might be expected primarily for strongly exothermic reactions with reactantlike transition states.]

In cases where isomeric cycloadducts are possible, the favored regioisomer is the one for whose reaction path the numerator in *(504)* has the highest possible value. This is demonstrated in Table 94 showing the favored formation of *"para"* isomers in the reaction of isoprene with acrylonitrile or vinylidene cyanide.[431]

The data in Table 94 demonstrate the important fact mentioned briefly on page 369 that in FMO-controlled reactions an *increase* in

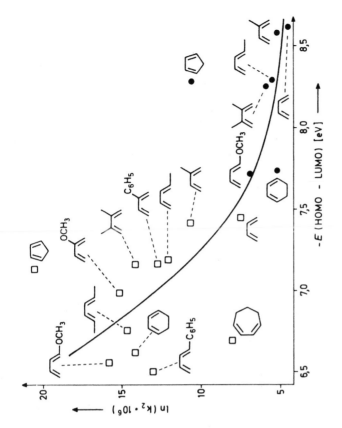

Fig. 116 Frontier-orbital dependence of rates of Diels Alder reactions of tetracyanoethylene (□) and maleic anhydride (●) with a series of acyclic butadienes and three cyclic dienes. The strong deviation for cyclopentadiene is probably due to the fixed cisoid conformation and minimization of steric effects. [From: R. Sustmann and R. Schubert, *Angew. Chem., Int. Ed.* **11**, 840 (1972) (Ref. 433)].

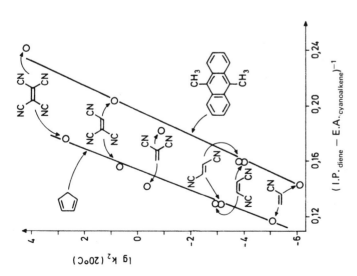

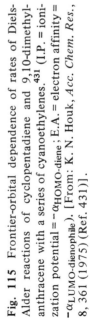

Fig. 115 Frontier-orbital dependence of rates of Diels-Alder reactions of cyclopentadiene and 9,10-dimethyl-anthracene with a series of cyanoethylenes.[431] (I.P. = ionization potential = $-\alpha_{HOMO\text{-diene}}$; E.A. = electron affinity = $-\alpha_{LUMO\text{-dienophile}}$). [From: K. N. Houk, *Acc. Chem. Res.*, **8**, 361 (1975) (Ref. 431)].

393

TABLE 94 Regioselectivities and Frontier Orbital Properties in Diels-Alder Reactions

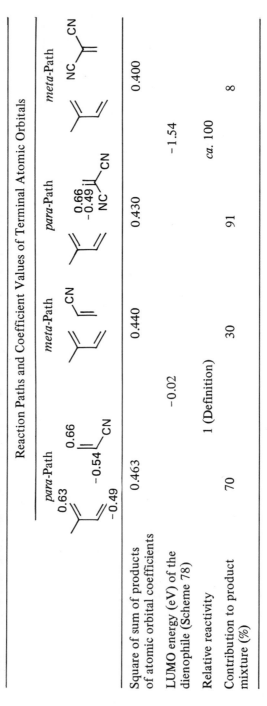

	Reaction Paths and Coefficient Values of Terminal Atomic Orbitals			
	para-Path	*meta*-Path	*para*-Path	*meta*-Path
Square of sum of products of atomic orbital coefficients	0.463	0.440	0.430	0.400
LUMO energy (eV) of the dienophile (Scheme 78)	−0.02		−1.54	
Relative reactivity	1 (Definition)		*ca.* 100	
Contribution to product mixture (%)	70	30	91	8

394

reactivity results in an *increase* in selectivity: low values of the HOMO-LUMO energy difference (leading to high reactivity) also accentuate differences in the numerator in (*504*) for competing reaction paths. This is further illustrated by Tables 95 and 96 which show that the more reactive dienophiles surpass the less reactive ones both in their ability to discriminate between various dienes (Table 95) and in the predominant formation of *endo* adducts (Table 96). The general relationships between reactivity, selectivity, and position of the transition state on the reaction coordinate of frontier orbital-controlled reactions are depicted in Fig. 116A. Higher reactivity (lower values of $\Delta G^{\ddagger}$) is reflected by a smaller slope of the line signifying the reactant changes. Since the properties of the products are considered to be uninfluential, the same product line applies to different pairs of reactants. Is is seen from Fig. 116A that the more reactive systems (dashed lines) have both higher selectivity ($\Delta\Delta G^{\ddagger}$) and later transition states.*

Values for crucial coefficients and for HOMO-LUMO energies of some dienes and dienophiles are given in ref. 435.

By means of the frontier orbital method Fukui was able to derive the stereochemistry of other pericyclic reaction types.[429] For instance, the electrocyclic reaction of hexatriene (*505*) is considered to be an intramolecular Diels-Alder reaction, where one part of the molecule functions as *dienophile* and the other as *diene*; the doubly suprafacial interaction results in disrotation.

TABLE 95 Kinetics of Reactions of Conjugated Dienes with Maleic Anhydride and Tetracyanoethylene in Dioxane[410]

Diene	Maleic Anhydride $10^8 \times k_2, 30°C$ (l/mol s)	Tetracyanoethylene $10^5 \times k_2, 20°C$ (l/mol s)
Cyclopentadiene	9,210,000	≈43,000,000
9,10-Dimethylanthracene	1,600,000	≈1,300,000,000
1,3-Cyclohexadiene	13,200	7,290
Hexachlorocyclopentadiene	1.14	—
1,2-Bismethylenecyclohexane	755,000	1,230,000
1,1'-Biscyclopentenyl	118,000	1,900,000
1-Methoxybutadiene	84,100	598,000
2,3-Dimethylbutadiene	33,600	24,300
trans-1-Methylbutadiene	22,700	2,060
2-Methylbutadiene	15,400	1,130
Butadiene	6,830	519
2-Chlorobutadiene	690	1.0

*This is plausible because there will be more charge transfer between reaction partners whose front orbitals are close to each other in energy.

TABLE 96   Reactivity and *exo/endo* Selectivity in Reaction[411]

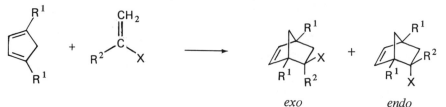

Diene	Dienophile			
R^1	R^2	X	k_{rel}	% *endo* Adduct
H	*N*-Phenylmaleinimide		2.45×10^3	>99.5
H	Maleic anhydride		2.30×10^3	99.5
H	H	COBr	5.83×10^2	92.2
H	H	COCl	3.77×10^2	90.7
H	H	$CO\text{-}i\text{-}C_3H_7$	3.20	82.1
H	H	$CO\text{--}C_3H_7$	2.40	80.5
H	H	COC_2H_5	2.28	82.8
H	H	$COCH_3$	2.28	82.3
H	H	CHO	2.00	75.0
H	H	CO_2CH_3	1.00 (Definition)	73.7
C_6H_5	H	CO_2CH_3	1.30×10^{-1}	51
p-Cl-C_6H_4	H	CO_2CH_3	4.81×10^{-2}	50
H	CH_3	CO_2CH_3	1.65×10^{-2}	35.3
H	C_2H_5	CO_2CH_3	1.62×10^{-2}	37.0
H	C_4H_9	CO_2CH_3	7.78×10^{-3}	30.6

The theoretical justification of this procedure is doubtful.

5.2.4.4   *Aromatic Transition States*

The most general treatment, initiated by Evans[436] and recently emphasized mostly by Zimmerman[437] and Dewar,[438] considers the transition
states of stereospecific pericyclic reactions to be *aromatic* and thus

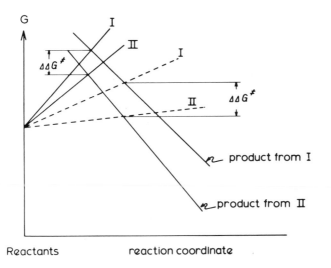

Fig. 116A Effects of reactivity on selectivities and on positions of transition states on the reaction coordinate of orbital controlled reactions. ——————— Energy changes on interaction of a less reactive reagent (e.g. maleic anhydride) with reaction partners (e.g. 1,3-dienes) I and II, respectively. — — — — — Energy changes of a more reactive reagent (e.g. tetracyanoethylene) with the same reaction partners.

endowed with special stability, lacking in both reactants and products. The disrotatory electrocyclic reaction of hexatriene is instructive (*506*):

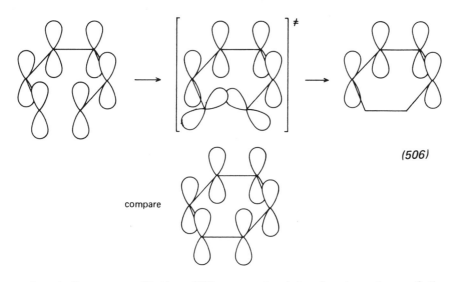

(506)

compare

Apart from quantitative differences (mainly due to values of the resonance integral between individual atomic orbitals), it is clear that in the activated complex of the disrotatory ring closure the mobile electrons find themselves in a situation similar to that of π electrons in benzene. In both cases six electrons can be delocalized over cyclic molecular orbitals obtained by linear combination of a cyclic basis set of six atomic orbitals, each of which can enter into a *bonding* overlap with *both* its neighbors.* Aromatic stabilization is to be expected to become operative in the transition state of the disrotatory cyclohexatriene cyclization and its contribution should lie between that of the stabilization in hexatriene, or cyclohexadiene (12.8-25.1 kJ/mol) and that in benzene (150.8 kJ/mol).

In accord with Hückel's rule the same should be true for all other *disrotatory* ring closures or ring openings involving *4n + 2* electrons (see examples in Table 98). On the other hand, disrotatory reactions in systems with *4n* mobile electrons should involve *antiaromatic* transition states (in the case of butadiene, "cyclobutadiene-like") and should be difficult to achieve. Favoring of *conrotatory* reactions in *4n* systems is due to the "*Möbius aromaticity*" of the corresponding transition states.

"*Möbius systems*" occur when in constructing molecular orbitals of a monocyclic conjugated system one starts with a basis set of *n 2p* orbitals in which each orbital is rotated by π/n degrees out of the plane of the preceding orbital, so that a nodal plane (sign inversion, phase inversion) is formed between the *n*th and the first atomic orbital (Fig. 117). In such systems stable (aromatic) states with closed shells are obtained when the total number of π electrons is *4n* (Fig. 118). (Basis sets with even numbers of sign inversions give rise to "Hückel situations.")

Stereospecificity is the result of preferential development of that topology whose basis set allows the formation of the aromatic system in accord with the number of mobile electrons (Hückel or Möbius).

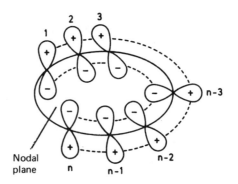

Fig. 117 Möbius basis set.

*This is the typical starting situation in the traditional Hückel MO theory.

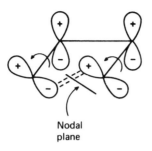

Nodal
plane

Fig. 118 Transition state of conrotatory cyclobutene formation as a Möbius system.

Transition states of other pericyclic reactions can also be considered to be "Hückel" or "Möbius systems." Substantial deviations from the situation in classical aromatic compounds may occur. However, the essential point is whether the number of overlaps occuring with sign inversion is even (e.g., 0) or odd (e.g., 1). (Sign inversions within the same atomic orbital may not be counted. See Figs. 119 and 120).

In this context we should recall the analysis of the σ^+-dependence of the ratio of *"para"* to *"meta"* adducts in the uncatalyzed and $AlCl_3$-

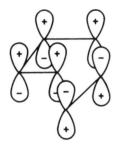

Fig. 119 Hückel basis set for the Diels-Alder reaction.

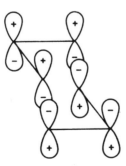

Fig. 120 Hückel basis set for the $[3_s,3_s]$-sigmatropic Cope (or Claisen) rearrangement.

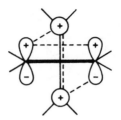

Fig. 121 Möbius basis set for the $(2_s + 2_a)$-ethylene dimerization.

catalyzed Diels-Alder reaction (p. 268 ff), implying the occurrence of an aromatic transition state.

With normal olefins the aromatic suprafacial-antarafacial dimerization to a cyclobutane (Fig. 121) is not observed because there are severe steric interactions between the ligands of the reaction partners. Strained *trans*-cycloalkenes do show (2s + 2a)- dimerization (507)[439]; however, it seems that a diradical process is operating in these cases[440a] as well as in cyclobutane formation from closely related bridgehead olefins.[440b]

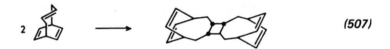

$$(507)$$

Cyclobutane rings are formed relatively easily from olefins and ketenes or allenes.[440c] In these cases steric hindrance must be less. It has also been hypothesized that in the case of ketenes acting as the 2a-component overlap of the additional (electrophilic) $2p$ lobe of C-2 with the ethylene π bond aids in putting the reaction partners together (Fig. 122). An alternative view is shown in (508).

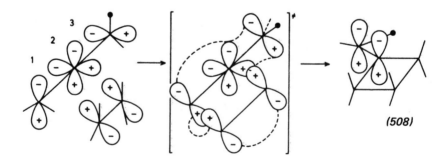

$$(508)$$

Atom 3 (and atom 1) of the cumulene partner must rotate 90° during the transformation and in the transition state (after ± 45° rota-

Fig. 122 Möbius basis set for a ketene-olefin cycloaddition.

tion) its $2p$ orbital can overlap with *both* $2p$ orbitals of C-2; this results in a Hückel basis set. The reaction effectively involves six π electrons and thus possesses an aromatic transition state.

The theory of aromatic transition states is particularly suitable for the analysis of sigmatropic reactions, since lack of symmetry elements in reactant and product rules out the analysis involving symmetry correlation diagrams.

Basis Sets for Stereospecific Allowed Sigmatropic Rearrangements. In constructing these sets one should start by assigning the orbital lobes formed from the original σ bond (which is broken during the transformation) the same sign.

1   Thermal Hydrogen Shifts.

[1_s ,5_s] (Hückel)

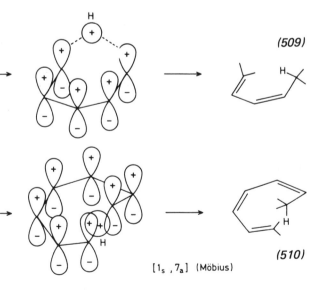

(509)

[1_s , 7_a] (Möbius)

(510)

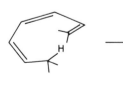

2 Shifts of Atoms Able to Use *p* Orbitals; Thermal Carbon Shifts.

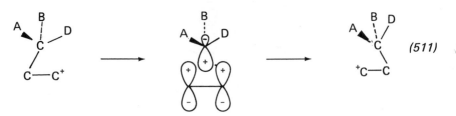

$[1_s,2_s]$ (Wagner-Meerwein rearrangement, Hückel)

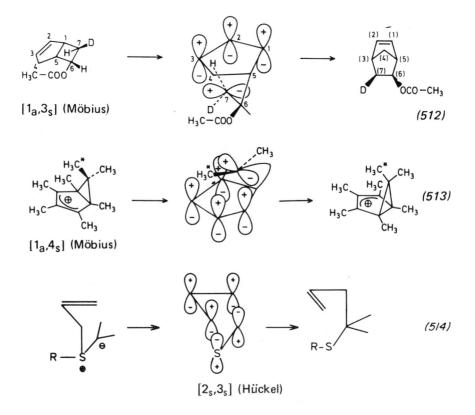

$[1_a,3_s]$ (Möbius)   (512)

$[1_a,4_s]$ (Möbius)   (513)

$[2_s,3_s]$ (Hückel)   (514)

5.2.4.5 *General Formulation of the Woodward-Hoffmann Rules*

In conclusion we present the general formulation of the Woodward-Hoffmann rules in a form easy to use as a mnemonic.[441] (See Table 97.) One starts with the total number of rearranging *electron pairs*. For reactions that can be formally split into two components (cyclo-additions, sigmatropic rearrangements, group transfer reactions, and

TABLE 97 Simplest Formulation of the Woodward-Hoffmann Rules

	Total Number of Shifting Electron Pairs	Stereochemistry of the Second Component (when the first reacts suprafacially with retention)	Electrocyclic Reactions
Thermal reactions	Even	Antara facial (or inversion)	Conrotatory
	Odd	Suprafacial (or retention)	Disrotatory
Photochemical reactions	Even	Suprafacial (or retention)	Disrotatory
	Odd	Antarafacial (or inversion)	Conrotatory

TABLE 98 Examples of Thermal and Photochemical Electrocyclic Reactions

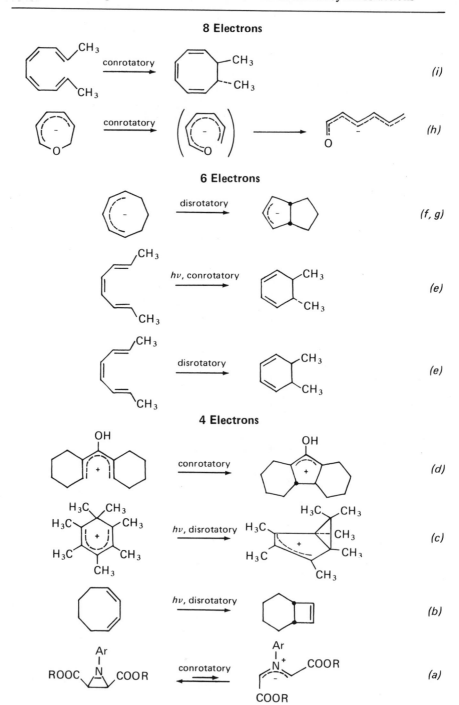

8 Electrons

(i)

(h)

6 Electrons

(f, g)

(e)

(e)

4 Electrons

(d)

(c)

(b)

(a)

cheletropic reactions)* one component reacts suprafacially and the stereochemistry of the other is governed by Table 97. Examples of electrocyclic and cycloaddition reactions are given in Tables 98 and 99.

5.2.4.6 Secondary Orbital Effects

The favoring of *endo* adducts in $(4_s + 2_s)$-cycloadditions (*"endo* rule"; the same is observed in *homo*-Diels-Alder reactions of bicyclo[2.2.1]-hepta-2,3-dienes[441a] mentioned on p. 386;) can be explained by the occurrence of *additional* bonding frontier orbital interactions (lacking in the isomeric *exo* transition states). Avoidance of antibonding interactions explains the favoring of the *exo* transition state in $(6_s + 4_s)$-cycloadditions (Fig. 123).

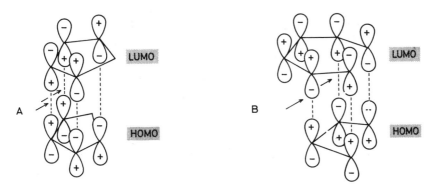

Fig. 123 (*a*) Additional bonding overlap ($\rightarrow$) in the *endo*-transition state for dimerization of cyclopentadiene. (*b*) Additional repulsions ($\rightarrow$) make the *endo*-transition state for $(6_s + 4_s)$-cycloaddition less favorable.

*In these reactions the rules are valid only if the leaving group moves away in a *linear* fashion: CX $\rightarrow$ C + |X. Opposite rules are valid when X rotates when departing: CX $\rightarrow$ C + X̲ (*"nonlinear* cheletropic reactions").[423]

TABLE 98 Footnotes

(*a*) R. Huisgen et al., *J. Amer. Chem. Soc.* **93**, 1779 (1971).
(*b*) R. S. H. Liu, *J. Amer. Chem. Soc.* **89**, 112 (1967).
(*c*) Ref. 446.
(*d*) R. B. Woodward, in "Aromaticity," Chem. Soc., Spec. Publ. No. 21, London 1967, p. 237.
(*e*) E. N. Marvell et al., *Tetrahedron Lett.* **1965**, 385; E. Vogel et al., *ibid.* **1965**, 391.
(*f*) P. R. Stapp and R. F. Kleinschmidt, *J. Org. Chem.* **30**, 3006 (1965).
(*g*) R. B. Bates and D. A. McCombs, *Tetrahedron Lett.* **1969**, 977.
(*h*) H. Kloosterziel and J. A. A. van Drunen, *Rec. Trav. Chim. Pays-Bas* **89**, 667 (1970).
(*i*) See Ref. 445.

TABLE 99 Some Recent Cycloadditions

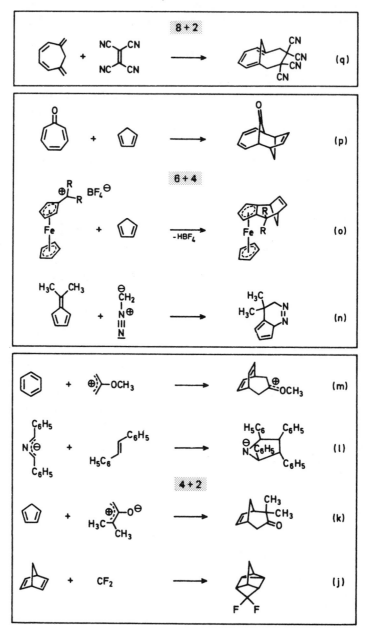

To understand the favoring of the chairform over the boatform transition state in sigmatropic $[3_s, 3_s]$ rearrangements (p. 236), one considers these two transition states to be points on the reaction coordinate connecting two allyl radicals to give the 1,4-cyclohexadiyl diradical and bicyclo[2.2.0]hexane, respectively [$(515a)$ and $(515b)$].

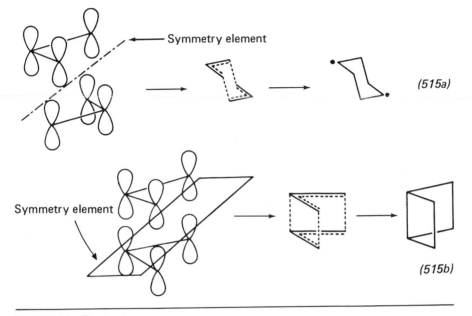

Symmetry element

Symmetry element

$(515a)$

$(515b)$

TABLE 99 Footnotes

(j) C. W. Jefford et al., *Tetrahedron Lett.* **1974**, 257. This reaction may be considered to be a linear cheletropic reaction.

(k) See N. J. Turro, *Acc. Chem. Res.* **2**, 25 (1967). The reacting dimethyloxyallyl cation is one of the tautomers of dimethylcyclopropanone.

(l) T. Kauffmann and E. Köppelmann, *Angew. Chem. Int. Ed.* **11**, 290 (1972).

(m) H. M. R. Hoffmann and A. E. Hill, *Angew. Chem. Int. Ed.* **13**, 136 (1974).

(n) K. N. Houk and L. J. Luskus, *Tetrahedron Lett.* **1970**, 4029.

(o) T. D. Turbitt and W. E. Watts, *J. Chem. Soc., Perkin Trans. 2*, **1974**, 195. The course of this cycloaddition becomes clear when one considers resonance structure a of the ferrocenyl cation.

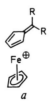

a

(p) R. C. Cookson et al., *J. Chem. Soc., Chem. Commun.* **1966**, 15; S. Ito et al., *Bull. Chem. Soc. J.* **39**, 1351 (1966).

(q) G. C. Farrant and R. Feldmann, *Tetrahedron Lett.* **1970**, 4979.

An orbital symmetry correlation of reactions (*515a*) and (*515b*) shows (Fig. 124) that because of the correlation of the second MO of the reactant with the fourth MO of the product, the energy rises faster for (*515b*) than for (*515a*); consequently, the boatlike transition state should be higher in energy than the chairlike.

However, such secondary effects are relatively weak and can be easily overcome by other factors, such as unfavorable steric interactions

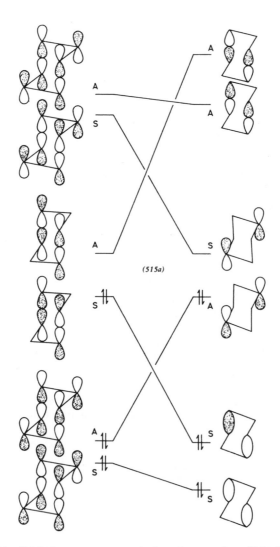

Fig. 124 Orbital symmetry correlations for reactions (*515a*) and (*515b*).

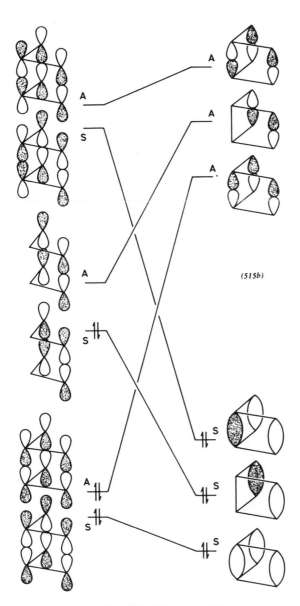

(515b)

Fig. 124 (Continued.)

of alkyl substituents of the dienophile and the CH_2 group of cyclopen-tadiene [442] (Scheme 79).

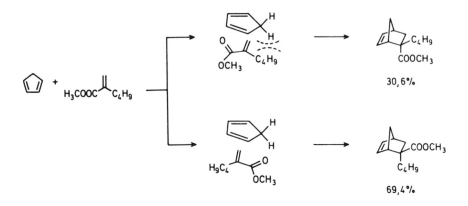

Scheme 79 Diminution of *endo* addition in a Diels-Alder reaction by steric inter-action of an alkyl group in the dienophile and the methylene group of cyclopenta-diene.[443]

Cope and Claisen rearrangements occur relatively easily even when the structure of the molecule allows only a boatlike transition state (pp. 301-432).

A stronger effect is observed in the *disrotatory* solvolysis of cyclo-propyl esters. In principle, opening of cyclopropyl cations can occur in two different ways (Scheme 80); however, it is observed that when

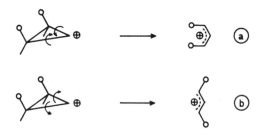

Scheme 80 Two possibilities for the disrotatory opening of a cyclopropyl cation.

ionization and ring opening are concerted, only the disrotation in which substituents *cis* to the leaving group rotate *inward* takes place. This can be understood if one attributes to the electron cloud of the breaking bond the role of a nucleophile in an intramolecular S_N2 reaction [back-side attack; (*516*)]. A different viewpoint is given on page 428.

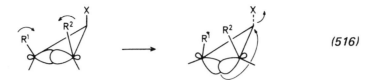

$$(516)$$

If R^1 and R^2 are ends of a (relatively short) bridge, only the leaving group in the *endo* position (X) is capable of concerted ionization; the leaving group Y in the *exo* position would lead to an exceedingly strained allyl cation (*517*).

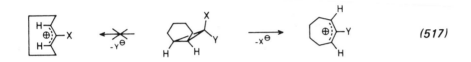

$$(517)$$

5.2.4.7   How Rigorous Are the Woodward-Hoffmann Rules?

The energetic favoring of a reaction path over its stereochemical alternative can be estimated by comparing the π-electron energies of the Hückel and Möbius transition states for the reaction under consideration.

The energies of π MOs of cyclic polyenes can be obtained in the following way: the appropriate polygon is inscribed into a circle of radius $2|\beta|$ (β = resonance integral). The ordinates of the intersections of polygon and circle give the MO energies. The ordinate of the center has the value α, where α is the Coulomb integral. *Hückel systems* are inscribed with one vertex down, *Möbius systems* are inscribed with one side down (Fig. 125).[444]

Fig. 125 Determination of orbital energies of cyclic Hückel and Möbius systems, illustrated for the benzene case. *Left*: π-electron energy, $6\alpha - 8|\beta|$; *right*: π-electron energy; $6\alpha - 6.93\ |\beta|$.

The thermal disrotatory ring closure of hexatriene is found to be more favorable than the conrotatory one by $-1.07\ |\beta|$ ($\beta = -75.5$ kJ/mol); this is the difference in π-electron energy between benzene and "Möbius benzene." Some experimental data are given in Table 100 (see also Scheme 72).

TABLE 100 Stereoselectivities in Some Pericyclic Reactions

Reaction	$\Delta\Delta G^{\ddagger}$ (kJ/mol), forbidden – allowed	Temperature K	Ref.
Item (*i*), Table 98	46.5	444	445
Equation (*513*)	$\geqslant 23.9$	264	446
Equation (*509*)	$\geqslant 33.5$	573	447

Pyrolysis of *cis*-3,4-dimethylcyclobutene (*518*)[447a] gave the products of the allowed and forbidden reactions in the ratio 99.9:0.005, corresponding to a value of $\Delta\Delta G^{\ddagger}$ at 280°C of 45 ± 2 kJ/mol. Inspection of other experimental data for allowed and forbidden cyclobutene thermolyses showed that $\Delta\Delta S^{\ddagger}$ is probably close to zero; consequently, $\Delta\Delta G^{\ddagger}$ could be equated to $\Delta\Delta H^{\ddagger}$. Taking into consideration the *steric* effect, which makes the formation of *cis*-1-methylbutadiene-1,3 less favorable by 20–40 kJ/mol (see also p. 287 and Table 79), it was concluded that the stabilization of the allowed over the forbidden transition state in this pericyclic reaction is $\geqslant 63$ kJ/mol.

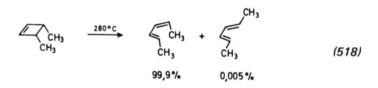

$$(518)$$

5.2.4.8 Forbidden Reactions[447b]

The Woodward-Hoffmann rules, like any other selection rules, are permissive but not prohibitive. If, for instance, the geometry demanded by the rules is incompatible with the steric properties of the substrates, or if it can be achieved only with difficulty, it *may* mean that the reaction will not take place [e.g., thermal ($2_s + 2_a$) cyclobutane formation from "normal" olefins].

A counterexample is the anionic 1,2-alkyl shift in the Wittig ether rearrangement[448] (*519*):

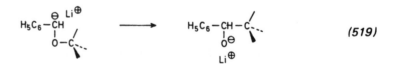

$$(519)$$

This reaction proceeds readily, although according to the rules it would involve a sterically rather unfavorable transition state (Fig. 126*a*).

Fig. 126 The Wittig ether rearrangement: (*a*) Möbius aromatic transition state and (*b*) intermediate radical pair.

It has been shown that this reaction (and similar anionic 1,2-rearrangements) chooses the two-step path, via a radical pair (Fig. 126*b*), rather than the unfavorable transition state of the concerted rearrangement.

The *relatively easy formation of intermediates* is also invoked in the interpretation of "forbidden" [2 + 2]-cycloadditions. Such reactions are observed when fluoroethylenes are heated with normal or, preferably, conjugated olefins[449] (Scheme 81), or when olefins strongly polarized in opposite directions react with each other (Scheme 82).[450]

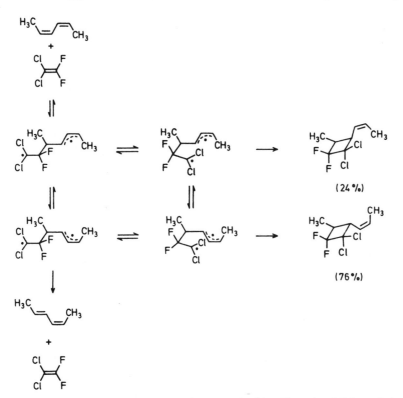

Scheme 81 Mechanism and products of a nonpolar [2 + 2]-cycloaddition of *cis,cis*-1,4-dimethylbutadiene-1,3.

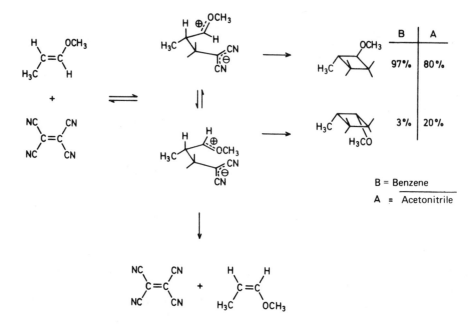

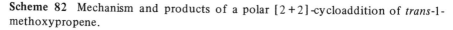

Scheme 82 Mechanism and products of a polar [2 + 2]-cycloaddition of *trans*-1-methoxypropene.

The driving force in the first case is the higher C—F bond energy in the tetrahedral intermediate and its additional stabilization by halogen atoms and allylic resonance; in the second case the driving force is the propensity of substituents to stabilize charges. The observed *cis-trans* isomerization of reactants in these reactions shows that the formation of the intermediates is reversible. [451]

Products of [2 + 2]-*cycloreversions* show that here, too, the reaction proceeds in two steps. In accord with the principle of microscopic reversibility, it should proceed along the same reaction coordinate as that of the corresponding (hypothetical) [2 + 2]-cycloadditions. A good example is the thermolysis of *cis*- and *trans*-6,7-dimethylbicyclo-[3.2.0]heptane (520),[452] which is practically nonstereoselective and gives in the case of the *trans* compound predominantly the "forbidden" butene-2.

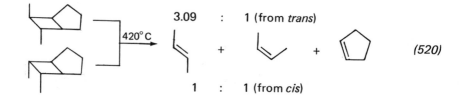

An alternative interpretation for nonstereospecific pericyclic transformations states that the reason for the formation of diastereomers is *not* the intermediate formation of a species capable of internal rotation, but rather *two simultaneously occurring competing one-step processes*, one of which is allowed and the other forbidden. This hypothesis was proposed primarily by Berson to explain the stereochemical outcome of thermal [1,3]-carbon migrations [e.g., *(521)*].[453]

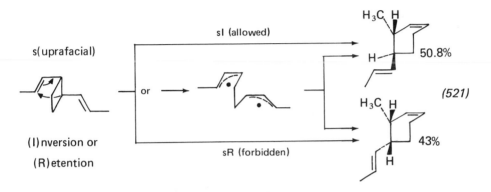

One of the arguments for the favoring of *synchronous*, Woodward-Hoffmann *forbidden* processes over *diradical* processes is that the interaction of a π system with the *second highest* occupied MO of an interaction partner ["subjacent orbital control," Fig. 127*b*] can make such a process still more favorable than the path involving noninteracting intermediate radicals (Fig. 127*a*).[454]

A clear choice between the two alternative interpretations is not yet possible.

The occurrence of reactions that violate the Woodward-Hoffmann rules may be a consequence of factors *not* encompassed by the rules. An example is the thermal *syn* elimination of HX from alkyl halides *(522)*, apparently a forbidden [2s + 2s]-transformation, but involving one of the unshared electron pairs of the halogen.[455] The rules do not allow unequivocal predictions for reactions involving an atom which contributes more than one orbital (see ref. 423b, p. 199).

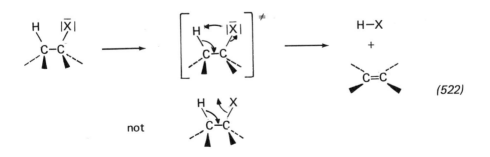

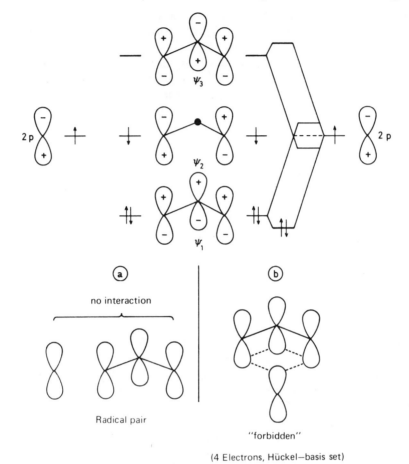

Fig. 127 Subjacent orbital control. Bonding interaction of a $2p$ orbital [e.g., of the migrating carbon atom in (521)] and the *second highest occupied* MO of the allyl fragment (ψ_1) in the geometry shown (b) (leading to retention of configuration at the migrating carbon) lowers the energy of ψ_1 (b). As a consequence, the Woodward-Hoffmann forbidden transition state (b) (see Table 97) becomes lower in energy than the radical pair (a) involved in the step-wise reaction. [From J. A. Berson, *Acc. Chem. Res.*, **5**, 406 (1972) (ref. 453).]

It was proposed to designate such reactions as *pseudopericyclic* [J. A. Ross, R. P. Seiders, and D. M. Lemal, *J. Amer. Chem. Soc.* **98**, 4325 (1976)]. Another example is possibly the hydroboration of olefins [see also K. Fukui, *Bull. Chem. Soc. Jap.* **39**, 498 (1966)].

The low activation enthalpy ($\Delta H^{\ddagger}$ = 94 kJ/mol) and the activation entropy ($\Delta S^{\ddagger}$ = −29 J/K mol) suggest that in the forbidden conrotatory cyclization of hexatriene **288** to **289** intermediate diradicals do not

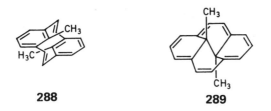

288   **289**

play any role.[456] The extended π system of these molecules contains empty molecular orbitals of relatively low energy, enabling mixing of electronic configurations as given by the *Aufbau* principle (configuration interaction). As a result, in a way analogous to mixing (interaction) of individual orbitals, the energy of the system is lowered even in the transition state of the forbidden reaction (Scheme 83). The isomerization of Dewar-benzene to benzene probably occurs largely via the forbidden disrotatory electrocyclic reaction path.[376a]

(*a*) Individual orbitals   (*b*) Individual electronic configurations

Scheme 83   (*a*) Lowering of electronic energy by interaction of an occupied orbital and an unoccupied orbital. (*b*) Lowering of electronic energy by interaction of two electronic configurations.

5.2.4.9   Acyclic Systems

The model of the dominant HOMO-LUMO interaction can also be applied to acyclic systems.[6a] For instance, the observed predominance of *trans eliminations* is in accord with calculations that show that the highest value of the square of atomic orbital coefficients of β-hydrogen atoms in the σ-LUMO of chloroalkanes (extending over all bonds) must be ascribed to the *trans-β*-hydrogen atom.[457a,457b]

The stereochemistry of the *S_N2 reaction* becomes apparent on inspection of the LUMO of methyl chloride (Fig. 128). Attack by the nucleophile from direction 1 is substantially more favorable than attacks from directions 2 (poor overlap, hindrance of departure of Cl⁻), 3 (no bonding interaction), and 3' (strong repulsion of the nucleophile by the electron pairs of Cl and the C—H bond).[457c]

The same results can be obtained in a qualitative treatment. It involves the following modification of the basic concept of the aromatic transition state, stating that any cyclic orbital system is qualita-

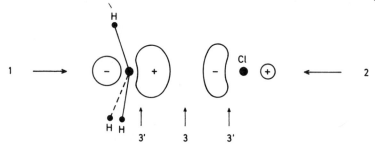

Fig. 128   LUMO (σ^*) of methyl chloride and possible directions (1-3′) of attack by the nucleophile. [From L. Salem, *Chem. Brit.* **5**, 449 (1969) (ref. 457c).]

tively equal to the basis sets of ideal Hückel or Möbius systems: the orbital systems of transition states of acyclic additions (and their reverse, eliminations) and of substitutions may be compared to orbital systems of *linear polyenes*.

In the analysis of the stereochemistry of the termolecular addition of A^+A^- to ethylene the process is dissected into two hypothetical consecutive steps: addition of the electrophile and addition of the nucleophile.[457d] In the first stage an allyloid system is formed (Fig. 129),

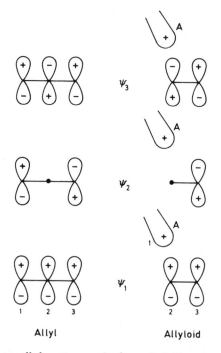

Allyl   Allyloid

Fig. 129   MOs of an allyl system and of an allyloid system. The latter is formed when reagent A interacts with one of the $2p$ orbitals of a double bond.

Fig. 130   Transfer of orbital signs from C-1 to C-2.

whose HOMO is ψ_1. The LUMO, ψ_2, controls the second step, the addition of A⁻. One considers the rehybridization of the carbon atoms of ethylene accompanying addition: if the C-1—A bond is formed via an sp^3 orbital endowed with a positive sign then the $2s$ orbital of C(1) and—because of the bond between C-1 and C-2—the $2s$ orbital of C-2 must also have a positive sign (Fig. 130). In the rehybridization of C-2, resulting from the addition of A⁻, superposition of the $2p$ and $2s$ orbitals (the sign of the former as in ψ_2) leads to the development of a bonding (positive) main lobe of the sp^3 orbital *trans* to the C-1—A bond.

Similarly, it can be shown that *1,4-additions* to butadiene, where one must consider *pentadienyloid* states **290**, proceed with *cis* stereochemistry.

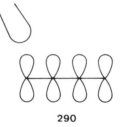

290

Several examples are known, for example,[458]

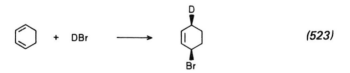

(523)

The stereochemistry of concerted eliminations must be the same as that of the corresponding additions. *syn* Elimination has been observed in a 1,4-elimination.[463]

The first of the two hypothetical steps in a S_N2' reaction[459] is the *breaking of the C—X bond*. The transmission of the sign of the orbital

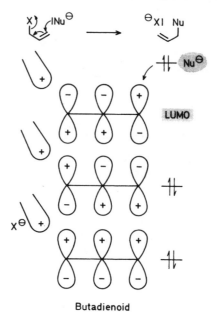

Butadienoid

Fig. 131 Derivation of *syn* stereochemistry for the $S_N 2'$ reaction.

of X through the chain of 2*s* orbitals causes in the LUMO of the *buta-dienoid* transition state the positive lobe of C-3 to develop into the main lobe of the sp^3 bond to Nu, which is in accord with the experimentally observed *syn* stereochemistry (Fig. 131).[457e]

In the transition state of the *syn* substitution, the nucleophile (Nu), the leaving group (X), and the allyl fragment can form a *Hückel aromatic* transition state, **291**, with six delocalized electrons. In the case of a neutral nucleophile the arrangement is also favored by electrostatic effects (**292**). With a negatively charged nucleophile the *anti* transition state **293** may be more favorable because of its lesser electrostatic destabilization.[460]

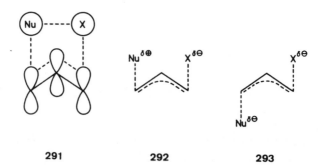

291   **292**   **293**

An elegant, stereospecific total synthesis of *d,l-prostaglandin* A_2 takes advantage of the *syn* stereochemistry of the S_N2' reaction (*522a*).[460a] The initial assumption was that the butadiene epoxide moiety of the β-ketoester anion will react in its *s-trans* conformation, both on steric and thermodynamic grounds. An intramolecular S_N2' reaction then leads not only to the five-membered ring and the side chain of the desired structure, but also to the natural configuration of the asymmetric carbon atoms 12 and 15 and the *trans* geometry of the double bond.

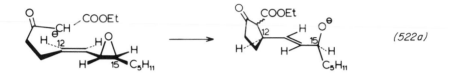

(*522a*)

The qualitative equality of MOs of delocalized acyclic transition states and those of acyclic polyenes is the basis for Zimmerman's "orbital following" method.[461] This method provides orbital correlation

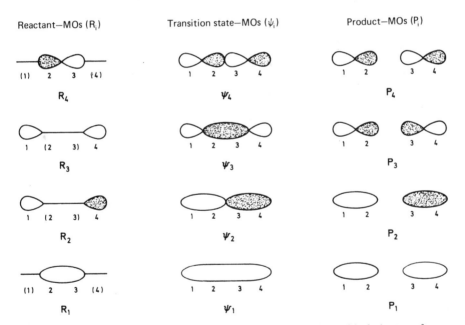

Fig. 132 Conservation of the node character of molecular orbitals in transformation 1,4 + 2-3 → (1-2-3-4)‡ → 1-2 + 3-4 taking place via a butadienoid transition state. [Schematic representation of electron clouds. Changes in sign are represented by the dark areas. The numbers represent atoms. As in Fig. 108, the reactant is considered to be the encounter complex of 1,4 and 2-3. Its MOs have the form of the MOs of the individual reactants.]

diagrams and thus allows the determination of favored reaction paths mainly when orbital symmetry correlations and analysis based on the aromaticity of transition states are impossible. The criterion for the correlation is the intuitive presumption that the *node character* of the individual MOs (number of nodes; symmetric arrangement of nodes with respect to the center of the chain) existing in the transition state (which is approximated by the corresponding polyene) is preserved in the net transformation reactant(s) → transition state → product(s). Figure 132 illustrates such behavior.

As an example the geometry of *addition of methylene to ethylene* is analyzed. The question is whether methylene attacks as in **294** or **295**. The butadienoid basis set of the transition state is represented in **296** (Scheme 84).

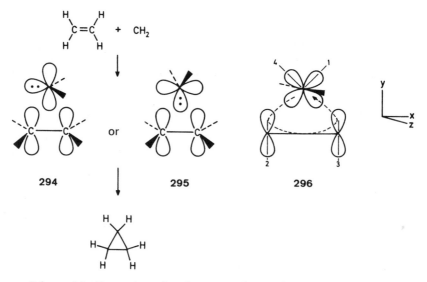

Scheme 84   Formation of cyclopropane from ethylene and methylene.

In **296** the sp^2-hybridization of methylene is lifted. What one has are two orthogonal $2p$ orbitals. The ligands lie on the z-axis of the coordinate system; the orbital correlation should tell whether at the beginning of the transformation they are in the z,x-plane (**294**), or in the z,y-plane (**295**). The correlation diagram is given in Fig. 133. The two $2p$ orbitals 1 and 4 give MOs $p_1 + p_4$ and $p_1 - p_4$. Starting with the sequence of butadiene MOs of the transition state, conservation of the node character (maintaining signs) gives the correlation shown (see also Fig. 132). The crucial result is that the $(p_1 - p_4)$ combination must function as the *occupied* MO of the reactant state. This MO should then correspond to the actual *occupied* nonbonding sp^2 orbital of methylene

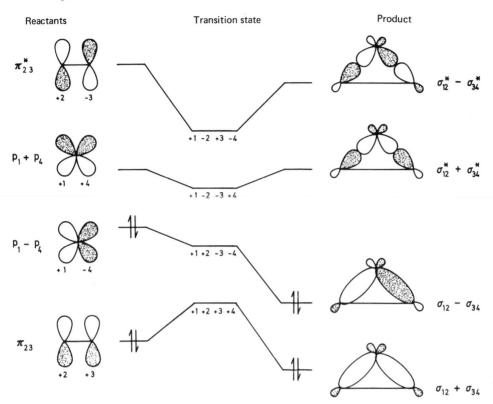

Fig. 133 Correlation of MOs of ethylene and methylene with those of cyclopropane as it results from the conservation of the node character of the butadienoid transition state. [Sign assignment to p_1 and p_4 is based on the basis set **296**: when all three overlaps (1,2, 2,3, and 3,4) are bonding, the orbital lobes involved must have the same sign (color).]

(which results from $p_1 - p_4$ by admixing of s-character from the sp-hybridized C—H bonds in **296**); the unoccupied $(p_1 + p_4)$ MO should represent the empty $2p$ orbital of methylene. This means (Scheme 85) that the reaction starts in arrangement **294** (the same as obtained by calculations, see Fig. 78, and which leads also to the most favorable HOMO-LUMO interaction).

Intramolecular frontier orbital interactions may be causes of stereoselective behavior. The preference for formation of equatorial cyclohexanols by axial addition of hydride or organometallics to cyclohexanones has been explained as follows.[462] The major interaction occurs between the nucleophile's HOMO and the substrate's LUMO. The most stable LUMO is the one resulting from interaction of $\pi^*_{C=O}$ and a σ^* orbital on a carbon atom adjacent to C=O (Fig. 134a). For optimum overlap the C-2—L bond has to be parallel to the π system. Nu$^-$

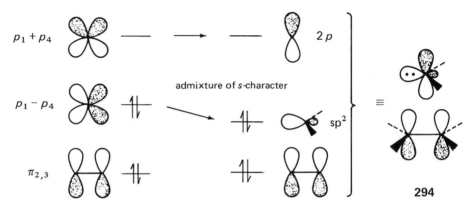

Scheme 85 Lifting the degeneracy (introduced in **296** and Fig. 133) of the non-bonding orbitals of methylene.

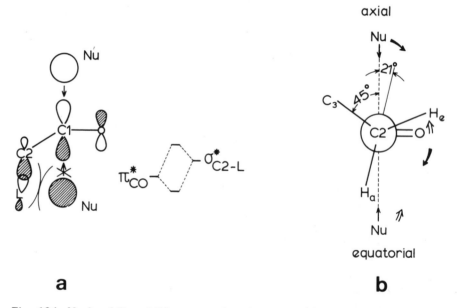

Fig. 134 Nucleophilic addition to carbonyl groups. (*a*) Interacting π^* and σ^* orbitals. (*b*) Newman projection showing antiperiplanarity in axial and equatorial addition, respectively, of a nucleophile to cyclohexanone.

approaches along the axis of the $2p$ lobe. Antiperiplanar attack is favored over synperiplanar attack for steric and overlap reasons.

Equatorial and axial attack on cyclohexanone are depicted in a Newman projection in Fig. 134*b*. It is seen that for axial attack the favorable antiperiplanarity of C-1 ⋯ Nu and C-2—H$_a$, (and C-6—H$_a$)

can be reached relatively easily by a 21° torsion of the segment Nu⋯C-1=O (heavy arrows). By contrast equatorial attack cannot be aided by antiperiplanarity of C-1⋯Nu and C-2—C-3 (and C-5—C-6) since this would require a 45° torsion of the Nu⋯C-1=O segment (opens arrows) which is impossible in a cyclohexane ring. The stereochemistry of hydride or methyllithium addition to cycloalkanones has also been thoroughly analyzed in terms of *steric approach control* and *product stability* effects.[462a]

The preferential occurrence of *trans* elimination can be understood if one considers the intramolecular HOMO-LUMO interaction between C—H and C—X bonds required for the electron flow (Scheme 86).[464]

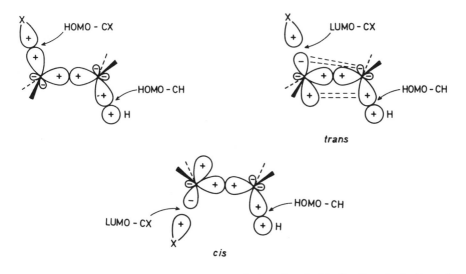

Scheme 86 Relationship of orbital phases in bonds suitable for the elimination of HX from a fragment X—C—C—H. In the *trans*-coplanar conformation there is the possibility of maximum interaction of HOMO—CH and LUMO—CX [and of HOMO—CX and LUMO—CH (not shown and insignificant because of the larger energy difference)]. Such an interaction is impossible in the *cis*-coplanar conformation.

The strongly enhanced reactivity of the "*α-effect*" *nucleophiles* [hydrazines, oxyanions (ClO^-, RO_2^-), hydroxylamine] toward electrophiles with π orbitals, primarily carbonyl compounds (e.g., phenyl acetates: Fig. 40 and Table 60), but also triarylmethyl cations and Michael acceptors was ascribed to the possibility of forming Hückel aromatic six-electron transition states,[465] (e.g., **297**). (Significantly, the α-effect should *not* appear in S_N2 reactions.)

The perturbation theory treatment of the reactivity of various organic reaction types was reviewed by Hudson.[466] In the examples

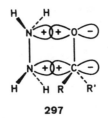

297

described, initial perturbation should determine the overall course of the reaction (see also Section 5.2.4.3).

5.2.5 The Principle of Least Nuclear Motion (PLNM)

An amazingly fruitful method for the elucidation of favorable reaction paths turned out to be the *principle of least nuclear motion*: among competing elementary reactions (subject to the same product stability and orbital symmetry factors) the one involving the least change in position of atoms should be the most favorable. For competing reactions $[\Delta G(1) = \Delta G(2)]$ Hine deduced this principle from a BEP analysis.[467]

$$
X - Y + Z \quad
\begin{cases}
\xrightarrow{(1)} X + Y - Z \text{ (small change in the geometry of Z)} \\[2mm]
\xrightarrow{(2)} X + Y - Z' \text{ (considerable change in the geometry of Z)}
\end{cases}
\qquad (523)
$$

If reaction (1) can be described by a pair of dissociation curves $x(X - Y \to X + Y)$ and $z[Y - Z \to Y + Z$ or $(Y - Z' \to Y + Z')]$, reaction (2) should be describable by the pair x' and z. The steeper path of x' corresponds to the additional energy required to overcome the most stable geometry of Z in the ground state.

The difference in potential energy of the two transition states (T, T'; Fig. 135) can be calculated, according to Hooke's law, from the difference in the squared changes of bond distances corresponding to the two reaction modes. In the analysis of mesomeric systems, changes in distances can be represented by changes in bond order (Δn). An example is given in Fig. 136.

The results given on pp. 348 and 350 can be interpreted in the same way, and from the principle of microscopic reversibility it is possible to draw conclusions with respect to the relative rates of deprotonation. The same is possible for the formation and reactions of mesomeric radicals and carbocations.[467b]

Tee investigated recently the predictive power of the PLNM[468] by calculating the geometric changes in competing reaction paths representing two stereochemical alternatives. Figure 136A is illustrative. The results are given in Table 101. A discussion can be found in ref. 467b.

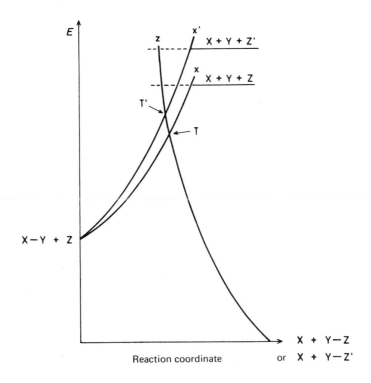

Fig. 135 Derivation of the principle of least nuclear motion. Curve x: energy change in the dissociation of X—Y, Z unchanged. Curve x′: dissociation of X—Y *plus* variation of Z. Curve z: X + Y + Z(Z′) → X + Y—Z(Y—Z′). [After J. Hine, *J. Org. Chem.* **31**, 1236 (1966) (ref. 467).]

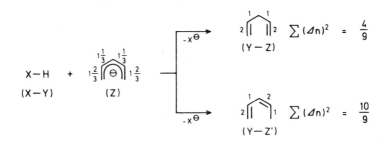

Fig. 136 Protonation of a pentadienyl anion, Z (notation as in Fig. 135) with formation of both isomeric pentadienes (Y—Z and Y—Z′). The sum of squares of bond distance changes (which can be expressed by $\Sigma(\Delta n)^2$, where n = bond order) is smaller for the formation of the nonconjugated hydrocarbon. This hydrocarbon is formed preferentially. (The numbers in the structural formulas indicate bond orders.)

427

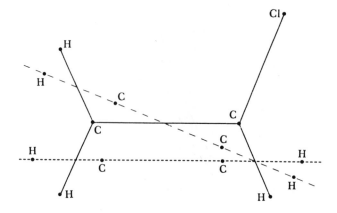

Fig 136A Changes in geometry in the dehydrochlorination of vinyl chloride (solid lines) to give acetylene via a *trans* elimination (long dashes) or a *cis* elimination (short dashes). Movements of the proton abstracted and the leaving chlorine are assumed to be the same for both modes of reaction. The long-dash line contains the four atoms of acetylene as formed by least motion *trans* elimination and the short-dash line shows the position of acetylene resulting from least motion *cis* elimination. The *trans* elimination moves the atoms by 0.320 Å^2 and the *cis* elimination moves them by 1.237 Å^2. According to the PLNM, *trans* elimination is therefore favored. [From J. Hine, *Adv. Phys. Org. Chem.* **15**, 1 (1977) (ref. 467b).]

TABLE 101 Predictions of the Principle of Least Nuclear Motion for Some Reactions

Reaction	Reaction Path of Least Nuclear Motion
1,2-Elimination	*trans* coplanar
1,4-Elimination	*syn* (from a 1,4-disubstituted *cis*-2-butene)
	anti (from a 1,4-disubstituted *trans*-2-butene)
Enolization	Abstraction of the proton whose C—H bond is parallel to the plane of the π bond. In 2-norbornanone *exo* abstraction is much more favorable than *endo* abstraction.
Epoxides from halohydrins	*trans* coplanar
Cyclopropyl $\xrightarrow{\Delta}$ allyl	
Cyclobutene $\xrightarrow{\Delta}$ butadiene	conrotation
Cyclohexadiene-1,3 $\longrightarrow$ hexatriene-1,3,5	conrotation

Only in the last reaction is there a discrepancy between prediction and experiment. The requirement for the conservation of orbital symmetry is here stronger than the tendency toward least geometric change. With respect to the principle of least nuclear motion, chair and boat conformations of the transition states of Cope and Claisen rearrangements turned out to be equivalent. It has been pointed out[468a] that the PLNM is applied most correctly if competing motions toward the transition state (not toward the product) are compared. In most cases this is not feasible, since too little is known about the detailed nature of the transition state.

Usually the PLNM-favored paths correspond to vibrational modes requiring less energy.[467b] If the four atoms of the acetylene molecule formed by *cis* or *trans* elimination from vinyl chloride were to remain in the positions they held in the vinyl chloride molecule, acetylene would be produced in two distorted conformations. It appears that the one formed by (favored) *trans* elimination is also obtained by a bending vibration of acetylene of lower frequency ($\bar{\nu} = 612$ cm^{-1}) and smaller force constant than the one ($\bar{\nu} = 730$ cm^{-1}) leading to the acetylene distorted in *cis* fashion.

5.2.6 The Energetic Criterion for Concerted Reactions

5.2.6.1 *General Considerations**

Stereospecificity conforming to the Woodward-Hoffmann rules or stereoelectronics derived in Section 5.2.4.9 for related and open chain systems constitute a *necessary* but not *sufficient* criterion for ascertaining that the reaction under consideration occurs in *one* step and involves a concerted (synchronous) change in *all* participating bonds, as assumed in the derivation of the rules. There remains the possibility of a *stepwise* reaction, where after a first step in which part of the changes in bonding required for the overall reaction have taken place, a second step sets in, before molecular motions in the intermediate (e.g., internal rotations) can eliminate the conformation derived from the starting material; the geometry of the final product will then have a specific relationship to the starting material. "Memory effects" in the intermediate can thus lead to stereospecificity and simulate a concerted reaction.

Similarly, noncompliance with the predictions of the rules ("forbidden reactions") does not necessarily mean that the reaction proceeds stepwise, as was pointed out in Section 5.2.4.8.

Even the appearance of mixtures of products may be compatible with concerted reactions if one assumes the involvement of concurrent one-step processes. The PLNM will ordinarily favor stepwise reactions:

*See also ref. 22.

a single step will then involve minimal motion (only that associated with *one* bond change). By contrast, concerted reactions will be favored (orbital symmetry allowing) when the formation of a sufficiently unstable intermediate may be avoided by combining several bond changes into one concerted step. This leads us to a second criterion, which *can* be indicative of concertedness: *the activation energy of a concerted reaction must be lower than that of any conceivable nonconcerted reaction path* leading to the same final state (see Section 5.2.2). Since in general the two modes of reaction cannot be realized independently of each other to allow a direct comparison, the application of this criterion requires the choice of a realistic *model* for the nonconcerted reaction to which the real reaction can then be compared. For *nonpolar reactions* (mainly unimolecular thermal gas phase reactions), which may involve *radicals* and *diradicals* as intermediates, such analyses are possible.

5.2.6.2 Gas Phase Reactions of Cyclobutane Derivatives; Mechanisms of the Cope Rearrangement and the Diels-Alder Reaction

The enthalpy of the transition state of the first, diradical-forming step in the generalized reaction (*524*) should be approximately equal to that of the diradical (Hammond's postulate); however, in three- and four-membered ring systems the enthalpy should be higher (by about 40 and 27 kJ/mol, respectively, see p. 340).*

$$A - B \ \rightleftharpoons \ A\cdot \ \cdot B \ \longrightarrow \ \text{products} \qquad (524)$$

　　Reactions involving transition states with enthalpies of formation *substantially lower*† than those calculated for reaction paths leading to diradicals *must* be concerted. If calculated and experimental activation energies are nearly equal, one may draw the conclusion that a reaction path involving diradicals (and leading to the observed reaction rate) is *possible*.

　　The energy difference between reactant and diradical can be calculated or estimated (e.g., from the dissociation energy of the A—B bond), taking into account all other factors that could influence the energies of starting material and diradical [steric energies (ring strain, nonbonded interactions), delocalization energies, stabilization (destabilization) by substituents]. Obviously, one can derive the energy of the

---

*The same should be true for the homolysis and formation of such other strained systems as bicyclo[2.2.1]heptane and was assumed in the formation of 1,5-cyclooctadiene from octa-1,7-diene-3,6-diyl diradicals.[469]

†About 12 kJ/mol; see W. v. E. Doering and K. Sachdev, *J. Amer. Chem. Soc.* **97**, 5512 (1975).

diradical from any other state (compound), as long as there are thermo-dynamic data for that state and for the transformations leading from it to the diradical. Finally, it is possible to compute the energy of the diradical from experimental values obtained for very similar diradicals.

In the following the use of the energetic criterion is demonstrated for some thermal reactions of cyclobutane derivatives.*

1 We begin with (525):

$$\begin{array}{c}H_2C-CH_2\\ |\qquad|\\ H_2C-CH_2\end{array} \rightleftharpoons \begin{array}{c}H_2C\cdot\ \ \cdot CH_2\\ |\qquad|\\ H_2C-CH_2\end{array} \rightleftharpoons \begin{array}{c}H_2C\\ \|\\ H_2C\end{array} + \begin{array}{c}CH_2\\ \|\\ CH_2\end{array} \quad (525)$$

Bond dissociation energy RH_2C-CH_2R	344 kJ/mol
Ring strain in reactant	110 kJ/mol
Reaction enthalpy (ΔH) cyclobutane → diradical	234 kJ/mol
Additional (four-membered ring) activation energy	27 kJ/mol
Calculated activation energy (E_A)	261 kJ/mol
Experimental activation energy	257 kJ/mol

The degree of agreement between the calculated and experimental value allows the assumption that the thermolysis of cyclobutane proceed stepwise, as shown in (525).

2 In the analogous formation of the diradical from *trans*-1,2-divinyl-cyclobutane (526) the activation energy should be lower by twice the allylic resonance energy.

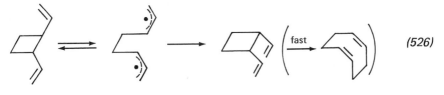

$$E_{A,calc} = 261 - (2 \times 52.5) = 156 \text{ kJ/mol}$$
$$E_{A,exp} = 148 \text{ kJ/mol}$$

The calculated and experimental values are still too close to conclude from the slightly lower value for $E_{A,exp}$ that the reaction is concerted.

3 The activation energy of the thermolysis of *cis*-1,2-divinylcyclobu-tane (527) is so low that this *Cope rearrangement*, although requiring the unfavorable boatlike transition state (see pp. 236 and

*The data are taken from refs. 43c, 369, 469, and 470.

407) must be strongly concerted. The *cis-trans* isomers of 1,2-divinylcyclopropane[479] behave very similarly to those of 1,2-divinyl-cyclobutane.

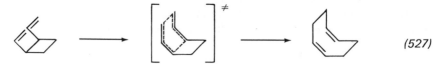

(527)

$$E_{A,exp} = 100 \text{ kJ/mol}$$
$$E_{A,calc} \text{ (diradical)} = 156 \text{ kJ/mol}$$

The analysis of the Cope rearrangement of *open chain* 1,5-hexa-dienes (*528*) illustrates the difficulties involved in choosing an appropriate model for the stepwise reaction. If one chooses two independent allyl radicals **299**, the concerted reaction is favored by 100 kJ/mol:

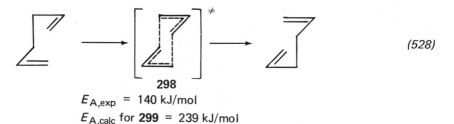

(528)

298

$$E_{A,exp} = 140 \text{ kJ/mol}$$
$$E_{A,calc} \text{ for } \textbf{299} = 239 \text{ kJ/mol}$$

If however, one considers that the first step is not the breaking of the original σ bond, but the formation of the new σ bond, leading to the 1,4-cyclohexadiyl diradical **300**, then ΔH for this step is roughly equal to E_A; that is, such a diradical path seems to become plausible under the given reaction conditions.

Calculation of the enthalpy of formation of **300** and of ΔH for the transformation 1,5-hexadiene → **300**:

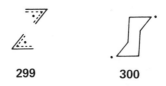

299   **300**

Enthalpy of formation of **300** = enthalpy of formation of cyclohexane *plus* twice the dissociation energy of $H\overset{\text{C}}{\underset{\text{C}}{\text{C}}}{-}H$

minus twice the enthalpy of

describe their reactivity in terms of the Hammett equation as applied to benzene derivatives if one considers them to be *substituted benzenes*. Replacement of one or more CH= groups or one or more —CH=CH— groups by a heteroatom or an annulated ring is expressed by the so called $\sigma_{replacement}$ or $\sigma^+_{replacement}$ constants.* For reactions of a β-side chain in naphthalene (using ρ values from the benzene series) one can write

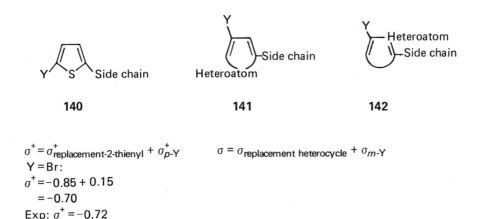

$$\log \frac{k \overset{\text{(naphthalene-R)}}{}}{k \overset{\text{(benzene-R)}}{}} = \rho \cdot \sigma_{replacement}(\text{2-naphthyl}) \qquad (139)$$

Assuming that substituent constants are additive (see above), one can account for the influence of additional substituents on the basic benzene ring by adding σ_p (σ^+_p) or σ_m for that substituent (140, 141).

Computation of substituent constants for side chain reactions (or electrophilic ring substitution) in substituted heterocycles:

Y—(thiophene)—Side chain	Y—(ring)—Side chain Heteroatom	Y—(ring, Heteroatom)—Side chain
140	**141**	**142**

$\sigma^+ = \sigma^+_{replacement\text{-}2\text{-thienyl}} + \sigma^+_{p\text{-}Y}$ $\sigma = \sigma_{replacement\ heterocycle} + \sigma_{m\text{-}Y}$

Y = Br:

$\sigma^+ = -0.85 + 0.15$

   $= -0.70$

Exp: $\sigma^+ = -0.72$

To treat structure type **142** one must use the Yukawa-Tsuno equation (with $\sigma_{p\text{-}Y}$ or $\sigma^+_{p\text{-}Y}$, although Y is formally in the *meta* position), since the heteroatom is particularly well suited to transmit the resonance effect.[124] Some $\sigma_{replacement}$ constants are given in Tables 24 and 25. (However, since experimental data are scarce, many of them should be used for orientation only.)

*Normally, the subscript "*replacement*" is not applied. It is used here to avoid confusion with "normal" σ's expressing the substituent effect of the respective polycycle or heterocycle when attached to a benzene ring.

TABLE 24 $\sigma_{\text{replacement}}$ Constants for Side Chain
Reactions of Naphthalene and Heterocycles: Arrows
Show Position of Chain[114 c]

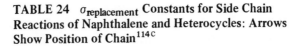

$\sigma = 0,042$ $(\sigma^+ = -0,145$, $\sigma^- = 0,166)$[a]

$0,96$[b]

$0,65$[b]

$0,75$[b]

$(\sigma^{*^1} = 0,65)$[d]

$0,33$[c] $(\sigma^{*^1} = 1,08)$[d]

0[c] $(\sigma^{*^1} = 0,65)$[d]

0[c] $(\sigma^{*^1} = 0,93)$[d]

[a] Ref. 125.   See pages 140 and 142.
[b] Ref. 126.
[c] Ref. 127.
[d] Ref. 128.

Using ρ values for corresponding reactions of benzene derivatives
and the substituent constants above one can at least approximately
describe:

1   Reactions of acids.[126-128]
2   Substitution reactions.

$$R-\overset{|}{\underset{|}{C}}-X \rightarrow R-\overset{|}{\underset{|}{C}}-Y \quad [129-132]$$

$$R-CH_2-CH_2-X \rightarrow R-CH_2-CH_2-Y \quad [127]$$

3   Electrophilic substitutions $R-H \rightarrow R-X$.[130]

More accurate results may be obtained by writing, for each ring system, *its own* correlation equations. For example, in Hammett equations
for side chain reactions and ring substitutions of substituted thiophenes
k_H is no longer the rate constant of the unsubstituted benzene (derivative) for the identical reaction [as in (*139*)], but rather that of the
unsubstituted thiophene (derivative); $\sigma_{\text{replacement}}$ constants drop out,
and the usual $\sigma(\sigma^+, \sigma^-)$ constants are used in equations like (*140*).

TABLE 25 $\sigma_{\text{replacement}}^{+}$ Constants: Arrows Show Position of Electrophilic Ring Substitution or of Side Chains Able to Interact with Ring by Resonance Effect[114c]

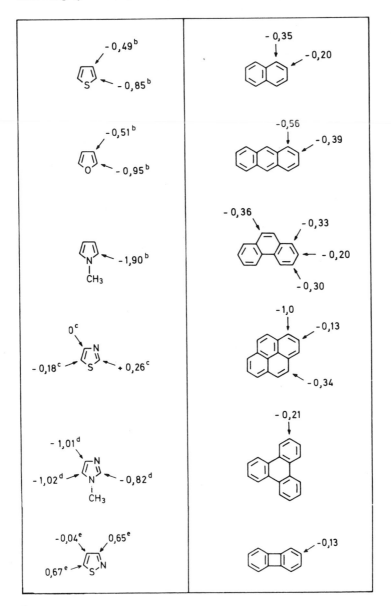

[a] Values for polynuclear aromatics from ref. 129.
[b] Ref. 130.
[c] Ref. 131.
[d] Ref. 132.
[e] Ref. 132a.

115

Values of ρ in such equations are different from those for benzene derivatives (Table 26).

TABLE 26   Electrophilic Substitution of Thiophene and Benzene Derivatives[133]

Electrophilic Substitution	ρ-Thiophenes	ρ-Benzenes
Bromination	-10.0	-12.1
Chlorination	-7.8	-10.0
Protodetritiation	-7.2	-8.2
Acetylation	-5.6	-9.1
Mercuration	-5.3	-4.0

One no longer has complete additivity of effects of the heteroatom and the substituent, because in this instance ρ values in (141), derived with respect to benzene, and those in (140), derived with respect to the unsubstituted heterocycle, should be equal:

$$\log \frac{k_{het-X}}{k_{het-H}} = \rho_{het} \cdot \sigma_X \tag{140}$$

$$\log \frac{k_{het-X}}{k_{benzene-H}} = \rho_{benzene} (\sigma_{replacement} + \sigma_X) \tag{141}$$

Combining (140) and (141) and rearranging one can write

$$\log k_{het-X} = \rho_{het} \cdot \sigma_X + \log k_{het-H} = \rho_{benzene} (\sigma_{replacement} + \sigma_X) +$$
$$\log k_{benzene-H} \tag{142}$$

and taking into consideration the definition of $\sigma_{replacement}$ [see (139)]:

$$\rho_{het} = \rho_{benzene} \tag{143}$$

For the small number of reactions of naphthalene derivatives investigated so far the four-parameter equation (134) gives a reasonable correlation.[134]

This equation also describes (in some cases quite well, in others only approximately) reactivities of side chains and double (and triple) bonds in olefins and acetylenes (and even ring reactivities of some cyclopropanes); all these compounds possess a rigid polarizable molecular structure (as do aromatic compounds), where steric substituent effects play only a subordinate role and inductive and resonance effects of X seem to act in a way similar to that in aromatic compounds.

Reactions of substrates

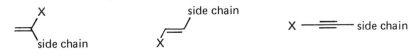

for which correlation equations are available[135] are the following:

1 Reactions of acids (dissociation, esterification, reaction with di-phenyldiazomethane, saponification of esters).
2 Solvolyses and nucleophilic substitutions of halides and reactive esters.
3 Electrophilic additions (halogenation, oxymercuration, hydrobora-tion, formation of complexes with silver ions).
4 Nucleophilic additions of amines.
5 Additions of radicals.
6 Diels-Alder and 1,3-dipolar additions.
7 H/D exchange in acetylenes.

3.1.1.3 Semiempirical Derivation of Substituent Constants According to Dewar and Grisdale

Dewar and Grisdale[136] proposed (*144*) for the computation of σ constants for the general situation **A**;

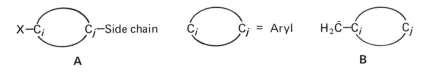

Dewar-Grisdale equation

$$\sigma_{ij} = \frac{F}{r_{ij}} + Mq_{ij} \qquad (144)$$

where F is a measure of the field effect of substituent X, M is its reso-nance effect (both are empirical values) (see below), r_{ij} is the distance between ring carbon atoms i and j, measured in units of the C–C distance in benzene. (The assumption of structures made of regular hexagons with C–C bond lengths as in benzene gives satisfactory results for polynuclear aromatics.) q_{ij} is the fraction of an elementary charge induced by a CH_2^- group at C_i (as a model for substitutent X) on C_j. It is equal to the square of coefficient a (see below) representing the contribution of the $2p$ orbital of C_j to the nonbonding molecular orbital (NBMO) of the arylcarbinyl system **B**.

TABLE 27 Values for r_{ij} and q_{ij}

	i	j	r_{ij}	q_{ij}
Benzene	1	3	$\sqrt{3}$	0
	1	4	2	1/7
Naphthalene	3	1	$\sqrt{3}$	0
	4	1	2	1/5
	5	1	$\sqrt{7}$	1/20
	6	1	3	0
	7	1	$\sqrt{7}$	1/20
Biphenyl	3'	4	$\sqrt{21}$	0
	4'	4	5	1/31

Values for r_{ij} and q_{ij} for benzene, naphthalene, and biphenyl are given in Table 27.

Directions for the calculation of NBMO Coefficients for Nonlinear, Alternant Hydrocarbons[137] *

1. In formulas representing the hydrocarbons carbon atoms are starred in such a way that starred positions alternate with nonstarred ones and the starred set is the more numerous one (see p. 119).

Only orbitals of the "active" starred atoms contribute to the NBMO.

2. The sum of the coefficients a of "active" carbon atoms adjacent to a nonstarred position must be zero.

3. Normalization. The sum of the squares of all coefficients must add up to unity (in our example $7a^2 = 1$, $a = 1/\sqrt{7}$). The NBMO wave function of the benzyl system is

$$\psi_{\text{NBMO,benzyl}} = -\frac{1}{\sqrt{7}} X_2 + \frac{1}{\sqrt{7}} X_4 - \frac{1}{\sqrt{7}} X_6 + \frac{2}{\sqrt{7}} X_7 \quad (145)$$

where X_i = wave function of atom i

*If in formulas representing *alternant conjugated hydrocarbons* (which may never contain odd-membered rings) carbon atoms are alternatively starred and unstarred, it can be seen that each starred atom has only unstarred neighbors, and vice versa. In *nonalternant hydrocarbons* (e.g., fulvene or azulene) the same starring procedure leads to formulas where a starred (unstarred) carbon atom has a neighbor of the same type. The terms *odd* and *even* indicate the number of carbon atoms in the conjugated system. An odd alternant hydrocarbon must exist as an ion or radical. The significance of the concept of alternant and nonalternant conjugated systems for molecular orbital theory is described, for example, in M. J. S. Dewar and R. C. Dougherty, *The PMO Theory of Organic Chemistry*, Plenum Press, New York and London, 1975.

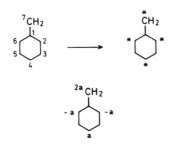

Values of F and M for individual substitutents are found by substituting their σ_m or σ_p and the corresponding values r_{ij} and q_{ij} for benzene in (144). They are given, along with other substituent constants in Table 21.

The values of F and M and (144) enable us to calculate constants for the various possibilities of orientation of substituents and side chain in polynuclear systems. Calculated σ values for sterically unhindered positions in naphthalene derivatives and in biphenyl derivatives with side chains at C-1 or C-4 are given in Tables 28 and 29 (see also Fig. 17).

The Dewar-Grisdale equation (144) applied to five-membered heterocyclic systems gives good agreement between σ_{25} and σ_{24} for the heterocyclic systems and σ_p and σ_m for the benzene system (see above).[138]

Calculation of σ^+ (σ^-) involves a similar procedure. In (144) q_{ij} is the charge generated at position j bearing substituent X, on addition of an

TABLE 28 Calculated σ Values for Sterically Unhindered Positions in Naphthalene Derivatives

Substituent	σ_{31}	σ_{41}	σ_{51}	σ_{61}	σ_{71}
NO_2	0.71	0.84	0.52	0.41	0.53
CN	0.56	0.73	0.43	0.32	0.43
Br	0.39	0.19	0.22	0.23	0.21
Cl	–	0.19	0.21	0.23	–
CH_3	–	−0.21	−0.07	−0.04	−0.08
OCH_3	–	−0.42	−0.05	0.07	−0.07
HO	0.12	−0.57	−0.09	0.07	−0.08
H_2N	–	−0.87	−0.29	–	–

TABLE 29 σ Values for Sterically Unhindered Positions in Biphenyl Derivatives

	$\sigma_{3'4}$		$\sigma_{4'4}$	
Substituent	Experimental	Calculated	Experimental	Calculated
NO$_2$	0.23	0.27	0.30	0.29
Br	0.12	0.15	0.13	0.15
Cl			0.13	0.15
CH$_3$			-0.02	-0.05
OCH$_3$			0.07	-0.04
HO			-0.19	-0.07
H$_2$N			-0.25	-0.19

electrophile (nucleophile) to atom i (see, e.g., **143**, **144**). The charge is again equal to the square of the corresponding NBMO coefficients of the odd alternant hydrocarbon formed upon addition (**143**, **144**, X = H).

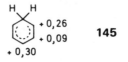

After standardization by means of σ^+ constants for benzene one can derive σ^+ constants for naphthalene[139]; they correlate reasonably well with the rates of protodeuteration of naphthalene derivatives via **143** and **144**.

Based on nuclear magnetic resonance (nmr) measurements, charge distribution in protonated benzene may be represented by **145**.

The ratio of charges in *ortho* and *para* positions of **145** is mirrored by the ratio of logarithms of partial rate factors (*148*) in protodetritiation of substituted benzenes [(*146*), (*147*)][140]:

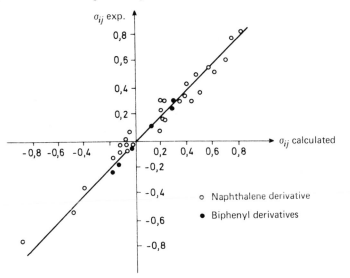

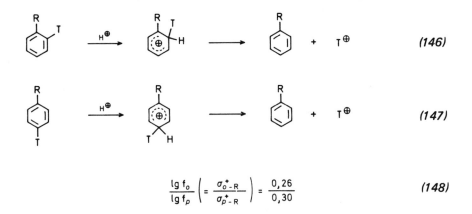

Fig. 17 Comparison of experimental and calculated σ constants. [From M. J. S. Dewar and P. J. Grisdale, *J. Amer. Chem. Soc.* **84**, 3548 (1962) (ref. 136).]

$$\frac{\lg f_o}{\lg f_p}\left(= \frac{\sigma^+_{o-R}}{\sigma^+_{p-R}}\right) = \frac{0,26}{0,30} \qquad\qquad (148)$$

This result may be understood in terms of the Dewar-Grisdale equation if one assumes that steric and inductive effects play *no* role in proto-detritiation, so that:

$$\frac{\sigma^+_{o-R}}{\sigma^+_{p-R}} = \frac{q_{ortho}}{q_{para}} \qquad\qquad (149)$$

Then σ^+_o constants valid for these special circumstances may be derived from σ^+_p constants via (*148*):

$$\sigma^+_{o-OCH_3}: \; -0,67;$$
$$\sigma^+_{o-SCH_3}: \; -0,515;$$
$$\sigma^+_{o-CH_3}: \; -0,27;$$
$$\sigma^+_{o-C_6H_5}: \; -0,155.$$

$\sigma^+_{\text{replacement}}$ constants of several polynuclear aromatics (p. 115) can be correlated with calculated charges on the exocyclic CH_2 group of the corresponding $Ar-CH^+_2$ system.[129]

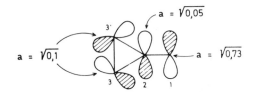

$$(150a)$$

$$(150b)$$

The substantial difference between stabilizing effects of a phenyl group in 1- and 3-positions of a cyclopropylcarbinyl cation [see ratio of solvolysis rate constants (150a) and (150b)] may be attributed in part to the relative values of NBMO coefficients a in this system[141] (Fig. 18).

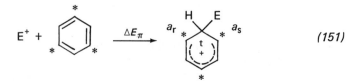

Fig. 18 NBMO coefficients a for the cyclopropylcarbinyl system. [From C. F. Wilcox, L. M. Loew, and R. Hoffmann, *J. Amer. Chem. Soc.* **95**, 8193 (1973) (ref. 141).]

3.1.1.4 Reactivity Indices N_t of Aromatic Systems

NBMO coefficients for odd alternant hydrocarbons play a role in empirical correlation equations for equilibria or rate constants of addition reactions of aromatic hydrocarbons. The change in π-electron energy (ΔE_π) on going from an even alternant hydrocarbon to an odd hydrocarbon, that is, the addition product (151)

$$E^+ + \quad \xrightarrow{\Delta E_\pi} \quad \qquad (151)$$

is given by[137]

$$\Delta E_\pi = -2\beta(a_r + a_s) = -\beta N_t \qquad (152a)$$

$$N_t = 2(a_r + a_s) \qquad (152b)$$

where β is the C—C resonance integral, a_r and a_s are the *NBMO coefficients* of the sp^2-hybridized carbon atoms r and s in the addition product next to carbon atom t, where additions took place (these coefficients are calculated according to rules given on p. 118). N_t is the *reactivity index* at position t (Fig. 19). *The lower the reactivity index, the more reactive is this position.*

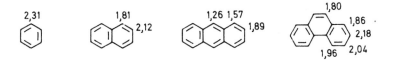

Fig. 19 Reactivity indices N_t.

Linear correlations of N_t (or similar parameters)[142] and log K or log f_t (f_t = partial rate factor for position t) exist for:

1 *Protonation* equilibria (assuming that protonation occurs at the most reactive position t) and *Diels-Alder reactions* with maleic anhydride.[144]
2 Rates of *electrophilic substitution*,[145] *free radical substitution*,[146] *hydroxylation with OsO$_4$*,[147] and *ozonolysis*[147] (Figs. 20 and 21).

Since the transformation shown in (*151*) is only *partially* achieved in the *transition states* of this reaction, only part of ΔE_π is operative, and so empirical values of β (Table 30) which plays here the role of the reaction constant are *lower* than the real value (-84 kJ/mol). Logarithms of *rates of formation* of arylmethyl cations, arylmethyl anions, and arylmethyl radicals from arylmethane derivatives (*154*) may be similarly correlated with the change in π-electron energy (ΔE_π).

$$R^{(+,\cdot,-)} + \text{aryl}-CH_2X \xrightarrow{\Delta E_\pi} \text{aryl}-CH_2^{(+,\cdot,-)} + RX \qquad (154)$$

where R^+ = polar solvent X = halogen[150]

$R^\cdot = \cdot CCl_3$ X = H[151]

$R^- = {}^-NHC_6H_{11}$ X = H[152]

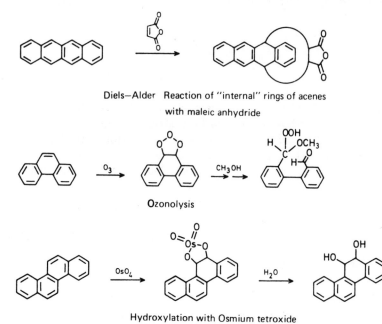

Diels–Alder Reaction of "internal" rings of acenes
with maleic anhydride

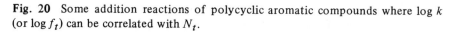

Ozonolysis

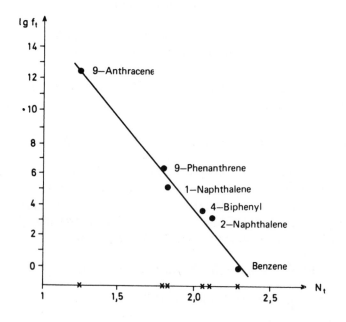

Hydroxylation with Osmium tetroxide

Fig. 20 Some addition reactions of polycyclic aromatic compounds where log k (or log f_t) can be correlated with N_t.

Fig. 21 Bromination of aromatic hydrocarbons[148]. $\log f_t = -67(N_t - N_{C_6H_6})/2.3RT$.[153]

124

TABLE 30 β Values for Aromatic Substitution Reactions[149] Obtained by Plotting $\log f_t$ versus $(N_t - N_{C_6H_6})/2.3RT$

Reaction	β (kJ/mol)
Phenylation (radicalic)	$-14, -13$
Methylation (radicalic)	-23
Trichloromethylation (radicalic)	-42
Nitration (electrophilic)	-20
Chlorination (electrophilic)	-61
Bromination (electrophilic)	-65

A simpler (less accurate) treatment is possible by means of equation

$$\Delta E_\pi = 2\beta(1 - a_{or}) \qquad (155)$$

using NBMO coefficients obtained as described on page 118. Here a_{or} is the NBMO coefficient of the exocyclic CH_2 group of the arylcarbinyl species formed in these reactions.

The basis for all these correlations is the following: within a series of aryl derivatives undergoing the same reaction (reaction series), it is assumed that all factors determining reactivity are roughly equal for all compounds except the changes in π-electron energy so that the latter alone are essentially responsible for differences between individual compounds.

3.1.1.5 The Inductive Parameter $1/r_{ij}$

Logarithms of rate constants of reactions where a *negative* charge is produced in the rate-determining step (e.g., lithiumcyclohexyl amide catalyzed exchange of deuterium attached to aromatic carbon atoms whose rate-determining step is shown below)[153]

and saponification of unhindered methylaryl acetates[154] may be satisfactorily represented as linear functions of parameter $\Sigma_j (1/r_{ij})$ [Figs. 22 and 23; see also (*144*)]. The terms r_{ij} are distances in angstrom units between the reacting carbon atom (*i*) and all the other carbon atoms (*j*) of the aromatic system (their nuclei attract the negative charge generated in the reaction) (see Fig. 24). The large discrepancies noticed

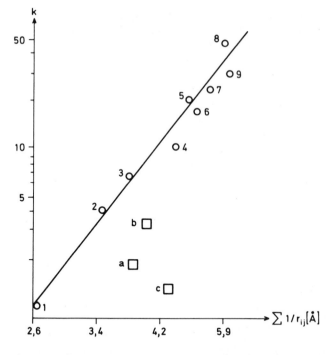

Fig. 22 Correlation of relative D exchange rates of aromatic hydrocarbons with lithium cyclohexylamide: 1, benzene ($k = 1$, definition); 2, 2-naphthalene; 3, 1-naphthalene; 4, 1-anthracene; 5, 2-pyrene; 6, 9-phenanthrene; 7, 1-pyrene; 8, 9-anthracene; 9, 4-pyrene; *a*, 4-biphenyl; *b*, 3-biphenyl; *c*, 2-biphenyl. [From A. Streitwieser, Jr., and R. G. Lawler, *J. Amer. Chem. Soc.* 87, 5388 (1970) (ref. 153).]

for biphenyl in the first case and for methyl esters of the 1-naphthyl acetate type in the second case are probably caused by lack of co-planarity of the two phenyl rings (see p. 112) and inhibition of OH⁻ attack by the *peri*-hydrogen atoms.

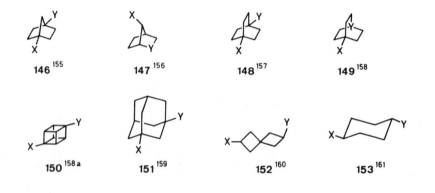

146 [155]   147 [156]   148 [157]   149 [158]

150 [158a]   151 [159]   152 [160]   153 [161]

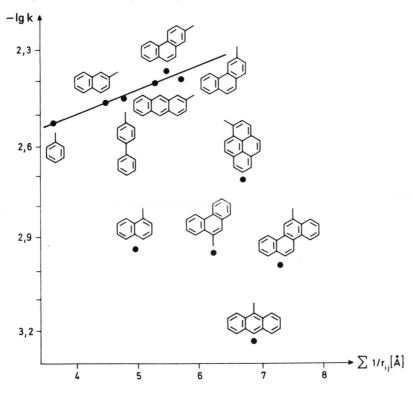

Fig. 23   Correlation of hydrolysis rates ($1 \cdot mol^{-1} s^{-1}$) of methyl aryl acetates ($-$ = $CH_2-COOCH_3$). [From N. Acton and E. Berliner, *J. Amer. Chem. Soc.* **86**, 3312 (1964) (ref. 154).]

Fig. 24   Distances that must be considered in reactions at the 2-position of pyrene.

3.1.2   Rigid Alicyclic Compounds; σ_I Correlations; Bridgehead Reactivities; Correlation of Reactivities and Spectroscopic Properties

Compounds **146-153** (X = substituent, Y = reaction center) correspond in a sense to aromatic compounds; steric effects and resonance effects between substituent and reaction center are out of the question; the rigidity of polycyclic compounds and the strong favoring of the di-

equatorial conformation in *trans*-1,4-disubstituted cyclohexanes gua-
rantee that within a reaction series the distance between X and Y will
remain constant. Consequently, for these and similar compounds, σ_m
constants are again a good measure for the *polar effect*, the only one
active here.

Even more appropriate are σ_I constants (Table 21) defined by means
of the dissociation constants K of 4-substituted bicyclo[2.2.2]octane-1-
carboxylic acids (148, Y = COOH)[157] (*156a*):

$$\log \frac{K}{K_0} = 1.65\ \sigma_I \quad \text{(ethanol-water, 1:1, 25°C)} \qquad (156a)$$

They reflect exclusively the polar effect of X, whereas σ_m reflects to a
small extent its influence on Y as well, via indirect resonance effects
[(*156b*)] **154** ↔ **155**. These correlations hold for common *reactions
of acids* (dissociation, esterification, rates of saponification of
esters),[155,157-161] also rates of S_N1 *reactions* (Y = Hal, OTos, C–Hal,
C–OTos)[162], S_N2 *reactions* (e.g., X–R–CH$_2$OTos + SR⁻),[163] and
hydrogen abstraction by radicals (X–R–H + ·CCl$_3$ → XR· + HCCl$_3$).[163]

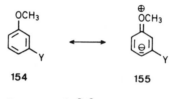

154   **155**

$$\sigma_m = \sigma_I + 0{,}3\ \sigma_R \qquad (156b)$$

3.1.2.1 Bridgehead Reactivities

Logarithms of rate constants of reactions at bridgehead atoms in many
alicyclic systems (e.g., *157a-157d*) may be correlated with strain differences between R–H and R⁺,[164] calculated by means of potential functions for bond lengths, bond angles, torsion, and van der Waals interactions.[163a]

$$
\begin{array}{llll}
\text{R–X} & \longrightarrow & \text{R}^{\oplus} + \text{X}^{\ominus} & (157a) \\
\text{R–N=N–R} & \longrightarrow & \text{R–N=N·} + \text{·R} & (157b) \\
\text{R–CO}_3\text{–}t\text{-C}_4\text{H}_9 & \longrightarrow & \text{R·} + \text{CO}_2 + \text{·O–}t\text{-C}_4\text{H}_9 & (157c) \\
\text{R–H} + \text{·X} & \longrightarrow & \text{R·} + \text{HX} & (157d)
\end{array}
$$

R =

The basis for this correlation is the special steric situation at the reaction center. Direction of attack by reagents, steric hindrance, stereochemistry of solvation, and electronic interactions with the rest of the molecule as well as entropy effects are the same for all compounds, so that the only variable is the *difference in strain*.

This method enables the correct prediction of regioselectivity in cationic bridgehead bromination at C-6 in protoadamantane[165] [(*158*)]:

(*158*)

The calculation predicts that ionization at the C-6 position will result in a smaller increase in the strain of the carbon skeleton than ionization at the other positions (C-1, C-3, and C-8).

Calculated differences in strain energy between ground states and models of transition states have also been correlated with relative reaction rates in the following cases: solvolysis of 2-alkyl-2-adamantyl esters,[165a], chromic acid oxidation of secondary alcohols,[165b] reduction of ketones by sodium borohydride,[165c] and trifluoroethanolysis of cycloalkyl tosylates.[165d]

3.1.2.2 The Foote-Schleyer Equation

Rates of S_N1 acetolysis of polycyclic and bicyclic secondary tosylates whose structure prevents S_N2 reactions with solvent as well as stabilization of the carbocation by neighboring group participation (see p. 30 and 207) are relatively uncomplicated and thus amenable to correlation. The prototype for these compounds is 2-adamantyl tosylate **156**. The axial hydrogen atoms shown inhibit the S_N2 reaction (Fig. 25) and the bridgehead nature of the CH groups surrounding C-2 makes hyperconjugation impossible.

Only *conformational* factors are responsible for reactivity; the most influential are bond angles. The larger the deviation of the $C-C^+-C$ angle in the cation from the ideal $120°$ value (toward smaller values) prompted by restrictions imposed by the molecular framework, the slower the ionization. A measure of these deviations are the C=O vibrational frequencies of ketones corresponding to the carbocations. (Car-

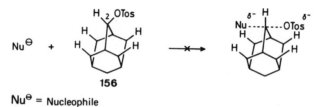

$Nu^{\ominus}$ = Nucleophile

Fig. 25 Steric hindrance in the S_N2 reaction of 2-adamantyl tosylate.

bonyl compounds are often chosen as models for carbocations, since their polar resonance structure is that of an oxidocarbocation). A smaller $>$C=O bond angle corresponds to a higher C=O frequency.[166] The correlation between rate of ionization k (s^{-1}) and carbonyl frequency $\nu_{C=O}$ (cm^{-1}) is given by:

$$\log k = -a\nu_{C=O} + C \tag{159}$$

The value of the constant C is obtained from the rate of acetolysis of cyclohexyl tosylate at 25°C:

$$\lg k_{rel} = \lg k - \lg k\langle\;\rangle\text{-OTos} = a\,(1715 - \nu_{C=O}) \tag{160}$$

The slope was determined by plotting the empirical values of $\log k_{rel}$ (160) for some bridge tosylates versus the carbonyl frequencies of the corresponding ketones [Fig. 26 and (161)].[167]

$$\log k_{rel} = \frac{1}{8}\,(1715 - \nu_{C=O}) \tag{161}$$

For higher precision one must take into consideration additional contributions from torsion effects,[163a] van der Waals interactions, and polar effects of functional groups[168]:

Foote-Schleyer equation

$$\log k_{rel} = \frac{1}{8}\,(1715 - \nu_{C=O}) + 1.32\sum_i (1 + \cos 3\phi_i) + \text{inductive term} + \frac{GS - TS}{5.6} \tag{162}$$

where ϕ_i = smallest torsion angle between the HCOTos group and its two neighboring bridgehead CH bonds (i)

GS – TS = difference in van der Waals energies (kJ/mol) between ground state and transition state

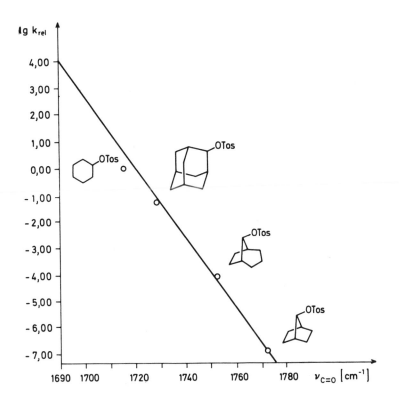

Fig. 26 Determination of the bond angle dependence of relative acetolysis rates of secondary tosylates; k_{rel} <cyclohexyl-OTos> = 1

The second term of the complete equation (*162*) is derived from the potential function for internal rotation in ethyl chloride and accounts for the fact that smaller values ($\phi_i < 60°$) of the angle of torsion between the HCOTos group and its two neighboring bridgehead CH groups (*i*) result in higher ground state energies. This is shown in Fig. 27 for the eclipsed rotamer **157** of ethyl chloride chosen as model. Transition from **157** to the ethyl cation **158**, which according to calculations[168a] must have the same energy in any conformation, requires less energy (and thus proceeds faster) than transition from the staggered rotamer **159**. This rotamer, just like the chair form of cyclohexyl toxylate, completely lacks torsional effects.

The inductive term due to *polar effects* of aryl groups or double bonds in the β-position is taken[168] to be -0.9;[168e] the estimated value for cyclopropane rings in β-position is -0.5; inductive effects of other

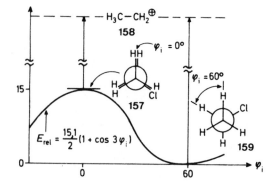

Fig. 27 Concerning the second term of the Foote-Schleyer equation. Torsion energies and ionization of rotamers of ethyl chloride.

substituents can be estimated from $\sigma\rho$ or $\sigma^*\rho^*$ correlations (p. 142). [If one uses $\sigma^*_{CH_2X}$ (X = phenyl, ethyl, see Table 21), then ρ^* for these acetolysis reactions estimated from the ratio $\log (k_{\beta\text{-phenyl}}/k_{\beta\text{-ethyl}}) \simeq -0.9$ turns out to be -3. These ρ^* values can then be used to estimate rate effects of other substituents by means of their $\sigma^*_{CH_2X}$].

The last term takes into account the possibility that van der Waals interactions in the ground state (GS) and in the transition state (TS) may contribute to differences in energy between those two states and thus increase or decrease the reaction rate as compared to the normal case (solid line in Fig. 28).

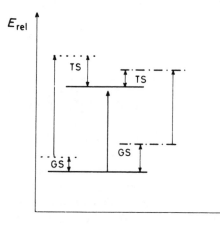

Fig. 28 Contributions (TS) of unfavorable van der Waals interactions to energies of transition states for acetolysis [relative to the normal case (solid lines)] can be larger (dots) or smaller (dots and dashes) than the corresponding contributions (GS) to the ground state energy of the reacting tosylates.

The energy contribution to GS can be estimated by inspection of models of the given molecules and their comparison with model compounds for which steric energies are known (p. 39). For instance, to evaluate steric interactions at the reaction site of aliphatic compounds listed in Table 31, one chooses as the basic model the corresponding cyclohexane derivative (generally the axial cyclohexyl tosylate with a conformational energy of 2.5 kJ/mol); corrections for specific situations must then be applied. (Van der Waals interactions in endo-2-norbornyltosylate are actually much stronger than in the axial cyclohexyl tosylate.[168b] A direction for the estimation of van der Waals energies can be found in the literature).[168c]

To estimate the energy contribution of TS is practically impossible, since this would require precise knowledge of the changes in position of the corresponding atoms in moving toward the transition state and of potential functions for bonding relationships in the transition state. Unequivocal interpretations are not possible, as can be seen for instance in the ionization of endo-2-norbornyl chloride (endo-2-chloro-bicyclo-[2.2.1]-heptane). If one starts from ground state 160, Scheme 38, the

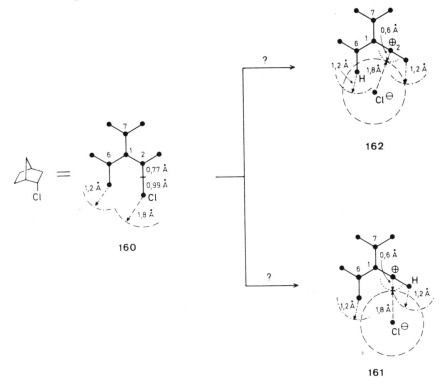

Scheme 38 Possible modes of bond breaking in the ionization of endo-2-norbornyl chloride. [From: H. C. Brown, I. Rothberg, P. v R. Schleyer, M. M. Donaldson, and J. J. Harper, Proc. Nat. Acad. Sci. (USA) 56, 1653 (1966) (ref. 170).]

TABLE 31 Relative Rate Constants for Acetolysis of Tosylates[168]

Tosylate	Ketone ν_{CO} (cm^{-1})	ϕ_i (degrees)	GS – TS (kJ/mol)	log k_{rel} Calculated[a]	Experimental
7-Norbornyl	1773	60.60	1.7	-7.0	-7.00
endo-8-Bicyclo[3.2.1]octyl	1750	60.60	2.5	-4.2	-4.11
endo-2-Benznorbornenyl	1756	0.40	1.7	-2.4	-2.22
endo-2-Norbornenyl	1745	0.40	1.7	-1.0	-1.48
2-Adamantyl	1727	60.60	2.5	-1.1	-1.18
Cyclohexyl	1716	60.60	0.00	-0.1	0.00
Cyclotetradecyl	1714	60.60?	0.00	+0.1	+0.08
Isopropyl	1718	60.60	0.00	-0.4	+0.15
endo-2-Norbornyl	1751	0.40	5.5	-0.2	+0.18
Cyclopentadecyl	1715	60.60?	0.00	0.0	+0.42
cis-4-t-Butylcyclohexyl	1716	60.60	2.5	+0.4	+0.42
Cyclododecyl	1713	60.60?	0.00	+0.3	+0.50
2-Butyl	1721	60.60	2.5	-0.3	+0.53
3,3-Dimethyl-2-butyl	1710	60.60	5.0	+1.5	+0.62
Cyclotridecyl	1713	60.60?	0.00	+0.3	+0.66
3-Methyl-2-butyl	1718	60.60	5.9	+0.6	+0.93
Cyclopentyl	1740	0.20?	0.00	+1.5	+1.51
Cycloheptyl	1705	45.45	0.00	+2.0	+1.78
trans-2-t-Butylcyclohexyl	1700	60.60	2.5	+2.2	+2.20
cis-2-t-butylcyclohexyl	1700	60.60	5.0	+2.6	+2.61

1,4-α-5,8-β-Dimethanoperhydro-9-anthracyl	1696	60.60	2.5	+2.67
Cyclopropyl	1815	0.00	0.00	-5.32
9-Bicyclo[3.3.1]nonyl	1726	60.60	2.5	+0.48
exo-Trimethylene-norborn-exo-2-yl	1751	0.40	1.3	+0.84
cis-3-Bicyclo[3.1.0]hexyl	1739	0.40	0.00	+1.14
axial-2-Bicyclo[3.2.1]octyl	1717	50.60	2.5	+1.62
2-Bicyclo[2.2.2]octyl	1731	0.60	1.7	+1.85
equat-2-Bicyclo[3.2.1]octyl	1717	50.60	0.8	+0.47
trans-3-Bicyclo[3.1.0]hexyl	1739	0.40	2.5	+0.17
syn-8-Bicyclo[3.2.1]oct-2-enyl	1758	60.60	4.6	-5.54
Cyclooctyl	1703	40.40?	0.00?	+2.76
Cyclononyl	1703	40.40?	0.00?	+2.70
Cyclodecyl	1704	40.40?	0.00?	+2.98
Cycloundecyl	1709	40.40?	0.00?	+2.05

[a]By means of the Foote-Schleyer equation.

C—Cl bond can be stretched in the direction of the original bond (**161**), in which case TS $\simeq$ 0 is a fair approximation. If, however, the leaving group moves in a direction that is at all times perpendicular to the front side of the carbocation being formed (so as to maximize the overlap between the leaving group and the $2p$ orbital at C-2), one must take into consideration considerable steric hindrance by H-6$_{endo}$ and GS – TS $<$ 0 (p. 32).

All data in Table 31 were obtained under the assumption TS = 0.[170] This is often a valid assumption ("evidently because leaving groups are generally able to find a propitious avenue for departure[99])[168] and has been proved to be correct for solvolyses of axial cyclohexane derivatives.[168d]

Experimental acetolysis rates for tosylates **163**[169] and **164-166**[170] are significantly lower than values calculated for TS = 0. Obviously, in this instance GS – TS $<$ 0. (In addition to the possibility of some physical barrier inhibiting the departure of the leaving group, one must take into consideration steric hindrance of solvation by alkyl groups; see, e.g., p. 254 and ref. 173a).

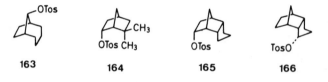

| 163 | 164 | 165 | 166 |

[It is surprising that the Foote-Schleyer equation correctly predicts reactivities of simple secondary cycloalkyl tosylates, even though the additional effect of nucleophilic solvent participation plays an important role in their solvolyses (see p. 201, 203, and 212)].[170a]

Relative epimerization rates of the following esters [(*163*) and Table 32] show that C=O stretching frequencies of the corresponding ketones correlate quite well with *rates of formation* of the *mesomeric* (ideally sp^2-hybridized) *carbanions*.[171]

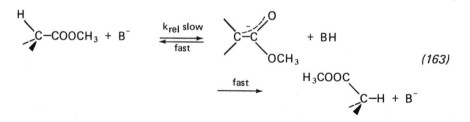

                                                                                                (*163*)

TABLE 32 Relative Epimerization Rates of Esters and C=O Stretching Frequencies of the Corresponding Ketones

Ketone	$\nu_{C=O}$ [cm^{-1}]	k_{rel} Epimerization		
(bicyclic ketone)	1780	(ester COOCH$_3$) 1(Def.)	(ester COOCH$_3$) 1.4
(bicyclic ketone)	1750	(ester COOCH$_3$) 66	(ester COOCH$_3$) 27
(bicyclic ketone)	1745	(ester COOCH$_3$) 26	(ester COOCH$_3$) 24
(H$_3$C cyclohexanone)	1715	(ester COOCH$_3$) 251	(ester COOCH$_3$) 46

3.1.2.3 CH Acidities and s-Character

We should mention here another correlation between spectroscopic data and reactivities investigated mainly in alicyclic compounds. Logarithms of rate constants for metallation of 3,3-dimethylcyclopropene and of bridgehead CH bonds in bicyclobutanes with an additional alkyl bridge by methyllithium (164) may be approximately correlated with the ^{13}C$-^1$H coupling constants (J) for these bonds[172] (Table 33).[172a]

The same is true for rate constants of tritium exchange between cycloalkanes and N$-^3$H$-$ cyclohexylamine, catalyzed by cesium cyclohexylamide[173] (Table 34 and Fig. 29).

$$\text{H}_3\text{C}-\square-\text{CH}_3 \ + \ \text{CH}_3\text{Li} \ \xrightarrow{k} \ \text{H}_3\text{C}-\square-\text{CH}_3 \ + \ \text{CH}_4 \qquad \textbf{(164)}$$

TABLE 33 Dependence of the Rate of Metallation of C–H by Methyllithium on Their $^{13}C-^{1}H$ Coupling Constants

Substrate	$J_{^{13}C-H}$ [Hz]	k/k_0	lg k/k_0
H₃C, CH₃ (structure) H, H	221 ± 0,5	2 500	3,4
H₃C—(structure)—CH₃	212 ± 2	65	1,8
(structure)—CH₃	206 ± 1	12	1,1
(structure)	200 ± 1	1 (k_0)	0,0

$$lg \frac{k}{k_0} \approx 0{,}16 \, J_{^{13}C-H} - 32$$

These correlations (which, depending on the chosen model compound and the base used differ mainly in their constant terms) are based on the fact that $J^{13}C-H$ is a *measure of the s-character* of the considered C–H bond and that the contribution of the unshared electron pair to the energy of the carbanion is the smaller the higher the

TABLE 34 Relative Rate Constants in the Isotope Exchange Reaction

$$R-H + H\bar{N}-\bigcirc \xrightarrow{slow,\, k_{rel}} R^- + H_2N-\bigcirc$$

$$R^- + TH\bar{N}-\bigcirc \xrightarrow{fast} R-T + H\bar{N}-\bigcirc$$

and $J^{13}C-H$ of cycloalkanes R–H

R–H	k/k_0	$J^{13}C-H$ [Hz]
Cyclopropane	$(7.0 \pm 0.9) \times 10^4$	161
Cyclobutane	28 ± 10	134
Cyclopentane	5.72 ± 0.27	128
Cyclohexane	$1.00\ (k_0)$	123
Cycloheptane	0.76 ± 0.09	123
Cyclooctane	0.64 ± 0.06	122

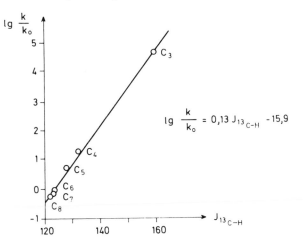

Fig. 29 Graphic representation of data from Table 34. [From A. Streitwieser, Jr., R. A. Caldwell, and W. Young, *J. Amer. Chem. Soc.* **91**, 529 (1969) (ref. 173).]

s-character of "its" orbital. Within a series of compounds the ratio between the *s*-character of the C—H bond of the hydrocarbon and the *s*-character of the carbanion formed from this hydrocarbon must remain constant and other factors should not have any influence on anion formation. (The simplest assumption is that the hydrocarbon and the carbanion have the same degree of hybridization, since their molecular skeletons should resist conformational changes caused by possible changes in hybridization on carbanion formation.)

3.1.3 Flexible Compounds; Separation of Steric and Polar Effects; The Taft Equations

Basic conditions for correlations involving substituent constants presented so far:

Absence of steric substituent effects,
Constant distance between substitutent and reaction center within a given reaction series

are no longer present in freely rotating aliphatic compounds. Their capacity to minimize substituent effects by conformational changes becomes evident if one compares the relative dissociation constants of 4-bromobicyclo[2.2.2]octane-1-carboxylic acid **167** and ω-bromo-valeric acid (**168**).
Such consequences of the flexibility of aliphatic chains can be taken into account if—starting with the reaction center—one considers the *whole* group bearing the substituent as substituent.

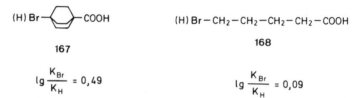

$$\lg \frac{K_{Br}}{K_H} = 0,49 \qquad\qquad \lg \frac{K_{Br}}{K_H} = 0,09$$

One would then describe the reactivity of ω-bromovaleric acid not by means of σ_I-Br, but by means of σ^*-(CH$_2$)$_4$-Br. Taft computed the σ^* constants for the *polar effect* of substituents X directly attached to the reaction center by means of rate constants k for acid- and base-catalyzed hydrolysis of esters X—COOR.[114,175]

$$\log \left(\frac{k_X}{k_{CH_3}}\right)_{basic} - \log \left(\frac{k_X}{k_{CH_3}}\right)_{acidic} = 2.48\, \sigma_X^* \qquad (165)$$

$$\sigma_{CH_3}^* = 0 \qquad\qquad (166)$$

The factor 2.48 was chosen to obtain for σ^* constants absolute values comparable to those of aromatic σ constants. The essence of Taft's analysis relies on the assumption that *steric* and *polar* effects of X act *independently*. The subsequent procedure is based on two postulates:

First Taft Postulate
Both *polar* and *steric* effects of X are operative in *base-catalyzed* hydrolysis; on the other hand, rates of *acid-catalyzed hydrolysis* are affected only by the *steric effect*.

The low values of ρ in the acid hydrolysis of esters of benzoic acid ($\rho = +0.14$) attest to the fact that polar effects have very little influence on this reaction. The reason is that the rate-determining step (*168*) consists of attack of water on the protonated ester **170**, formed in pre-equilibrium [(*167*)] :

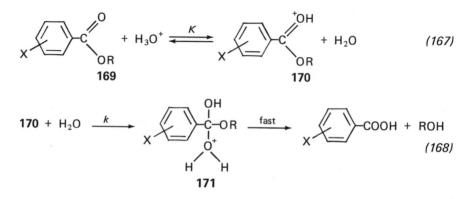

For the rate-determining formation of **171** one can write

$$\frac{d[171]}{dt} = k\,K\,[169][H_3\overset{\oplus}{O}] = k_{exp}\,[169][H_3\overset{\oplus}{O}] \tag{169}$$

Application of the Hammett equation to k_{exp} gives:

$$lg\left(\frac{k_X}{k_H}\right)_{exp} = lg\frac{k_X}{k_H} + lg\frac{K_X}{K_H} = (\rho_{(167)} + \rho_{(168)})\,\sigma_x = \rho_{exp}\cdot\sigma_x \tag{170}$$

Opposite signs for $\rho(167)$ and $\rho(168)$ [an electron-withdrawing substituent X favors reaction (*168*), but disfavors reaction (*167*)] are responsible for the low value of ρ_{exp}.

Second Taft Postulate
(Steric effect of X – steric effect of CH_3)$_{base}$ – (steric effect of X – steric effect of CH_3)$_{acid}$ = 0   (*171*)

Application of both postulates to (*165*), where the log k terms represent the sum of the corresponding polar (P) and steric (S) effects $[(P_X + S_X)_{base} - (P_{CH_3} + S_{CH_3})_{base} - (P_X + S_X)_{acid} - (P_{CH_3} + S_{CH_3})_{acid} = 2.48\,\sigma_X^*$ (*165a*)] gives:

$$\text{polar effect of X – polar effect of } CH_3 = 2.48\sigma^* \tag{172}$$

The assumption that differences in steric effects of two substituents in the acid hydrolysis of esters are the same as differences in base hydrolysis (independent of the nature of substituents) makes sense, since aside from solvation effects, the structures of the tetrahedral intermediates in both reactions (**172, 173**) differ only by two protons and thus by only a small steric effect, which remains constant in various systems.

(In general one must anticipate that even *differences* in steric effects will depend on the nature of the reacting systems. For instance, replacement of H by CH_3 in the 2-adamantyl system (**174**, R = H or CH_3, respectively) results in an *increase* in acetolysis rate, whereas in the *endo*-2-norbornyl system (**175**, R = H or CH_3, respectively) the rate decreases![173a])

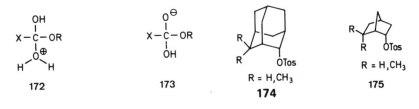

172   **173**   R = H,CH₃   **174**   R = H,CH₃   **175**

Neither structure **172** nor **173** allows for resonance between X and the reaction center. Since the initial states are the same in both reactions (with potential resonance interactions between X and COOR), resonance contributions to both terms of equation (*165*) must be the same and must cancel out.

The hypothesis that σ^* constants reflect only the *inductive effect of X* is consistent with the observation that σ_X^* is proportional to $\sigma_{I,X}$ and that insertion of a CH_2 group between X and the rest of the molecule diminishes the σ^* values of *all* substituents by the same factor.

$$\sigma_X^* = 6,23\,\sigma_{I,X} \tag{173}$$

$$\sigma_{CH_2X}^* = 0,36\,\sigma_X^* \tag{174}$$

Charton described correlations that allow the computation of inductive substituent constants from dissociation constants of substituted acetic acids.[174] $\sigma_{CH_2-X}^*$ constants can be found in Table 21.

1 First Taft equation

Reactions where *steric* and *conjugative effects* do not play any role may be described by (*175*) and (*176*):

$$\log \frac{k_X}{k_{CH_3}} = \rho^* \cdot \sigma_X^* \tag{175}$$

and for several substituents

$$\log \frac{k_X}{k_{CH_3}} = \rho^* \sum \sigma_X^* \tag{176}$$

where k = rate constant or equilibrium constant

Examples[175] are various acid-base dissociation equilibria (carboxylic acids, ammonium and phosphonium ions, thiols) and also rate constants of:

(*a*) Acid-catalyzed hydrolysis of acetals and epoxides.[176]
(*b*) Esterification of carboxylic acids with diphenyl diazomethane.
(*c*) S_N1 and S_N2 reactions[177a] (see, however, p. 144).
(*d*) Additions to olefins [e.g., Br_2, Hg^{2+} (oxymercuration), $\cdot CCl_3$].
(*e*) Oxidations with chromic acid[177b] and abstraction of α-protons from esters.[177c]

According to Taft's first postulate the relationship between structure and rates of acid hydrolysis of esters is exclusively a consequence of *steric factors*, which thus can be determined quantitatively by means of this reaction (*177*):

$$\log\left(\frac{k_X}{k_{CH_3}}\right)_{\text{acid hydrol. of esters}} = E_s \qquad (177)$$

where E_s = steric parameter; for values of E_s, see Table 21.

Charton[178] showed that the E_s parameter is a linear function of van der Waals radii r_X, and

$$E_{s,\ CH_2X} = -0{,}412\,r_X + 0{,}445 \qquad (178)$$

$$E_{s,\ CHX_2} = -2{,}88\,r_X + 3{,}49 \qquad (179)$$

$$E_{s,\ CX_3} = -3{,}49\,r_X + 4{,}14 \qquad (180)$$

[Values of ΔG for equilibria between axial and equatorial conformers of monosubstituted cyclohexanes (p. 39) do not correlate with van der Waals radii of substituents.[179] This shows that steric effects in organic molecules depend to a great extent on the specific character of a given system. In the cyclohexane case one is dealing with 1,3-interactions between C—X and axial C—H groups. The C—H bond is exceedingly short (1.09 Å); as the van der Waals radius of X becomes larger, so does the C—X bond length, and large substituents "grow" somehow "over" the C—H groups; therefore the increase in size of X does not result in stronger interaction with C—H groups. In interactions of C—X[1] with C—C or C—X[2] this situation obtains to a much lesser degree.]

2 Second Taft equation

When reaction rates are affected exclusively by *steric effects* of the type observed in reactions used to define E_s (additions to sp^2-hybridized carbon atoms) (*181*) (with the susceptibility constant δ) should be valid.

$$\log \frac{k_X}{k_{CH_3}} = \delta E_s \qquad (181)$$

Examples[175] are acid-catalyzed transesterification, catalytic hydrogenation of olefins, hydroboration with "disiamylborane" (bis[3-methylbutyl-(2)]-borane),

$$\begin{array}{cc} CH_3 & CH_3 \\ | & | \\ (H_3C-CH-CH)_2 & BH \end{array}$$

and quaternization of heterocycles [(*182*)] [180]:

$$\qquad \qquad (182)$$

Compounds in which free rotation of R is impossible (e.g., **176**) no longer obey the linear E_s correlation.

176

3   In a third group of reactions polar *and* steric effects act independently; they can be described by a combination of equations mentioned above (*183*):

$$\lg \frac{k_X}{k_{CH_3}} = \varrho^* \sigma_X^* + \delta E_{s,X} \qquad (183)$$

Examples are (within the limits of applicability of E_s) polar carbonyl additions (e.g., base-catalyzed transesterification and reduction with BH_4^-)[181] and additions to olefins[182] when substrates bear sterically demanding substituents (branched alkyl groups).

4   Reactions where polar and steric effects do *not* seem to operate *independently* or where steric interactions differ in character from those in the reaction used to define E_s cannot be described by means of Taft's σ^*-E_s correlations. Examples are S_N2 reactions of alkyl halides.[183] Their rates may be described by (*184*)

$$\lg \left(\frac{k_{RX}}{k_{C_2H_5X}} \right) = r \, \alpha_R \qquad (184)$$

where α_R is the "alkyl reactivity constant" of the alkyl group R and r is the reaction constant. The α_R values in Table 35 show very clearly the enhanced reactivity of methyl halides and the reduced reactivity of branched isomers.

Hudson's (*ibid.*, p. 222); also : G. Klopman, *J. Amer. Chem. Soc.* **90**, 223 (1968).

186a. Cf. *Organic Peroxides*, D. Swern (Eds.), Wiley-Interscience, New York-London-Sydney-Toronto, Vol. I, 1970, Vol. III, 1972; B. Plesuicar, in W. S. Trahanowsky (Ed.), *Oxidation in Organic Chemistry*, Academic Press, New York-San Francisco-London, 1978, p. 211.

187. M. Arbelot, J. Metzger, M. Chanon, C. Guimon, and G. Pfister-Guillonzo, *J. Amer. Chem. Soc.* **96**, 6217 (1974).

188. R. Pearson, in N. B. Chapman and J. Shorter (Ed.), *Advances in Linear Free Energy Relationships*, Plenum Press, London-New York, 1972, p. 281.

188a. A similar pattern is found for acyclic ketones and alcohols: increasing alkyl substitution at the carbon atoms adjacent to C—O makes them harder (I. A. Blair, J. H. Bowie, and V. C. Trenerry, *J. Chem. Soc., Chem. Commun.* **1979**, 230).

189. H. Kollmar and H. O. Smith, *Angew. Chem. Int. Ed.* **9**, 462 (1970); J. E. Williams, V. Buss, and L. C. Allen, *J. Amer. Chem. Soc.* **93**, 6870 (1971); J. R. Grunwell and J. F. Sebastian, *Tetrahedron* **27**, 4387 (1971).

190. D. J. DeFrees, J. E. Bartmess, J. K. Kim, R. T. McIver, Jr., and W. J. Hehre, *J. Amer. Chem. Soc.* **99**, 6451 (1977), and references therein.

191a. W. J. le Noble, *Synthesis* **2**, 1 (1970).

191b. G. J. Heiszwolf and H. Kloosterziel, *Rec. Trav. Chim. Pays-Bas* **89**, 1153 (1970).

191c. R. Gompper and H.-U. Wagner, *Angew. Chem. Int. Ed.* **15**, 321 (1976).

192. G. J. Heiszwolf and H. Kloosterziel, *Rec. Trav. Chim. Pays-Bas* **89**, 1217 (1970).

193. S. G. Smith and M. P. Hanson, *J. Org. Chem.* **36**, 1931 (1971).

194. F. G. Bordwell, *Organic Chemistry*, Macmillan Co., New York-London, 1963, p. 218, quoted in ref. 197.

195. K. Okamoto, H. Matsuda, H. Kawasaki, and H. Shingu, *Bull. Chem. Soc. Jap.* **40**, 1917 (1967).

195a. G. Modena, *Acc. Chem. Res.* **4**, 73 (1971).

195b. F. Théron, *Bull. Soc. Chim. Fr.* **1969**, 278.

195c. Y. Kobayashi, T. Taguchi, T. Morikawa, T. Takase, and T. Takanashi, *Tetrahedron Lett.* **1980**, 1047.

195d. J. E. G. Kemp et al., *Tetrahedron Lett.* **1980**, 2991.

196. N. Kornblum, R. A. Smiley, R. K. Blackwood, and D. C. Iffland, *J. Amer. Chem. Soc.* **77**, 6269 (1955).

197. R. G. Pearson and J. Songstad, *J. Amer. Chem. Soc.* **89**, 1827 (1967).

198. Ref. 186, p. 82.

199. R. K. Lustgarten, M. Brookhart, and S. Winstein, *J. Amer. Chem. Soc.* **94**, 2347 (1972).

200. A. Diaz, M. Brookhart, and S. Winstein, *J. Amer. Chem. Soc.* **88**, 3133 (1966).

201. W. Kirmse and F. Scheidt, *Chem. Ber.* **103**, 3711 (1970).

201a. See however: W. Kirmse and T. Olbricht, *Chem. Ber.* **108**, 2616 (1975).

202. D. T. Clark and G. Smale, *Tetrahedron* **25**, 13 (1969).

203. With respect to the 7-norbornenyl cation we point out in this context the work of R. D. Bach, N. A. LeBel, et al., *J. Amer. Chem. Soc.* **95**, 8182 (1973).

204.   O. Eisenstein, Y. M. Lefour, C. Minot, N. T. Anh, and G. Soussau, *C.R. Acad. Sci Paris C.* **274** 1310 (1972). For similar studies with dilithium carboxylates, see J. Mulzer, G. Hartz, U. Kühl, and G. Büntrup, *Tetrahedron Lett.* **1978**, 2949.

205.   J. Bottin, O. Eisenstein, C. Minot, and N. T. Anh, *Tetrahedron Lett.* **1972**, 3015.

205a.   In less extreme cases one finds deviation from this behavior. In organomagnesium halide systems the preferred state can be (depending on the nature of the halogen and the solvent), $R_2Mg + MgX_2$, RMgX, "dimers" $MgR_2 \cdot MgX_2$ and $RMgX \cdot RMgX$, but also ionic complexes such as $RMg^+ \cdot MgX_2R^-$. [See, e.g., K. Nützel, in E. Müller (Publ. Ed.), *Methoden der Organischen Chemie (Houben-Weyl)*, Vol XIII/2a, Georg Thieme Verlag, Stuttgart, 1973, p. 508.

        In the system $2RHgX \rightleftarrows HgR_2 + HgX_2$ (X = halogen) the *left* side is preferred. [see, e.g., K. P. Zeller, H. Straub, and H. Leditschke, in E. Müller (Publ. Ed.), *Methoden der Organischen Chemie (Houben-Weyl)*, Vol. XIII/2b, George Thieme Verlag, Stuttgart, 1974, p. 247.

205b.   High preference for 1,2-reduction is also achieved with $LiAlH_4$ if the reaction is carried at $-78°C$: J. T. Groves and C. A. Bernhardt, *J. Org. Chem.* **40**, 2806 (1975); H. C. Brown and H. M. Hess, *ibid.* **34**, 2206 (1969).

205c.   W. C. Still and A. Mitra, *Tetrahedron Lett.* **1978**, 2659.

206.   R. G. Pearson and J. Songstad, *J. Org. Chem.* **32**, 2899 (1967).

207.   (a) A. Streitwieser, Jr., *Solvolytic Displacement Reactions*, McGraw-Hill Book Co., New York, 1962, p. 30. (b) More data regarding several types of reactions can be found in: C. J. M. Stirling, *Acc. Chem. Res.* **12**, 198 (1979).

208.   N. Kornblum, R. E. Michel, and R. C. Kerber, *J. Amer. Chem. Soc.* **88**, 5660 (1966).

208a.   A. J. Kresge, *Chem. Soc. Rev.* **2**, 475 (1973).

209.   A. Streitwieser, Jr., W. B. Hollyhead, G. Sonnichsen, A. H. Pudjaatmaka, C. J. Chang, and T. L. Kruger, *J. Amer. Chem. Soc.* **93**, 5096 (1971).

210.   Data from ref. 184a, p. 181.

211.   F. G. Bordwell, W. J. Boyle, Jr., J. A. Hautala, and K. C. Yee, *J. Amer. Chem. Soc.* **91**, 4002 (1969).

212.   J. N. Brönsted and K. J. Pederson, *Z. Phys. Chem.* **108**, 195 (1924).

213.   F. Hibbert, F. A. Long, and E. A. Walters, *J. Amer. Chem. Soc.* **93**, 2829 (1971).

214.   Because of the pyramidal form of H_3O^+, H_2O is assigned two spatially different equivalent basic positions, so that $q = 2$ (see V. Gold and D. C. A. Waterman, *J. Chem. Soc. (B)* **1968**, 839 [M. C. R. Symons, *J. Amer. Chem. Soc.* **102**, 3982 (1980)].

215.   A. J. Kresge, H. L. Chen, Y. Chiang, E. Murrill, M. A. Payne, and D. S. Sagatys, *J. Amer. Chem. Soc.* **93**, 413 (1971).

216.   A. J. Kresge, S. Slae, and D. W. Taylor, *J. Amer. Chem. Soc.* **92**, 6312 (1970).

217.   R. P. Bell and W. C. E. Higginson, *Proc. R. Soc. London, Ser. A* **197**, 141 (1949).

218.   T. H. Fife, *Acc. Chem. Res.* **5**, 264 (1972).

219a.   The dehydration of 1-(2,4,6-trimethoxyphenyl)-2-phenylethanol tàkes place under specific acid catalysis and general base catalysis (see also p. 173 and 176):

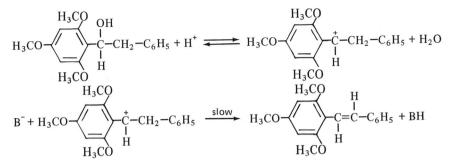

This proves that the reverse reaction, the hydration of *trans*-2,4,6-trimethoxystilbene, takes place under general acid catalysis and that the often quoted reversible formation of a π complex

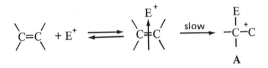

does not occur in this instance.

π Complexes (and specific catalysis because of a preequilibrium) are most plausible when **A** is relatively unstable and thus is formed very slowly (see p. 339), for example, in addition in nonpolar media, when stabilization by resonance is low, or when E = I, Br. G. M. Loudon and D. S. Noyes, *J. Amer. Chem. Soc.* **91**, 1433 (1969).

219b. Because of the logarithmic form of the *Brönsted law*, experimental limitations make it impossible to prove general catalysis for extremely small or extremely large values of α (or β). When $\alpha = 1$ the enormous acidity of H_3O^+ manifests itself so strongly that the contributions of other acid catalysts vanish and the reaction appears to be specifically catalyzed. Conversely, when $\alpha = 0$ all pK differences vanish and only relative concentrations of various acids are decisive. Effects of added acids are by far surpassed by the catalytic effect of the solvent water and the reaction appears to be uncatalyzed. The following example may serve as illustration: a reaction takes place in a buffered aqueous solution of acetic acid containing 0.1 mol/l acetic acid and 0.1 mol/l sodium acetate. Depending on the value of α the contributions of the individual acids to catalysis are as follows:

Contributions to Catalysis by

α	H_3O^+	% CH_3COOH	H_2O
0.1	0.002%	2%	98%
0.5	3.6%	96.4%	0.001%
1.0	99.8%	0.2%	$5 \times 10^{-12}\%$

220. A. Streitwieser, Jr., P. J. Scannon, and H. M. Niemeyer, *J. Amer. Chem. Soc.* **94**, 7936 (1972).

220a. A. Streitwieser, Jr., and L. L. Nebenzahl, *J. Amer. Chem. Soc.* **98**, 2188 (1976).

221. W. P. Jencks and M. Gilchrist, *J. Amer. Chem. Soc.* **90**, 2622 (1968).

222.   W. P. Jencks and Y. Carriuolo, *J. Amer. Chem. Soc.* **82**, 1778 (1960).

223.   M. J. Cho and I. H. Pitman, *J. Amer. Chem. Soc.* **96**, 1843 (1974).

224.   R. F. Hudson and G. Loveday, *J. Chem. Soc.* **1962**, 1068.

225.   From ref. 184a, p. 196.

226.   C. G. Swain and C. B. Scott, *J. Amer. Chem. Soc.* **75**, 141 (1953).

227a. J. Koskikallio, *Acta Chem. Scand.* **23**, 1477 (1969).

227b. J. Koskikallio, *Acta Chem. Scand.* **23**, 1490 (1969).

228.   H. K. Hall, Jr., *J. Org. Chem.* **29**, 3540 (1964).

229.   J. O. Edwards, *J. Amer. Chem. Soc.* **76**, 1540 (1954).

230.   R. E. Davis, in M. J. Janssen (Ed.), *Organosulfur Chemistry*, Interscience Publishers, New York-London-Sydney, 1967, p. 311.

231.   H. D. Zook and J. A. Miller, *J. Org. Chem.* **36**, 1113 (1971).

232.   C. D. Ritchie and P. O. I. Virtanen, *J. Amer. Chem. Soc.* **95**, 1882 (1973).

232a. C. D. Ritchie, *J. Amer. Chem. Soc.* **97**, 1170 (1975).

232b. J. L. Kice and L. F. Mullan, *J. Amer. Chem. Soc.* **98**, 4259 (1976).

232c. C. D. Ritchie and M. Sawada, *J. Amer. Chem. Soc.* **99**, 3754 (1977).

232d. K. Hillier, J. M. W. Scott, D. J. Barnes, and F. J. P. Steele, *Can. J. Chem.* **54**, 3312 (1976).

232e. P. O. I. Virtanen and R. Korhonen, *Acta Chem. Scand.* **27**, 2650 (1973).

232f. S. Hoz and D. Speizman, *Tetrahedron Lett.* **1978**, 1775.

233.   R. G. Fuchs and L. L. Cole, *J. Amer. Chem. Soc.* **95**, 3194 (1973).

233a. Gas phase studies [D. K. Bohme and L. B. Young, *J. Amer. Chem. Soc.* **92**, 7354 (1970)] have shown that MeO^----$HOCH_3$ reacts with CH_3Cl much more slowly than MeO^- to produce Cl^-.

234.   M. S. Puar, *J. Chem. Educ.* **47**, 473 (1970).

235.   Reference 58 in ref. 197.

236a. W. N. Olmstead and J. I. Brauman, *J. Amer. Chem. Soc.* **99**, 4219 (1977).

236b. O. I. Asubiojo and J. I. Brauman, *J. Amer. Chem. Soc.* **101**, 3715 (1979).

236c. E. K. Fukuda and R. T. McIver, Jr., *J. Amer. Chem. Soc.* **101**, 2498 (1979).

236d. K. Takashima and J. M. Riveros, *J. Amer. Chem. Soc.* **100**, 6128 (1978); M. Comisarow, *Can. J. Chem.* **55**, 171 (1977).

236e. See J. E. Bartmess and R. T. McIver, Jr., in M. T. Bowers (Ed.), *Gas Phase Ion Chemistry*, Vol. 2, Academic Press, New York-San Francisco-London, 1979, p. 88.

237.   C. L. Liotta, H. P. Harris, M. McDermott, T. Gonzalez, and K. Smith, *Tetrahedron Lett.* **1974**, 2417; H. D. Durst, *Tetrahedron Lett.* **1974**, 2421.

237a. For a description of the properties of crown ethers see: C. J. Pedersen and H. K. Frensdorff, *Angew. Chem. Int. Ed.* **11**, 16 (1972) and G. W. Gokel and H. D. Durst, *Synthesis* **1976**, 168.

238.   C. L. Liotta and H. P. Harris, *J. Amer. Chem. Soc.* **96**, 2250 (1974); L. A. Carpino and A. C. Sau, *J. Chem. Soc., Chem. Commun.* **1979**, 514.

239.   D. Y. Sam and H. E. Simmons, *J. Amer. Chem. Soc.* **96**, 2252 (1974).

240.   R. A. Moss and F. G. Pilkiewicz, *J. Amer. Chem. Soc.* **96**, 5632 (1974).

241.   Ref. 245; B. G. Cox and A. J. Parker, *J. Amer. Chem. Soc.* **95**, 402 (1973); ref. 243; I. M. Kolthoff and M. K. Chantooni, Jr., *J. Amer. Chem. Soc.* **93**, 7104 (1971).

242.   I. M. Kolthoff and M. K. Chantooni, Jr., *J. Phys. Chem.* **76**, 2024 (1972).

243. R. Alexander, A. J. Parker, J. H. Sharp, and W. E. Waghorne, *J. Amer. Chem. Soc.* **94**, 1148 (1972).

244. P. Müller and B. Siegfried, *Helv. Chim. Acta* **55**, 2400 (1972).

245. A. J. Parker, *Chem. Rev.* **69**, 1 (1969).

246. S. Winstein, E. Grunwald, and H. W. Jones, *J. Amer. Chem. Soc.* **73**, 2700 (1951).

247. E. Grunwald and S. Winstein, *J. Amer. Chem. Soc.* **70**, 846 (1948).

248. Data from ref. 247, and from A. H. Fainberg, and S. Winstein, *J. Amer. Chem. Soc.* **79**, 1597 (1957); S. G. Smith, A. H. Fainberg, and S. Winstein, *J. Amer. Chem. Soc.* **83**, 618 (1961); D. J. Raber, R. C. Bingham, J. M. Harris, J. L. Fry, and P. von Ragué Schleyer, *J. Amer. Chem. Soc.* **92**, 5977 (1970); J. E. Nordlander, R. R. Gruetzmacher, W. J. Kelly, and S. P. Jindal, *J. Amer. Chem. Soc.* **96**, 184 (1974).

249. A. Streitwieser, Jr., *Solvolytic Displacement Reactions*, McGraw-Hill Book Co., New York, 1962, p. 45.

250. A. H. Fainberg and S. Winstein, *J. Amer. Chem. Soc.* **79**, 1608 (1957). The same was found for 1-adamantyl bromide: D. J. Raber et al. *J. Amer. Chem. Soc.* **92**, 5977 (1970).

251. K. Okamoto, K. Matsubara, and T. Kinoshita, *Bull. Chem. Soc. Jap.* **45**, 1191 (1972).

251a. M. F. Ruasse and J. E. Dubois, *J. Amer. Chem. Soc.* **97**, 1977 (1975).

252. T. W. Bentley and P. v. R. Schleyer, *J. Amer. Chem. Soc.* **98**, 7658 (1976); F. L. Schadt, T. W. Bentley, and P. v. R. Schleyer, *ibid.* **98**, 7667 (1976).

252a. D. N. Kevill and G. M. L. Lin, *J. Amer. Chem. Soc.* **101**, 3916 (1979).

253. See D. J. Raber and J. M. Harris, *J. Chem. Educ.* **49**, 60 (1972); J. M. Harris, A. Becker, J. F. Fogan, and F. A. Walden, *J. Amer. Chem. Soc.* **96**, 4484 (1974), discussed details of the solvolysis mechanism of 2-adamantyl derivatives.

254. J. Slutsky, R. C. Bingham, P. von Ragué Schleyer, W. C. Dickason, and H. C. Brown, *J. Amer. Chem. Soc.* **96**, 1970 (1974).

255. V. J. Shiner, Jr., R. D. Fischer, and W. Dowd, *J. Amer. Chem. Soc.* **91**, 7748 (1969).

256. A. J. Kresge and R. J. Preto, *J. Amer. Chem. Soc.* **87**, 4593 (1965).

257. H. C. Brown, C. J. Kim, C. J. Lancelot, and P. von Ragué Schleyer, *J. Amer. Chem. Soc.* **92**, 5244 (1970).

258. C. J. Lancelot and P. von Ragué Schleyer, *J. Amer. Chem. Soc.* **91** 4297 (1969).

259. D. Fain, J. Toullec, and J.-E. Dubois, *Tetrahedron Lett.* **1974**, 1725.

259a. M. F. Ruasse and J. E. Dubois, *J. Org. Chem.* **39**, 2441 (1974).

260. P. G. Gassman and A. F. Fentiman, Jr., *J. Amer. Chem. Soc.* **92**, 2549 (1970).

260a. See also: H. G. Richey, Jr., et al., *J. Amer. Chem. Soc.* **92**, 3783 (1970).

261. P. G. Gassman and A. F. Fentiman, Jr., *J. Amer. Chem. Soc.* **92**, 2551 (1970).

262. R. Sustmann, *Tetrahedron Lett.* **1974**, 963; [see also *Pure Appl. Chem.* **40**, 576 (1974)]. W. Bihlmaier, R. Huisgen, H.-U. Reissig, and S. Voss, *Tetrahedron Lett.* **1979**, 2621.

262a. P. Beltrame, P. Sartina, and C. Vintani, *J. Chem. Soc. (B)* **1971**, 814.

262b. R. H. Wollenberg, J. S. Nimitz, and D. Y. Gokcek, *Tetrahedron Lett.* **1980**, 2791.

263. J. O. Schreck, *J. Chem. Educ.* **48**, 103 (1971).

264. See also refs. 114 for further examples.

265. Regarding the stabilization of positive charges by neighboring cyclopropane rings see: G. A. Olah and R. J. Spear, *J. Amer. Chem. Soc.* **97**, 1539 (1975).

266. J. M. Harris and S. P. McManus, *J. Amer. Chem. Soc.* **96**, 4693 (1974).

267. P. Zuman, *Collect. Czech. Chem. Commun.* **25**, 3225 (1960).

268. Ref. 227a.

269. D. J. Raber, J. M. Harris, R. E. Hall, and P. von Ragué Schleyer, *J. Amer. Chem. Soc.* **93**, 4821 (1971); see also J. M. Harris, D. C. Clark, A. Becker, and J. F. Fagan, *J. Amer. Chem. Soc.* **96**, 4478 (1974).

270. W. Kirmse, *Carbene, Carbenoide und Carbenanaloge*, Verlag Chemie, Weinheim/Bergstr., 1969, p. 48.

270a. N. G. Rondan, K. N. Houk, and R. A. Moss, *J. Amer. Chem. Soc.* **102**, 1770 (1980).

270b. R. A. Moss and C. B. Mallon, *J. Amer. Chem. Soc.* **97**, 344 (1975); R. A. Moss, C. B. Mallon, and C.-T. Ho, *ibid.* **99**, 4105 (1977).

271. G. A. Russell, in J. K. Kochi (Ed.), *Free Radicals*, Vol. 1, John Wiley & Sons, New York, 1973, p. 289.

272. From: W. F. Sliwinski, T. M. Su, and P. von Ragué Schleyer, *J. Amer. Chem. Soc.* **94**, 133 (1972).

272a. D. J. McLennan and R. J. Wong, *J. Chem. Soc., Perkin Trans.* 2, **1974**, 1373.

272b. W. H. Saunders, Jr., D. G. Bushman, and A. F. Cokerill, *J. Amer. Chem. Soc.* **90**, 1775 (1968).

273. Ref. 114b, p. 25.

274. A. Streitwieser, Jr., M. A. Hammond, R. H. Jagow, R. M. Williams, R. G. Jesaitis, C. J. Chang, and R. Wolf, *J. Amer. Chem. Soc.* **92**, 5141 (1970).

275. P. von Ragué Schleyer, J. L. Fry, L. K. M. Lam, and C. J. Lancelot, *J. Amer. Chem. Soc.* **92**, 2542 (1970).

276. A detailed discussion is given in: K. Schwetlick, *Kinetische Methoden zur Untersuchung von Reaktionsmechanismen*, VEB, Deutscher Verlag der Wissenschaften, Berlin, 1971. L. P. Hammett, *Physical Organic Chemistry*, 2nd ed., McGraw-Hill Book Co., New York, 1970. A. A. Frost and R. G. Pearson, *Kinetics and Mechanism*, 2nd ed., John Wiley & Sons, London-New York, 1961. K. B. Wiberg, *Physical Organic Chemistry*, John Wiley & Sons, London-New York, 1964. G. L. Pratt, *Gas Kinetics*, John Wiley & Sons, London-New York, 1969. See also R. W. Hoffmann, *Aufklärung von Reaktionsmechanismen*, Georg Thieme Verlag, Stuttgart, 1976.

277. N. L. Avery, III, dissertation, Harvard University, 1972, quoted and described by W. v. E. Doering and K. Sachdev, *J. Amer. Chem. Soc.* **96**, 1168 (1974), footnote 43.

278. B. S. Rabinovitch and M. C. Flowers, *Quart. Rev. Chem. Soc.* **18**, 122 (1964).

279. A. M. Mansoor and I. D. R. Stevens, *Tetrahedron Lett.* **1966**, 1733; H. M. Frey and I. D. R. Stevens, *J. Amer. Chem. Soc.* **84**, 2647 (1962).

280. D. Y. Curtin, *Rec. Chem. Prog. (Kresge-Hooker Sci. Lib.)* **15**, 111 (1954).

281. E. P. Kohler and H. M. Chadwell, *Org. Synth.* **1**, 178 (1944).

282. D. R. Brown, R. Lygo, J. McKenna, J. M. McKenna, and B. G. Hutley, *J. Chem. Soc. (B)* **1967**, 1184; J. McKenna, *Top. Stereochem.* **5**, 275 (1970).

282a. For a recent example see: E. Vedejs, M. J. Arco, and J. M. Renga, *Tetrahedron Lett.* **1978**, 523.

283. W. von E. Doering and W. R. Roth, *Tetrahedron* **18**, 67 (1962).

284. C. L. Perrin and D. J. Faulkner, *Tetrahedron Lett.* **1969**, 2783. H.-J. Hansen and H. Schmid, *Tetrahedron* **30**, 1959 (1974).

285. D. J. Faulkner, M. R. Petersen, W. S. Johnson, L. Werthemann, W. R. Bartlett, T. J. Brocksom, and Tsung-tee Li, *J. Amer. Chem. Soc.* **92**, 741 (1970).

286. W. Sucrow and W. Richter, *Chem. Ber.* **104**, 3679 (1971); W. Sucrow, B. Schubert, W. Richter, and M. Slopianka, *ibid.* **104**, 3689 (1971).

287. V. Rautenstrauch, *J. Chem. Soc., Chem. Commun.* **1970**, 526; see also: J. E. Baldwin and J. E. Patrick, *J. Amer. Chem. Soc.* **93**, 3556 (1971); P. A. Grieco and R. S. Finkelhor, *J. Org. Chem.* **38**, 2245 (1973); D. A. Evans and G. C. Andrews, *Acc. Chem. Res.* **7**, 147 (1974).

288. W. D. Closson and D. Gray, Jr., *J. Org. Chem.* **35**, 3737 (1970) and references therein.

289. Review: G. Maier, *Angew. Chem. Int. Ed.* **6**, 402 (1967).

290. K. Alder, *Experientia, Suppl. II*, p. 86, Birkhäuser Verlag, Basel, 1955; quoted by R. Huisgen, R. Grashey, and J. Sauer, in S. Patai (Ed.), *The Chemistry of Alkenes*, Interscience Publishers, New York, 1964, p. 887.

291. R. Huisgen and F. Mietzsch, *Angew. Chem. Int. Ed.* **3**, 83 (1964).

292. A. P. Johnson and V. Vajs, *J. Chem. Soc., Chem. Commun.* **1979**, 817.

293. S. Winstein and N. J. Holness, *J. Amer. Chem. Soc.* **77**, 5562 (1955).

294. E. L. Eliel and R. S. Ro, *J. Amer. Chem. Soc.* **79**, 5995 (1957).

295. S. Winstein and R. Heck, *J. Amer. Chem. Soc.* **78**, 4804 (1956).

296. S. P. Acharya and H. C. Brown, *J. Org. Chem.* **35**, 3874 (1970).

297. W. G. Dauben and R. E. Wolf, *J. Org. Chem.* **35**, 2361 (1970).

298. E. Havinga, *Experientia* **29**, 1181 (1973).

298a. I. Fleming and J. P. Michael, *J. Chem. Soc., Chem. Commun.* **1978**, 245.

299. H. Plieninger and W. Maier-Borst, *Chem. Ber.* **98**, 2504 (1965).

300. See ref. 200, and J. Lhomme, A. Diaz, and S. Winstein, *J. Amer. Chem. Soc.* **91**, 1548 (1969).

301. E. M. Arnett, *Prog. Phys. Org. Chem.* **1**, 223 (1963).

301a. C. Rüchardt, *Top. Curr. Chem.* **88**, 1 (1980).

302. P. D. Bartlett and T. T. Tidwell, *J. Amer. Chem. Soc.* **90**, 4421 (1968).

303. J. E. Dubois and J. S. Lomas, *Tetrahedron Lett.* **1973**, 1791.

304. W. Duismann and C. Rüchardt, *Chem. Ber.* **106**, 1083 (1973).

305. E. L. Eliel, N. L. Allinger, S. J. Angyal, and G. A. Morrison, *Conformational Analysis*, Interscience Publishers New York, 1965, pp. 81-84.

305a. P. Müller and J.-C. Perlberger, *Helv. Chim. Acta* **59**, 2335 (1976).

306. See *Isotope Effects in Chemical Reactions*, American Chemical Society Monograph, C. J. Collins and N. S. Bowman (Eds.), Van Nostrand Reinhold, New York, 1970; and R. W. Hoffmann *Aufklärung von Reaktionsmechanismen*, Georg Thieme Verlag, Stuttgart, 1976; and ref. 313.

307. This simplification corresponds to reality for partition functions of motions of parts of the molecule more than two bonds removed from the site of the isotopic substitution. For other motions (vibrations) it corresponds to the *predominance* of the zero-point energy effect when there are *considerable* primary isotope effects (ref. 308, p. 519).

308. R. P. Bell, *Chem. Soc. Rev.* **3**, 513 (1974).

309. H. Kwart, J. Slutsky, and S. F. Sarner, *J. Amer. Chem. Soc.* **95**, 5242 (1973).

309a. L. B. Sims, A. Fry, L. T. Netherton, J. C. Wilson, K. D. Reppond, and S. W. Crook, *J. Amer. Chem. Soc.* **94**, 1364 (1972).

310. R. P. Bell and D. M. Goodall, *Proc. R. Soc., Ser. A* **294**, 273 (1966), and further examples in ref. 308; see also H. Wilson, J. D. Caldwell, and E. S. Lewis, *J. Org. Chem.* **38**, 564 (1973).

311. E. F. Caldin, *Chem. Rev.* **69**, 135 (1969); and R. W. Hoffmann, *Aufklärung von Reaktionscmechanismen*, Georg Thieme Verlag, Stuttgart, 1976.

312. R. A. More O'Ferrall, *J. Chem. Soc. (B)* **1970**, 785.

312a. M. M. Green, G. J. Mayotte, L. Meites, and D. Forsyth, *J. Amer. Chem. Soc.* **102**, 1464 (1980).

312b. E. L. Motell, A. W. Boone, and W. H. Fink, *Tetrahedron* **34**, 1619 (1978).

312c. F. Hibbert and H. J. Robbins, *J. Chem. Soc., Chem. Commun.* **1980**, 141.

313. L. Melander, *Isotope Effects on Reaction Rates*, Ronald Press, New York, 1960.

314. A. Koeberg-Telder and H. Cerfontain, *Rec. Trav. Chim. Pays-Bas* **91**, 22 (1972).

315. R. P. Bell, *The Proton in Chemistry*, Methuen and Co. Ltd., London, 1959, p. 201.

316. A. Fry, *Chem. Soc. Rev.* **1**, 163 (1972).

317. B. M. Benjamin and C. J. Collins, *J. Amer. Chem. Soc.* **95**, 6145 (1973).

318. G. Müller-Hagen and W. Pritzkow, *J. Prakt. Chem.* **311**, 874 (1969); see also R. D. Bach and R. F. Richter, *J. Org. Chem.* **38**, 3442 (1973); *Tetrahedron Lett.*, **1973**, 4099; A. Lewis and J. Azoro, *Tetrahedron Lett.* **1979**, 3627.

319. P. D. Bartlett, *Quart. Rev.* **24**, 473 (1970).

320. E. J. Dewitt, C. T. Lester, and G. A. Ropp, *J. Amer. Chem. Soc.* **78**, 2101 (1958).

321. R. W. Henderson and R. D. Ward, Jr., *J. Amer. Chem. Soc.* **96**, 7556 (1974); R. W. Henderson, *ibid.* **97**, 213 (1975); W. A. Pryor and W. H. Davis, Jr., *ibid.* **96**, 7557 (1974).

322. G. A. Russell, in J. K. Kochi (Ed.), *Free Radicals*, Vol. 1, John Wiley & Sons, London-New York, 1973, p. 295.

323. L. W. Christensen, E. E. Waali, and W. M. Jones, *J. Amer. Chem. Soc.* **94**, 2118 (1972).

324. T. Inukai and T. Kojima, *J. Org. Chem.* **36**, 924 (1971).

325. A. Schriesheim, C. A. Rowe, Jr., and L. Naslund, *J. Amer. Chem. Soc.* **85**, 2111 (1963), and earlier work quoted therein.

326. K. M. Ibne-Rasa, R. H. Pater, J. Ciabattoni, and J. O. Edwards, *J. Amer. Chem. Soc.* **95**, 7894 (1973).

327. J. E. Leffler, *J. Org. Chem.* **20**, 1202 (1955).

328. O. Exner, *Prog. Phys. Org. Chem.* **10**, 411 (1973).

329. For examples, see ref. 328. It was assumed earlier that many more reaction series obey the isokinetic relationship. However, it turned out[328] that these were artifacts due to the methods used in determining errors and the statistical analysis.

330. T. Matsui, H. C. Ko, and L. G. Hepler, *Can. J. Chem.* **52**, 2906 (1974).

331. S. Winstein and A. H. Fainberg, *J. Amer. Chem. Soc.* **79**, 5937 (1957).

332. J.-E. Dubois and M. Marie de Ficquelmont-Loizos, *Tetrahedron Lett.* **1976**, 635.

333. L. G. Hepler, *J. Amer. Chem. Soc.* **85**, 3089 (1963).

334. T. M. Krygowski and W. R. Fawcett, *Can J. Chem.* **53**, 3622 (1975).

335. R. Yamdagni, T. B. McMahon, and P. Kebarle, *J. Amer. Chem. Soc.* **96**, 4035 (1974).

335a. L. D. Hansen and L. G. Hepler, *Can. J. Chem.* **50**, 1030 (1972).

336a. T. Mitsuhashi and O. Simamura, *J. Chem. Soc. (B)* **1970**, 705.

336b. T. Mitsuhashi, O. Simamura, and Y. Tezuka, *J. Chem. Soc., Chem. Commun.* **1970**, 1300.

336c. T. Mitsuhashi, H. Miyadera, and O. Simamura, *J. Chem. Soc., Chem. Commun.* **1970**, 1301.

336d. J. Hinz, A. Oberlinner, and C. Rüchardt, *Tetrahedron Lett.* **1973**, 1975.

337. H. M. Frey, B. M. Pope, and R. F. Skinner, *J. Chem. Soc., Faraday Trans.* **63**, 1166 (1967).

338. H. M. Frey, J. Metcalfe, and B. M. Pope, *J. Chem. Soc., Faraday Trans.* **67**, 750 (1971).

339. P. D. Bartlett and G. D. Sargent, *J. Amer. Chem. Soc.* **87**, 1297 (1965).

340. P. v. R. Schleyer and G. W. van Dine, *J. Amer. Chem. Soc.* **88**, 2321 (1966).

341. P. G. Gassman and D. S. Patton, *J. Amer. Chem. Soc.* **91**, 2160 (1969).

342. M. A. McKinney and S. H. Smith, *Tetrahedron Lett.* **1971**, 3657.

343. W. M. Schubert and B. Lamm, *J. Amer. Chem. Soc.* **88**, 120 (1966).

344. Ref. 342, footnote 11.

344a. A. de Meijere, O. Schallner, C. Weitemeyer, and W. Spielmann, *Chem. Ber.* **112**, 908 (1979).

345. J. Hine, N. W. Burske, M. Hine, and P. B. Langford, *J. Amer. Chem. Soc.* **79**, 1406 (1957).

346. K. E. Richards, A. L. Wilkinson, and G. I. Wright, *Aust. J. Chem.* **25**, 2369 (1972).

346a. F. G. Bordwell and G. J. McCollum, *J. Org. Chem.* **41**, 2391 (1976).

347. M. E. H. Howden and J. D. Roberts, *Tetrahedron* **19**, Suppl. 2, 403 (1963).

348. O. L. Chapman and P. Fitton, *J. Amer. Chem. Soc.* **85**, 41 (1963). The products of this reaction were not analyzed, but by analogy to similar transformations they should consist of mixtures of isomeric cycloheptatrienes and acetates [e.g., $(CH_3)_2(C_7H_7)OCOCH_3$]. N. A. Nelson, J. H. Fassnacht, and J. U. Piper, *J. Amer. Chem. Soc.* **83**, 206 (1961).

349. P. R. Story, L. C. Snyder, D. C. Douglass, E. W. Anderson, and R. L. Kornegay, *J. Amer. Chem. Soc.* **85**, 3630 (1963).

349a. Cf. H. Hart and M. Kuzuya, *J. Amer. Chem. Soc.* **98**, 1545 (1976), and L. M. Loew and C. F. Wilcox, *ibid.* **97**, 2296 (1975).

350. S. Winstein and C. Ordronneau, *J. Amer. Chem. Soc.* **82**, 2084 (1960).

351. H. G. Richey, Jr., and N. C. Buckley, *J. Amer. Chem. Soc.* **85**, 3057 (1963); P. R. Story and S. R. Fahrenholtz, *ibid.* **86**, 527 (1964).

351a. E. N. Peters and H. C. Brown, *J. Amer. Chem. Soc.* **95**, 2397 (1973).

352. C. J. Lancelot and P. von R. Schleyer, *J. Amer. Chem. Soc.* **91**, 4296 (1969).

353. Examples. M. Simonetta, *Top. Curr. Chem.* **42**, 1, (1973). See also: Jahres-rückblick 1975—Theoretische Chemie, *Nachr. Chem. Techn.* **24**, No. 3, 51 (1976). K. Müller, *Angew. Chem. Int. Ed.* **19**, 1 (1980).

354. I. Shavitt, R. M. Stevens, F. L. Minn, and M. Karplus, *J. Chem. Phys.* **48**, 2700 (1968).

355. J. A. Horsley, Y. Jean, C. Moser, L. Salem, R. M. Stevens, and J. S. Wright, *J. Amer. Chem. Soc.* **94**, 279 (1972).

356. A. Rastelli, A. S. Pozzoli, and G. Del Re, *J. Chem. Soc.*, Perkin *Trans. 2* **1972**, 1571.

357. N. Bodor, M. J. S. Dewar, and J. S. Wasson, *J. Amer. Chem. Soc.* **94**, 9100 (1972).

358a. H. B. Bürgi, J. D. Dunitz, and E. Shefter, *J. Amer. Chem. Soc.* **95**, 5065 (1973).

358b. H. B. Bürgi, J. D. Dunitz, J. M. Lehn, and G. Wipf, *Tetrhaedron* **30**, 1563 (1974).

358c. P. G. Jones and A. J. Kirby, *J. Chem. Soc., Chem. Commun.* **1979**, 288.

358d. G.-A. Craze and I. Watt, *J. Chem. Soc., Chem. Commun.* **1980**, 147.

358e. F. M. Menger and L. E. Glass, *J. Amer. Chem. Soc.* **102**, 5404 (1980).

358f. H.B. Bürgí, E. Shefter u. J.D. Dunitz, Tetrahedron 31. 3089 (1975).

359. R. C. Bingham and M. J. S. Dewar, *J. Amer. Chem. Soc.* **94**, 9107 (1972).

360. W. T. A. M. van der Lugt and P. Ros, Chem. *Phys. Lett.* **4**, 389 (1969).

361. M. J. S. Dewar and D. H. Lo, *J. Amer. Chem. Soc.* **93**, 7201 (1971); A. K. Cheng, F. A. L. Anet, J. Mioduski, and J. Meinwald, *ibid.* **96**, 2887 (1974).

362. See E. N. Peters and H. C. Brown, *J. Amer. Chem. Soc.* **96**, 265 (1974).

363. Such a plot was first given by H. L. Goering and C. B. Schewene, *J. Amer. Chem. Soc.* **87**, 3516 (1965).

364. See: G. D. Sargent, in G. Olah and P. v. R. Schleyer (Eds.), *Carbonium Ions*, Vol. III, John Wiley & Sons, New York, 1972, Chapter 24.

365. R. L. Burwell, Jr., and R. G. Pearson, *J. Phys. Chem.* **70**, 300 (1966).

366. S. W. Benson, D. M. Golden, and K. W. Egger, *J. Chem. Phys.* **42**, 4265 (1965); see also R. M. Noyes, D. E. Applequist, S. W. Benson, D. M. Golden, and P. S. Skell, *ibid.* **46**, 1221 (1967) and references therein.

367. W. P. Jencks, *Prog. Phys. Org. Chem.* **2**, 63, 86 (1964).

368. Jencks formulated a "rule" stating under what conditions one can expect the occurrence of such a general acid-base catalysis: W. P. Jencks, *J. Amer. Chem. Soc.* **94**, 4731 (1972).

369. H. M. Frey, and R. Walsh, *Chem. Rev.* **69**, 103 (1969).

370a. If a transition state were to belong to two reaction coordinates, it would re-semble a hilltop rather than a pass (col). A hilltop cannot be the lower barrier between any two points, since one could always cross a lower barrier by going around the hill. J. N. Murrell and K. J. Laidler, *J. Chem. Soc., Faraday Trans.* **64**, 371 (1968); J. N. Murrell, *J. Chem. Soc., Chem. Commun.* **1972**, 1044; see also P. G. Wright and M. R. Wright, *J. Phys. Chem.* **74**, 4398 (1970).

370b. A counterexample: the calculation of the system $(CH_2)_4$ [R. Hoffmann et al., *J. Amer. Chem. Soc.* **92**, 7091 (1970)] showed minima only for cyclo-butane and two molecules of ethylene along with an *extended energetically flat region* on the energy hypersurface. Because of the considerable exten-sion of this *region, which corresponds to a transition state*, the reacting sys-tem should spend a relatively long time ($> 10^{-12}$ s) in this state before it can find a path to one of the minima. The relatively long lifetime of such species (called "twixtyls") should make them behave like intermediates (be trappable and capable of forming *several* products). A similar situation for $(CH_2)_3$ was discussed by J. P. Freeman et al. [*J. Org. Chem.* 37, 1894 (1972)].

371. L. Salem, *Acc. Chem. Res.* 4, 322 (1971).

371a. J. W. McIver, Jr., *Acc. Chem. Res.* 7, 72 (1974).

372a. H. M. Frey and R. Walsh, *Chem. Rev.* 69, 103 (1969).

372b. H. M. Frey, A. M. Lamont, and R. Walsh, *J. Chem. Soc. (A)* 1971, 2642.

373. D. F. DeTar, *J. Amer. Chem. Soc.* 96, 1254 (1974); 1255 (1974); see also: F. Becker, *Z. Naturforsch.* 166, 236 (1961), and refs. 165b and 165c.

374. I. Tarvainen and J. Koskikallio, *Acta Chem. Scand.* 24, 1129 (1970).

375. J. E. Leffler, *Science* 117, 340 (1953).

376. G. S. Hammond, *J. Amer. Chem. Soc.* 77, 334 (1955).

376a. P. Lechtken, R. Breslow, A. H. Schmidt, and N. J. Turro, *J. Amer. Chem. Soc.* 95, 3025 (1973); M. J. Goldstein and R. S. Leight, *ibid.* 99, 8112 (1977).

377. R. P. Bell, *Proc. R. Soc. London, Ser. A* 154, 414 (1936); *Acid-Base Catalysis*, Oxford University Press, New York, 1941.

378. M. G. Evans and M. Polanyi, *J. Chem. Soc., Faraday Trans.* 32, 1340 (1936).

378a. R. A. More O'Ferral, *J. Chem. Soc. (B)* 1970, 274.

379. J. E. Critchlow, *J. Chem. Soc., Faraday Trans. 1,* 68, 1774 (1972).

379a. W. P. Jencks, *Chem. Rev.* 72, 705 (1972).

379b. J. J. Gajewski, *J. Amer. Chem. Soc.* 101, 4393 (1979).

379c. This conclusion has also been reached by a quantum mechanical treatment of (*414*): R. D. Gandour, G. M. Maggiora, and R. L. Schowen, *J. Amer. Chem. Soc.* 96, 6967 (1974).

380. J. Coburn-Harris and J. L. Kurz, *J. Amer. Chem. Soc.* 92, 349 (1970).

381. E. R. Thornton, *J. Amer. Chem. Soc.* 89, 2915 (1967).

382. Data from Ref. 372a.

383a. See also R. Walsh, D. M. Golden, and S. W. Benson [*J. Amer. Chem. Soc.* 88, 650 (1966)], D. M. Golden, A. S. Rodgers, and S. W. Benson [*ibid.* 88, 3196 (1966)], W. v. E. Doering and G. H. Beasley [*Tetrahedron* 29, 2231 (1973)], and W. R. Roth, G. Ruf, and P. W. Ford [*Chem. Ber.* 107, 48 (1974)]. In the last two studies the allylic resonance energy is measured as a lowering of the activation energy of the *cis-trans* isomerization of C=C double bond systems. Since the occurrence of intermediates is less likely in such reactions (the state of ethylene where the two groups lie in orthogonal planes should be a transition state), the validity of Hammond's postulate $E_{A(CH_2=CH-CH_2\cdot + \cdot R)} = 0$ is a less important prerequisite than it is in the former reactions.

383b. ESR studies of allyl radicals [P. J. Krusic, P. Meakin, and B. E. Smart, *J. Amer. Chem. Soc.* 96, 6211 (1974)] showed that even at temperatures of $280°C$ ($R^1 = R^2 = H$), resp. $180°C$ ($R^1 = R^2 = CH_3$) there is *no* indication of interconversion of R^1 and R^2.

Consequently, this exchange requires *free activation energies* higher than 71 and 59 kJ/mol, respectively.

383c. For a very low (34 kJ/mol) estimate of the resonance stabilization of the benzyl radical, see: M. J. S. Dewar and L. E. Wade in ref. 471. Resonance energies of many other radicals or ions are collected in ref. 43g.

384. M. J. S. Dewar and H. N. Schmeising, *Tetrahedron* **5**, 166 (1959).

385a. See H. E. O'Neal and S. W. Benson, *Int. J. Chem. Kin.* **2**, 423 (1970), and earlier references therein.

385b. Other authors [L. M. Stephenson, T. A. Gibson, and J. I. Brauman, *J. Amer. Chem. Soc.* **95**, 2849 (1973)] have pointed out that trimethylene, which can be *formally* obtained by removal of two hydrogen atoms from propane under the assumption that the C—H bond dissociation energy is the *same* for both bonds ("classic" scheme), represents only a classical *model* for reality and need not be identical to it. A *destabilization* of 1,3- and 1,4-diradicals by about 40 (27) kJ/mol (so far unrecognized and not considered in the "classic" scheme) could increase, for example, the dissociation energy of the second C—H bond by this amount and make the enthalpy of formation of trimethylene equal to that of the transition state from cyclopropane.

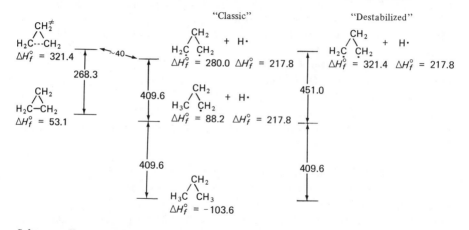

Scheme. Two models ("classic" and "destabilized") for the thermochemistry (kJ/mol) of trimethylene.

Analogously, one should take into account that also on homolysis of other strained systems the total strain energy of reagents is *not* released in the transition state, so that $\Delta H^{\ddagger} > \Delta H$. The activation energy for ring closure of trimethylene-methane was measured to be 30 kJ/mol, and this value was also set as a maximum for the energy of activation of ring closure of trimethylene [P. Dowd and M. Chow, *J. Amer. Chem. Soc.* **99**, 6438 (1977); see also: R. G. Bergman in J. K. Kochi (Ed.), *Free Radicals*, Vol. I, John Wiley & Sons, New York, 1973, and P. S. Engel et al., *J. Amer. Chem. Soc.* **100**, 1876 (1978)]. A different view is given in ref. 496.

386. E. W. Bittner, E. M. Arnett, and M. Saunders, *J. Amer. Chem. Soc.* **98**, 3734 (1976).

387. V. Gold, *J. Chem. Soc., Faraday Trans. 1*, **68**, 1611 (1972).

387a. A critical analysis of the application of Hammond's postulate is given by Farcaşiu, [D. Farcaşiu, *J. Chem. Educ.* **52**, 76 (1975)].

387b. P. C. Martino and P. B. Shevlin, *J. Amer. Chem. Soc.* **102**, 5429 (1980).

387c. B. Franzus, M. L. Scheinbaum, D. L. Waters, and H. B. Bowlin, *J. Amer. Chem. Soc.* **98** 1241 (1976).

387d. D. A. Evans and J. V. Nelson, *J. Amer. Chem. Soc.* **102** 774 (1980).

387e. O. Papies and W. Grimme, *Tetrahedron Lett.* **1980**, 2799.

387f. D. A. Evans and D. J. Baillargeon, *Tetrahedron Lett.* **1978**, 3315, 3319.

387g. M. E. Jung and J. P. Hudspeth, *J. Amer. Chem. Soc.* **100**, 4309 (1978).

387h. M. J. Goldstein and S. A. Kline, *Tetrahedron Lett.* **1973**, 1085.

387i. R. Breslow and J. M. Hoffmann, Jr., *J. Amer. Chem. Soc.* **94**, 2111 (1972); see also K. N. Houk in ref. 423e for a discussion of substituent effects in sigmatropic rearrangements.

388. A. Krantz, *J. Amer. Chem. Soc.* **94**, 4020 (1972).

389. J. Haywood-Farmer, *Chem. Rev.* **74**, 315 (1974).

390. D. F. Eaton and T. G. Traylor [*J. Amer. Chem. Soc.* **96**, 1226 (1974)] claim to have demonstrated for cyclopropylcarbinyl systems the *"vertical"* participation of cyclopropane rings, that is, the formation of carbocations with conservation of geometry and strain of the starting material (the designation is meant to stress the analogy with Franck-Condon processes). In such instances the stabilizing effect of the cyclopropyl group is due to hyperconjugation.

390a. W. C. Still and T. L. Macdonald, *J. Org. Chem.* **41**, 3620 (1976).

391. M. W. Rathke and D. Sullivan, *Tetrahedron Lett.* **1972**, 4249.

392. E. J. Corey and D. E. Cane, *J. Org. Chem.* **34**, 3053 (1969).

393. W. S. Murphy, R. Boyce, and E. A. O'Riordan, *Tetrahedron Lett.* **1971**, 4157.

394. R. B. Bates, S. Brenner, C. M. Cole, E. W. Davidson, G. D. Forsythe, D. A. McCombs, and A. S. Roth, *J. Amer. Chem. Soc.* **95**, 926 (1973).

395. B. J. L. Huff, F. N. Norman Tuller, and D. Caine, *J. Org. Chem.* **34**, 3070 (1969).

395a. E. Haselbach, *Tetrahedron Lett.* **1970**, 1543.

395b. R. Hoffmann, C. C. Levin, and R. A. Moss, *J. Amer. Chem. Soc.* **95**, 629 (1973).

395c. B. Giese, *Angew. Chem. Int. Ed.* **16**, 125 (1977).

395d. A. Pross, *Adv. Phys. Org. Chem.* **14**, 69 (1977).

396. Data from: R. Huisgen, *Angew. Chem. Int. Ed.* **9**, 751 (1970).

396a. See also: Z. Rappoport, *Tetrahedron Lett.* **1979**, 2559.

397. B. Giese, *Angew. Chem. Int. Ed.* **15**, 174 (1976).

398. B. Giese, *Angew. Chem. Int. Ed.* **15**, 173 (1976).

398a. B. Giese and J. Meister, *Angew. Chem. Int. Ed.* **17**, 594 (1978).

399. J. R. Knowles, R. O. C. Norman, and J. H. Prosser, *Proc. Chem. Soc.* **1961**, 341.

400. J. F. Kirsch, W. Clewell, and A. Simon, *J. Org. Chem.* **33**, 127 (1968).

401. See C. D. Johnson, *Chem. Rev.* **75**, 755 (1975); *Tetrahedron* **36**, 3461 (1980).

401a. D. S. Kemp and M. L. Casey, *J. Amer. Chem. Soc.* **95**, 6670 (1973).

402. C. D. Johnson and K. Schofield, *J. Amer. Chem. Soc.* **95**, 270 (1973).

403. J. R. Murdoch, *J. Amer. Chem. Soc.* **94**, 4410 (1972); *ibid.*, **102**, 71 (1980).

404. R. A. Marcus, *J. Phys. Chem.* **72**, 891 (1968).

405. G. W. Koeppl and A. J. Kresge, *J. Chem. Soc., Chem. Commun.* 371 (1973).
406. M. Eigen, *Angew. Chem. Int. Ed.* **3**, 1 (1964).
407. See: A. J. Kresge, *Acc. Chem. Res.* **8**, 354 (1975).
407a. R. A. Benkeser, J. Hooz, T. V. Liston, and A. E. Trevillyan, *J. Amer. Chem. Soc.* **85**, 3984 (1963).
408. R. A. Marcus, *J. Amer. Chem. Soc.* **91**, 7224 (1969).
408a. F. G. Bordwell and W. J. Boyle, Jr., *J. Amer. Chem. Soc.* **97**, 3447 (1975).
409. A. Pross, *J. Amer. Chem. Soc.* **98**, 776 (1976).
409a. A. Argile and M.-F. Ruasse, *Tetrahedron Lett.* **1980**, 1327.
410. Data from: J. Sauer, *Angew. Chem. Int. Ed.* **6**, 16 (1967).
411. K. Seguchi, A. Sera, Y. Otsuki, and K. Maruyama, *Bull Chem. Soc. Jap.* **48**, 3641 (1975).
412. A refinement of the zero-point energy interpretation (neglecting all other possible contributions) deprives the ratio k_H/k_D of its status as a measure for the "symmetry" of the transition state. In the simple interpretation it is assumed that constant a in the equation $k_A k_C - \beta^2 = a$ for the coupled oscillator system $A \cdots H \cdots C$ is zero, where $k_A(k_C)$ = force constant for the change in position of A(C) and β = coupling constant. More realistic, negative values of a lead to $k_H k/_D$ as a function of the extent of bond making (bond breaking) x,[413] as shown in the figure below.

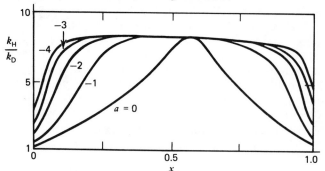

Calculated[413] change of k_H/k_D as a function of bond formation (bond breaking), x ($0 \leqslant x \leqslant 1$) and of the value of a.

413. A. V. Willi and M. Wolfsberg, *Chem. Ind. (London)* **1964**, 2097.
414. F. G. Bordwell and W. J. Boyle, Jr., *J. Amer. Chem. Soc.* **93**, 512 (1971).
415. P. J. Smith and S. K. Tsui, *Tetrahedron Lett.* **1973**, 61.
416. See R. C. Fort and P. von R. Schleyer, *Adv. Alicyclic Chem.* **1**, 284, 312 (1966). Many observations indicate that solvolysis and deamination differ not only in the position of their transition states on the reaction coordinate, but also in the nature of the carbocation formed in these processes. Deamination of *exo*-2-norbornylamines, unlike the solvolysis of reactive esters does not lead to the nonclassical 2-norbornyl cation but to a "classical" one [E. J. Corey, J. Casanova, Jr., P. A. Vatakencherry, and R. Winter, *J. Amer. Chem. Soc.*, **85**, 169 (1963); C. J. Collins and B. M. Benjamin, *ibid.* **92**, 3182 (1970) and B. M. Benjamin and C. J. Collins, *ibid.* **92**, 3183 (1970)]. An elegant demonstration of the *vertical* character (see ref. 390) in a deami-

nation and of a nonvertical character in the corresponding solvolysis is given by W. Kirmse and G. Voigt [*J. Amer. Chem. Soc.* **96**, 7598 (1974)].

417. P. Geneste, G. Lamaty, C. Moreau, and J.-P. Roque, *Tetrahedron Lett.* **1970**, 5011.

418. P. Geneste, G. Lamaty, and J.-P. Roque, *Tetrahedron Lett.* **1970**, 5007.

419. P. Haberfield, *J. Amer. Chem. Soc.* **93**, 2091 (1971).

420. J. L. Kurz, *Acc. Chem. Res.* **5**, 1 (1972).

421. R. A. Sneen and J. W. Larsen, *J. Amer. Chem. Soc.* **91**, 6031 (1969); see also: R. A. Sneen, *Acc. Chem. Res.* **6**, 46 (1973).

422. J. L. Kurz and J. C. Harris, *J. Amer. Chem. Soc.* **92**, 4117 (1970); see also: J. L. Kurz and Y.-N. Lee, *J. Amer. Chem. Soc.* **97**, 3841 (1975) and J. L. Kurz and J. Lee, *ibid.* **102**, 5427 (1980).

423a. R. B. Woodward and R. Hoffmann, *Conservation of Orbital Symmetry*, Verlag Chemie, Weinheim, 1970. T. L. Gilchrist and R. C. Storr, *Organic Reactions and Orbital Symmetry*, 2nd ed., Cambridge University Press, Cambridge, 1979.

423b. Nguyen Trong Anh, *Die Woodward-Hoffmann Regeln und ihre Anwendung*, Verlag Chemie, Weinheim Bergstrasse, 1972.

423c. R. E. Lehr and A. P. Marchand, *Orbital Symmetry; A Problem-Solving Approach*, Academic Press, New York-London, 1972.

423d. *Pericyclic Reactions*, Vol. I, A. P. Marchand and R. E. Lehr (Eds.), Academic Press, New York-San Franscisco-London, 1977.

423e. *Pericyclic Reactions*, Vol. II, A. P. Marchand and R. E. Lehr (Eds.), Academic Press, New York-San Francisco-London, 1977.

424. R. B. Woodward and R. Hoffmann, *J. Amer. Chem. Soc.* **87**, 395 (1965).

425. The symmetry correlation of individual MOs is a simpler version of the fundamental treatment of pericyclic reactions demonstrated for the first time by H. C. Longuet-Higgins and E. W. Abrahamson [*J. Amer. Chem. Soc.* **87**, 2045 (1965); see also ref. 447b]. They established symmetry correlations for *electronic states* (synonym: electronic configurations) of reactant(s) and product(s). Symmetries of electronic configurations are obtained by multiplying the symmetries of all electrons involved in that configuration (S · S = S, A · A = S, and A · S = A). With respect to conrotation, the symmetry of the ground state of butadiene ($\psi_1^2 \psi_2^2$) is $A^2 S^2 = S$; the symmetry characteristics of individual MOs are taken from Fig. 106. The ground state of cyclobutene ($\sigma^2 \pi^2$) is also symmetric with respect to conrotation. The symmetries of excited states of butadiene ($\psi_1^2 \psi_2 \psi_3$, $\psi_1 \psi_2^2 \psi_4$, and $\psi_1^2 \psi_3^2$) and of cyclobutene ($\sigma^2 \pi \pi^*$, $\sigma \pi^2 \sigma^*$, and $\sigma^2 \pi^{*2}$) can also be easily derived from the symmetries of individual MOs on page 381. (In setting up excited states one must take into account that only electronic transitions between MOs of opposite symmetry are allowed.) In correlating electronic states one must consider their "composition" in addition to their symmetry: only states constructed from MOs that themselves can be correlated may be linked by provisional (dashed) correlation lines. The final correlation (solid lines) is governed by the quantum mechanical noncrossing rule: provisional correlation lines between states of equal symmetry may not cross. This leads to the deflection of the pairs of lines in the following diagrams:

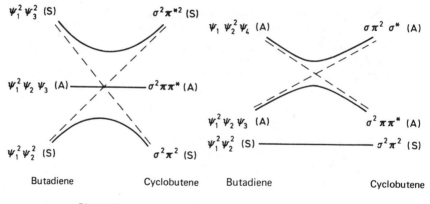

The diagrams show that on disrotatory reaction the energies of the ground states increase; on conrotatory reaction no such quantum mechanical barrier must be overcome. In reactions of the first excited state conrotation shows a barrier and disrotation is now easier to achieve.

425a. For cations see: T. S. Sorensen and A. Rauk, ref. 423e, p. 1 for carbanions see: S. W. Staley, ref. 423d, p. 199.

426. E. Eckel, R. Huisgen, R. Sustmann, G. Walbillich, D. Grashey, and E. Spindler, *Chem. Ber.* **100**, 2192 (1967).

427. E. Heilbronner and H.-D. Martin, *Helv. Chim. Acta* **55**, 1490 (1972).

428. J. C. Dalton and L. E. Friedrich, *J. Chem. Educ.* **52**, 721 (1975).

428a. K. N. Houk in ref. 423e, p. 182; see I. Fleming, in ref. 6a.

429. K. Fukui, *Acc. Chem. Res.* **4**, 57 (1971).

430. See N. D. Epiotis, *J. Amer. Chem. Soc.* **94**, 1924 (1972).

431. K. N. Houk, *Acc. Chem. Res.* **8**, 316 (1975).

432. R. Sustmann, *Pure Appl. Chem.* **40**, 569 (1974).

433. R. Sustmann and R. Schubert, *Angew. Chem. Int. Ed.* **11**, 840 (1972).

434. T. Kojima and T. Inukai, *J. Org. Chem.* **35**, 1342 (1970).

435. K. N. Houk, *J. Amer. Chem. Soc.* **95**, 4092 (1973).

436. M. G. Evans, *Trans. Faraday Soc.* **35**, 824 (1939).

437. H. E. Zimmerman, *Acc. Chem. Res.* **4**, 272 (1971); ref. 423d, p. 53.

438. M. J. S. Dewar, *Angew. Chem. Int. Ed.* **10**, 761 (1971); see also: C. L. Perrin, *Chem. in Brit.* **8**, 163 (1972).

439. K. Kraft and G. Koltzenburg, *Tetrahedron Lett.* **1967**, 4357, 4723.

440a. A. Padwa and S. Clough, *J. Amer. Chem. Soc.* **92**, 5308 (1970); J. Leitich, *Angew. Chem. Int. Ed.* **8**, 909 (1969); see also corresponding behavior of benzyne: J. Leitich, *Tetrahedron Lett.* **1980**, 3025.

440b. P. Warner, S.-C. Chang, D. R. Powell, and R. A. Jacobson, *J. Amer. Chem. Soc.* **102**, 5125 (1980).

440c. L. Ghosez and M. J. O'Donnell, in ref. 423e, p. 79.

441. R. H. Wollenberg and R. Belloli, *Chem. Brit.* **10**, 95 (1974).

441a. R. C. Cookson, J. Dance, and J. Hudec, *J. Chem. Soc.* **1964**, 5416.

442. For discussion of factors that could possibly influence *exo-endo* selectivity, see J. M. Mellor and C. F. Webb, *J. Chem. Soc., Perkin Trans 2*, **1974**, 26, and previous work.

443. K. Seguchi, A. Sera, and K. Maruyama, *Tetrahedron Lett.* **1973**, 1585.

444. This mnemonic was first formulated for Hückel systems by A. A. Frost and B. Mussulin [*J. Chem. Phys.* **21**, 572 (1953)]. The modification for Möbius systems is due to H. E. Zimmerman [*J. Amer. Chem. Soc.* **88**, 1564 (1966)].

445. A. Dahmen and R. Huisgen, *Tetrahedron Lett.* **1969**, 1465.

446. R. F. Childs and S. Winstein, *J. Amer. Chem. Soc.* **90**, 7146 (1968).

447. W. R. Roth, J. König, and K. Stein, *Chem. Ber.* **103**, 426 (1970).

447a. J. I. Brauman and W. C. Archie, Jr., *J. Amer. Chem. Soc.* **94**, 4262 (1972).

447b. J. E. Baldwin, in ref. 423e, p. 273.

448. See: U. Schöllkopf, *Angew Chem. Int. Ed.* **9**, 763 (1970).

449. See: D. Seebach, in *Methoden der Organischen Chemie (Houben-Weyl)*, Vol. IV/4, Georg Thieme Verlag, Stuttgart 1971. P. D. Bartlett, *Science* **159**, 833 (1968).

450. R. Huisgen, *Acc. Chem. Res.* **10**, 117 (1977); R. Huisgen and R. Schug, *J. Amer. Chem. Soc.* **98**, 7819 (1976).

451. For new quantum mechanical studies of such reactions see: N. D. Epiotis, R. L. Yates, D. Carlberg, and F. Bernardi, *J. Amer. Chem. Soc.* **98**, 453 (1976).

452. A. T. Cocks, H. M. Frey, and I. D. R. Stevens, *J. Chem. Soc., Chem. Commun.* 458 (1969).

453. J. A. Berson, *Acc. Chem. Res.* **5**, 406 (1972).

453a. J. A. Berson and P. B. Derwan, *J. Amer. Chem. Soc.* **95**, 269 (1973).

454. J. A. Berson and L. Salem, *J. Amer. Chem. Soc.* **94**, 8917 (1972).

455. P. C. Hiberty, *J. Amer. Chem. Soc.* **97**, 5975 (1975).

456. W. Schmidt, *Tetrahedron Lett.* **1972**, 581. The role of configuration interactions in other pericyclic reactions was discussed by N. D. Epiotis [*Angew. Chem. Int. Ed.* **13**, 751 (1974)].

457a. K. Fukui and H. Fujimoto, *Tetrahedron Lett.* **1965**, 4303.

457b. K. Fukui, H. Hao, and H. Fujimoto, *Bull Chem. Soc. Jap.* **42**, 348 (1969).

457c. L. Salem, *Chem. Brit.* **5**, 449 (1969); see also ref. 457b.

457d. K. Fukui, *Tetrahedron Lett.* **1965**, 2427.

457e. W. Drenth, *Rec. Trav. Chim. Pays-Bas* **86**, 318 (1967).

458. G. S. Hammond and J. Warkentin, *J. Amer. Chem. Soc.* **83**, 2554 (1961).

459. Cf. R. M. Magid, *Tetrahedron* **36**, 1901 (1980).

460. R. L. Yates, N. D. Epiotis, and F. Bernadi, *J. Amer. Chem. Soc.* **97**, 6615 (1975).

460a. J. Martel, E. Toromanoff, J. Mathieu, and G. Nomine, *Tetrahedron Lett.* **1972**, 1491.

461. H. E. Zimmerman, *Acc. Chem. Res.* **5**, 393 (1972); ref. 423d, p. 53.

462. N. T. Anh, *Top. Curr. Chem.* **88**, 145 (1980); see also: J. Klein, *Tetrahedron Lett.* **1973**, 4307.

462a. Min-hon Rei, *J. Org. Chem.* **44**, 2760 (1979).

463. R. K. Hill and M. G. Bock, *J. Amer. Chem. Soc.* **100**, 637 (1978).

464. G. Baddeley, *Tetrahedron Lett.* **1973**, 1645; a refinement of considerations presented in Scheme 86 can also explain the occurrence of *gauche* rotamers as favored conformations of 1,2-dihaloalkanes and similar compounds.

465. J. F. Liebman and R. M. Pollak, *J. Org. Chem.* **38**, 3444 (1973).

466. R. F. Hudson, *Angew. Chem. Int. Ed.* **12**, 36 (1973).

467. (a) J. Hine, *J. Org. Chem.* **31**, 1236 (1966); (b) *Adv. Phys. Org. Chem.* **15**, 1 (1977).

468. J. A. Altmann, O. S. Tee, and K. Yates, *J. Amer. Chem. Soc.* **98**, 7132 (1976); O. S. Tee, J. A. Altmann, and K. Yates, *ibid.* **96**, 3141 (1974), and earlier papers.

468a. S. Ehrenson, *J. Amer. Chem. Soc.* **96**, 3778, 3784 (1974).

469. W. v. E. Doering, M. Franck-Neumann, D. Hasselmann, and R. L. Kaye, *J. Amer. Chem. Soc.* **94**, 3833 (1972).

470. M. R. Willcott, R. L. Cargill, and A. B. Sears, *Prog. Phys. Org. Chem.* **9**, 25 (1972).

471. M. J. S. Dewar and L. E. Wade, *J. Amer. Chem. Soc.* **95**, 290 (1973); *ibid.* **99**, 4417 (1977), see also ref. 495.

472. See also: J. J. Gajewski and N. D. Conrad, *J. Amer. Chem. Soc.* **100**, 6268 (1978).

472a. A. Padwa and T. J. Blacklock, *J. Amer. Chem. Soc.* **102**, 2797 (1980).

473. K. B. Wiberg and R. A. Fenoglio, *J. Amer. Chem. Soc.* **90**, 3395 (1968).

474. C. R. Criegee and H. G. Reinhart, *Chem. Ber.* **101**, 102 (1968).

475. L. M. Stephenson, R. V. Gemmer, and S. Current, *J. Amer. Chem. Soc.* **97**, 5909 (1975).

476. J. A. Berson and R. Malherbe, *J. Amer. Chem. Soc.* **97**, 5910 (1975).

477. R. E. Townshend, G. Ramunni, G. Segal, W. J. Hehre, and L. Salem, *J. Amer. Chem. Soc.* **98**, 2190 (1976).

477a. G. Jenner and J. Rimmelin, *Tetrahedron Lett.* **1980**, 3039.

478. P. D. Bartlett and K. E. Schueller, *J. Amer. Chem. Soc.* **90**, 6071 (1968).

479. M. Schneider, *Angew Chem. Int. Ed.* **14**, 707 (1975).

480. F. G. Bordwell, *Acc. Chem. Res.* **3**, 281 (1970); *ibid.* **5**, 374 (1972).

480a. K. N. Houk, *Top. Curr. Chem.* **79**, 1 (1979), and ref. 423e, p. 218.

480b. H. E. O'Neal and S. W. Benson, *J. Phys. Chem.* **71**, 2903 (1967).

481. D. J. McLennan and R. J. Wong, *J. Chem. Soc., Perkin Trans. 2*, **1974**, 1373.

482. See W. H. Saunders, Jr., *Acc. Chem. Res.* **9**, 19 (1976).

483. C. H. De Puy and D. H. Froemsdorf, *J. Amer. Chem. Soc.* **79**, 3710 (1957); C. H. De Puy and C. A. Bishop, *ibid.* **82**, 2532, 2535 (1960).

484. Similar behavior was found for other systems; review: R. A. Sneen, *Acc. Chem. Res.* **6**, 46 (1973).

484a. Cf. D. J. McLennan, *Acc. Chem. Res.* **9**, 281 (1976).

485. R. A. Sneen and H. M. Robbins, *J. Amer. Chem. Soc.* **94**, 7868 (1972).

486. G. A. Gregoriou, *Tetrahedron Lett.* **1974**, 233; P. J. Dais and G. A. Gregoriou, *ibid.* **1974**, 3827.

487. D. J. McLennan, *J. Chem. Soc., Perkin Trans. 2*, **1974**, 481; see also D. J. McLennan, *Tetrhaedron Lett.* **1971**, 2317.

488. D. J. McLennan, *Tetrahedron Lett.* **1975**, 4689.

489. J. M. W. Scott, *Can. J. Chem.* **48**, 3807 (1970).

490. M. H. Abraham, *J. Chem. Soc., Perkin Trans. 2*, **1973**, 1893.
491. Taken from more detailed data in ref. 490.
492. C. D. Ritchie, *J. Amer. Chem. Soc.* **94**, 3275 (1972).
493. A. Pross and R. Koren, *Tetrahedron Lett.* **1975**, 3613.
494. V. R. Raaen, T. Juhlke, F. J. Brown, and C. J. Collins, *J. Amer. Chem. Soc.* **96**, 5928 (1974).
495. R. Wehrli, H. Schmid, D. Belluš, and H.-J. Hansen, *Helv. Chim. Acta* **60**, 1325 (1977).
496. W. v. E. Doering, *Proc. Natl. Acad. Sci. USA* **78**, 5279 (1981).

Index

488

Index

Index